Sound Synthesis and Sampling

Second Edition

Martin Russ

ELSEVIER

AMSTERDAM   BOSTON   HEIDELBERG   LONDON   NEW YORK
OXFORD   PARIS   SAN DIEGO   SAN FRANCISCO   SINGAPORE
SYDNEY   TOKYO

Focal Press is an imprint of Elsevier

Focal Press

Focal Press
An imprint of Elsevier
Linacre House, Jordan Hill, Oxford OX2 8DP
30 Corporate Drive, Burlington, MA 01803

First published 1996
Reprinted 1998, 1999, 2000 (twice), 2002 (twice)
Second edition 2004
Reprinted 2005

British Library Cataloguing in Publication Data
Russ, Martin
    Sound synthesis and sampling. – 2nd ed. – (Music technology series)
    1. Computer sound processing 2. Synthesizer (Musical instrument)
    I. Title
    786.7′4

Library of Congress Cataloguing in Publication Data
A catalogue record for this book is available from the Library of Congress

ISBN 0 240 51692 3

Typeset by Charon Tec Pvt Ltd., Chennai, India
www.charontec.com
Printed and bound in Meppel, The Netherlands by Krips bv.

Sound Synthesis and Sampling

Second Edition

Music Technology Series

Acoustics and Psychoacoustics, 2nd edition (with web site)
David M. Howard and James Angus

Composing Music with Computers (with CD-ROM)
Eduardo Reck Miranda

**Computer Sound Design: Synthesis Techniques and Programming,
2nd edition** (with CD-ROM)
Eduardo Reck Miranda

Desktop Audio Technology
Francis Rumsey

Digital Sound Processing for Music and Multimedia (with web site)
Ross Kirk and Andy Hunt

Network Technology for Digital Audio
Andrew Bailey

Sound and Recording: An introduction, 4th edition
Francis Rumsey and Tim McCormick

Sound Synthesis and Sampling, 2nd edition
Martin Russ

Spatial Audio
Francis Rumsey

Contents

Series introduction

The Focal Press Music Technology Series is intended to fill a growing need for authoritative books to support college and university courses in music technology, sound recording, multimedia and their related fields. The books will also be of value to professionals already working in these areas and who want either to update their knowledge or to familiarise themselves with topics that have not been part of their mainstream occupations.

Information technology and digital systems are now widely used in the production of sound and in the composition of music for a wide range of end users. Those working in these fields need to understand the principles of sound, musical acoustics, sound synthesis, digital audio, video and computer systems. This is a tall order, but people with this breadth of knowledge are increasingly sought after by employers. The series will explain the technology and techniques in a manner which is both readable and factually concise, avoiding the chattiness, informality and technical woolliness of many books on music technology. The authors are all experts in their fields and many come from teaching and research backgrounds.

Dr Francis Rumsey
Series Consultant

Preface to first edition

This is a book about sound synthesis and sampling. It is intended to provide a reference guide to the many techniques and approaches that are used in both commercial and research sound synthesizers. The coverage is more concerned with the underlying principles, so this is not a 'build your own synthesizer' type of book, nor is it a guide to producing specific synthesised sounds. Instead it aims to provide a solid source of information on the diverse and complex field of sound synthesis. As well as the details of the techniques of synthesis, some practical applications are described to show how synthesis can be used to make sounds.

It is designed to meet the requirements of a wide range of readers, from enthusiasts to undergraduate level students. Wherever possible, a non-mathematical approach has been taken, and the book is intended to be accessible to readers without a strong scientific background.

This book brings together information from a wealth of material which I have been collecting and compiling for many years. Since the early 1970s I have been involved in the design, construction and use of synthesizers. More recently this has included the reviewing of electronic musical instruments for *Sound on Sound*, the leading hi-tech music recording magazine in the UK.

The initial prompting for this book came from Francis Rumsey of the University of Surrey's Music Department, with support from Margaret Riley at Focal Press. I would like to thank them for their enthusiasm, time and encouragement throughout the project. I would also like to thank my wife and children for putting up with my disappearance for long periods over the last year.

Martin Russ
February 1996

Preface to second edition

This second edition has revised and updated all of the material in the first edition, correcting a few minor errors, and adding a completely new chapter on performance aspects (Chapter 8), which shows how synthesizers have become embedded within more sophisticated musical performance instruments, rather than always being stand-alone synthesizers *per se*. This theme is also explored further in the extended 'Future of synthesis' chapter.

I have strived to maintain the abstraction of the techniques away from specific manufacturers, and with only a few exceptions, the only place where details of individual instruments or software will be found is in the 'Examples' sections at the end of each chapter.

Taking a cue from other books in the Focal Press Music Technology series, I have added additional notes alongside the main text, which are intended to reinforce significant points.

I must thank Beth Howard and others at Focal Press who have helped me to finish this edition. Their patience and support has been invaluable.

I would also like to thank the many readers, reviewers and other sources of feedback for their suggestions, as many as possible of these have been incorporated in this edition. I welcome additional suggestions for improvement, as well as corrections: please send these to me via Focal Press.

Martin Russ
October 2003

Visual map

About this book

This book is divided into nine chapters, followed by References, Glossary, Jargon and finally, an Index. The Jargon section is designed to try and prevent the confusion that often results from the wide variation in the terminology which is used in the field of synthesizers. Each entry consists of the term which is used in this book, followed by the alternative names which can be used for that term.

Book guide

The chapters can be divided into five major divisions:

- Background
- Techniques
- Applications
- Analysis
- Reference

Background

Chapter 1 sets the background, and places synthesis in a historical perspective.

Techniques

Chapters 2–5 describe the main methods of producing and manipulating sound.

Applications

Chapters 6–8 show how the techniques described can be used to synthesise sound and music, in the studio and in live performance.

Analysis

Chapter 9 provides analysis of the development of sound synthesis and some speculation on future developments.

Reference

References, Glossary, Jargon and Index.

Chapter guide

Chapter 1   Background

This chapter introduces the concept of synthesis, and briefly describes the history. It includes brief overviews of acoustics, electronics, digital sampling and musical instrument digital interface (MIDI).

Chapter 2   Analogue synthesis

This chapter describes the main methods which are used for analogue sound synthesis: subtractive, additive, AM, FM, ring modulation, ringing oscillators and others.

Chapter 3   Hybrid synthesis

This chapter shows the way that synthesis techniques changed from the primarily analogue electronic circuit designs of the 1960s and 1970s, to the predominantly digital circuitry of the 1980s and 1990s. Synthesizers whose design incorporates a mixture of both design techniques are included.

Chapter 4   Sampling

This chapter describes the three types of sampling technology: tape, analogue and digital. It also looks at the continuing convergence between sampling and synthesis by looking at sampling and synthesis (S&S) synthesizers.

Chapter 5   Digital synthesis

This chapter looks at the major techniques which are used for digital sound synthesis: FM, waveshaping, physical modelling, granular, FOF, analysis–synthesis and resynthesis.

Chapter 6   Using synthesis

This chapter deals with the use of synthesis to make music and other sounds.

Chapter 7   Controllers

This chapter looks in detail at how synthesizers are controlled, including using MIDI.

Chapter 8   Performance

This chapter looks at the ways that synthesizers can be used in live performance.

Chapter 9   The future of synthesis

This chapter attempts to place sound synthesis in a wider context, by describing the probable development of music, MIDI and computing in the future.

Chapter section guide

Within each chapter, there are sections which deal with specific topics. The format and intention of some of these may be unfamiliar to the reader, and thus they deserve special mention.

Examples

These sections are illustrated with block diagrams of the internal function and front panel controls of some representative example instruments or software, together with some notes on their main features. This should provide a more useful idea of their operation than just black and white photographs. Further information and photographs of a wide range of synthesizers and other electronic musical instruments can be found in Julian Colbeck's comprehensive *Keyfax* books (Colbeck, 1985). Details on some specific instruments can be found in Mark Vail's *Vintage Synthesizers* (Vail, 1993) book, which is a collection of articles from the American magazine *Keyboard*.

Time line

The Time lines are intended to show the development of a topic in a historical context. Reading text which contains lots of references to dates and people can be confusing. By presenting the major events in time order, the developments and relationships can be seen more clearly. The time lines are deliberately split up so that only entries relevant to each chapter are shown. This keeps the material in each individual time line succinct.

Overall time line

Chapters 2–5 of this book do not represent a linear historical record, even though the apparent progression from analogue, via hybrid, to digital synthesis methods is a compelling metaphor. Synthesis techniques, like fashion, regularly recycle the old and mostly forgotten with 'retro' revivals of buzzwords like FM, analogue, valves, FETs, modular, resynthesis and more. The overall timeline shown overleaf is intended to show just some of the complex flow of the synthesis timeline.

Questions

Each chapter ends with a few questions, which can be used as either a quick comprehension test, or as a guide to the major topics covered in that chapter.

Event	Year		Category
Metronome patented	1815		Performance
First magnetic tape recorder		1920	Sampling
First true commercial magnetic tape recorder	1937		Sampling
RCA mark II synthesizer		1955	Analogue
Buchla black box, early analogue synth	1963		Analogue
Mellotron	1965		Sampling
Wendy Carlos's 'switched on bach'	1968		Performance
Mini Moog launched	1969		Analogue
Ralph Deutsch electronic piano		1970	Hybrid
Roland TR-33 Rhythm Unit		1972	Performance
Roland MC-4 Sequencer		1978	Digital
Fairlight CMI		1979	Hybrid
Fairlight CMI		1979	Sampling
Wasp synthesizer uses hybrid of analogue and digital	1980		Digital
First dedicated sampler: emulator	1980		Sampling
CD launched	1982		Digital
PPG 2.2 polyphonic hybrid synth	1982		Hybrid
Yamaha DX7 first commercial all-digital synth	1983		Digital
TR-909 First MIDI drum machine	1983		Performance
Roland MC-202 Micro-composer	1984		Performance
Ensoniq mirage: affordable sampler	1984		Sampling
Roland D50 digital synth	1987		Digital
Waldorf MicroWave wavetable synth	1989		Hybrid
Roland JD-800 analogue polysynth		1991	Analogue
Roland DJ-70 sampling workstation		1992	Performance
Yamaha VL1 physical modelling synth		1994	Digital
Digidesign ProTools free		1995	Sampling
Roland MC-303 Groovebox		1996	Performance
Yamaha AN1X analogue modelling synth		1997	Digital
Yamaha DJ-X mass-market sampling groovebox		1998	Performance
Reason virtual studio in a rack software	2001		Digital
Yamaha RS7000 Music Production Studio	2001		Performance
Hard disk recorder, mixer and CD-writer in one box	2001		Digital

Time

Background

1   Background

This chapter introduces the concept of synthesis, and briefly describes the history. It includes brief overviews of acoustics, electronics, digital sampling and musical instrument digital interface (MIDI).

1.1   What is synthesis?

'Synthesis' is defined in the 2003 edition of the *Chambers 21st Century Dictionary* as 'building up; putting together; making a whole out of parts'. The process of synthesis is thus a bringing together, and the 'making a whole' is significant because it implies more than just a random assembly: synthesis should be a creative process. It is this artistic aspect, which is often overlooked in favour of the more technical aspects of the subject. Although a synthesizer may be capable of producing almost infinite varieties of output, controlling and choosing them requires human intervention and skill.

The word 'synthesis' is frequently used in just two major contexts: the creation of chemical compounds and the production of electronic sounds. But there are a large number of other types of synthesis.

1.1.1   Types

All synthesizers are very similar in their concept – the major differences are in their output formats and the way that they produce that output. Just some of the types of synthesizers are as follows:

- **Texture synthesizers**, used in the graphics industry, especially in 3D graphics.
- **Video synthesizers**, used to produce and process video signals.
- **Colour synthesizers**, used as part of 'son et lumiere' presentations.
- **Speech synthesizers**, used in computer and telecommunications applications.
- **Sound synthesizers**, used to create and process sounds and music.
- **Word synthesizers**, more commonly known as authors using 'word processor' software!

Synthesizers have two basic functional blocks: a *control interface*, which is how the parameters which define the end product are set; and a *'synthesis engine'*, which interprets the parameter values and produces the output. In most cases there is a degree of abstraction involved between the control interface and the synthesis engine itself. This is because the complexity of the synthesis process is often very high, and it is often necessary to reduce the apparent complexity of the control by using some sort of simpler conceptual model. This enables the user of the synthesizer to use it without requiring a detailed knowledge of the inner workings. This idea of models and abstraction of interface is a recurring theme which will be explored many times in this book (Figure 1.1.1).

Figure 1.1.1 The user uses a metaphor in order to access the functions of the synthesizer. The synthesizer provides a model to the user and maps this model to internal functionality. This type of abstraction is used in a wide variety of electronic devices, particularly those employing digital circuitry.

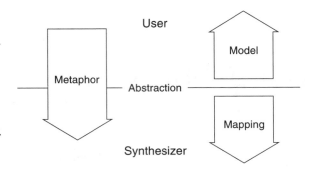

1.1.2 Sound synthesis

Sound synthesis is the process of producing sound. It can reuse existing sounds by processing them, or it can generate sound electronically or mechanically. It may use mathematics, physics or even biology; and it brings together art and science in a mix of musical skill and technical expertise. Used carefully, it can produce emotional performances, which paint sonic landscapes with a rich and huge set of timbres, limited only by the imagination and knowledge of the creator.

Many members of the general public have unrealistic expectations of the capabilities of synthesizers. The author has encountered feedback comments such as 'I thought it did it all by itself!' when he has shown that he can indeed 'play' a synthesizer.

Sounds can be simple or complex, and the methods used to create them are diverse. Sound synthesis is not solely concerned with sophisticated computer-generated timbres, although this is often the most publicised aspect. The wide availability of high quality recording and synthesis technology has made the generation of sounds much easier for musicians and technicians, and future developments promise even easier access to ever more powerful techniques. But the technology is nothing more than a set of tools which can be used to make sounds: the creative skills of the performer, musician or technician are still essential to avoid music becoming mundane.

1.1.3 Synthesizers

Although synthesizer can be spelt with a 'zer' or 'ser' ending, the 'zer' ending will be used in this book. Also, the single word 'synthesizer' is used here to imply 'sound synthesizer', rather than a generic synthesizer.

Sounds are synthesised using a sound synthesizer. The synthesis of sounds has a long history. The first synthesizer might have been an early ancestor of *Homo sapiens* hitting a hollow log, or perhaps learning to whistle. Singing uses a sophisticated synthesizer whose capabilities are often forgotten: the human vocal tract. All musical instruments can be thought of as being 'synthesizers', although few people would think of them in this context. A violin or clarinet is viewed as being 'natural', whereas a synthesizer is seen as 'artificial', even though all of these instruments produce a sound by essentially synthetic methods.

Note that the production of a wide range of sounds by a synthesizer can be very significant. A 'synthesizer' which produces a restricted range of sounds can often be viewed as being more musically acceptable.

Recently, the word 'synthesizer' has come to only mean an electronic instrument that is capable of producing a wide range of different sounds. The actual categories of sounds which qualify for this label of synthesizer are also very specific: purely imitative sounds are frequently regarded as nothing other than recordings of the actual instrument, in which case the synthesizer is seen as little more than a replay device. In other words, the general public seems to expect synthesizers

The electronic piano is one example, where the same synthesis capability could be packaged in two different ways, and would consequently be sold separately to synthesists and piano players.

to produce 'synthetic' sounds. This can be readily seen in many low-cost keyboard instruments which are intended for home usage: they typically have a number of familiar instrument sounds with names like 'piano', 'strings', 'guitar', etc. But they also have sounds labelled 'synth' for sounds which do not fit into the 'naturalistic' description scheme. As synthesizers become better at fusing elements of real and synthetic sounds, the boundaries of what is regarded as 'synthetic' and what is 'real' are becoming increasingly diffuse.

This blurred perception has resulted in broad acceptance of a number of 'hyper-real' instrument sounds, where the distinctive characteristics of an instrument are exaggerated. Fret buzz and finger noise on an acoustic guitar and breath noise on a flute are just two examples. Drum sounds are frequently enhanced and altered considerably, and yet, unless they cross that boundary line between 'real' and 'synthetic', their generation is not questioned – it is assumed to be 'real' and 'natural'. This can cause considerable difficulties for performers who are expected to reproduce the same sound as the compact disk (CD) in a live environment. The actual sound of many live instruments may be very different from the sound that is 'expected' from the familiar recording that was painstakingly created in a studio. Drummers are one example: they may have a physical drum kit where many parts of the kit are present merely to give a visual cue or 'home' to the electronically generated sounds that are being controlled via triggers connected to the drums, and where the true sound of the real drums is an unwanted distraction.

Forms

Synthesizers come in several different varieties, although many of the constituent parts are common to all of the types. Most synthesizers have one or more audio outputs; one or more control inputs; some sort of display; and buttons or dials to select and control the operation of the unit. The major split is into *performance* and *modular* forms:

- **Performance synthesizers** have a standard interconnection of their internal synthesis modules already built-in. It is usually not possible to change this significantly, and so the signal flow always follows a set path through the synthesizer. This enables the rapid patching of commonly used configurations, but does limit the flexibility. Performance synthesizers form the vast majority of commercial synthesizer products.
- Conversely, **modular synthesizers** have no fixed interconnections, and the synthesis modules can be connected together in any way. Changes can be made to the connections whilst the synthesizer is making a sound, although the usual practice is to set up and test the interconnections in advance. Because more connections need to be made, modular synthesizers are harder and more time-consuming to set up, but they do have much greater flexibility. Modular synthesizers are much rarer than performance synthesizers, and are often used for academic or research purposes.

Non-ideal interfaces are actually very common. The 'qwerty' typewriter keyboard was originally intended to slow down typing speeds and so help prevent the jamming of early mechanical typewriters. It has become dominant (and commercially, virtually essential!) despite many attempts to replace it with more ergonomically efficient alternatives. The music keyboard has also seen several carefully human-engineered improvements which have also failed to gain widespread acceptance. It is also significant that the qwerty and music keyboards have both become metaphors for computers/information and music in general.

Both performance and modular synthesizers can come with or without a music keyboard. The keyboard has become the most dominant method of controlling the performance aspect of a synthesizer, although it is not necessarily the ideal controller. Synthesizers which do not have a keyboard (or any other type of controller device) are often referred to as *expanders* or *modules*, and these can be controlled either by a synthesizer which does have a keyboard or from a variety of other controllers. It has been said that the choice of a keyboard as the controller was probably the biggest setback to the wide acceptance of synthesizers as a musical instrument. Chapter 7 describes some of the alternatives to a keyboard.

1.1.4 Sounds

Synthesised sounds can be split into simple categories like 'imitative' or 'synthetic'. Some sounds will not be easy to place in a definite category, and this is especially true for sounds, which contain elements of both real and synthetic sounds.

Imitative sounds often sound like real instruments, and they might be familiar orchestral or band instruments. In addition, imitative sounds may be more literal in nature, the sound effects. In contrast, synthetic sounds will often be unfamiliar to anyone who is used to hearing only real instruments, but over time a number of clichés have been developed: the 'string synth' and 'synth brass' being just two examples.

Synthetic sounds can be divided into various types, depending on their purpose.

'Imitations' and 'emulations'

'Imitations' and 'emulations' are intended to provide many of the characteristics of real instruments, but in a sympathetic way where the synthesis is frequently providing additional control or emphasis of significant features of the sound. Sometimes an emulation may be used because of tuning problems, or difficulties in locating a rare instrument. The many 'electronic' piano sounds are one example of an emulated sound.

'Suggestions' and 'hints'

'Suggestions' and 'hints' are sounds where the resulting sound has only a slight connection with any real instrument. The 'synth brass' sound produced by analogue polyphonic synthesizers in the 1970s is an example of a sound where just enough of the characteristics of the real instrument are present and so strongly suggest a 'brass'-type instrument to an uncritical listener, but where detailed comparison immediately highlights the difference to a real brass instrument.

'Alien' and 'off-the-wall'

'Alien' and 'off-the-wall' sounds are usually entirely synthetic in nature. The cues which enable a listener to determine if a sound is synthetic are complex, but are often related to characteristics that are connected with the physics of real instruments: unusual or unfamiliar harmonic structures and their changes over time; constancy of timbre over a wide range; and pitch change without timbre change. By deliberately moving outside of the physical limitations of conventional instruments, then it is noise-like.

Noise-like

Of course, most synthesizers can also produce variations on 'noise', of which 'white noise' is perhaps the most unpitched and unusual sound of all, since it has the same sound energy in linear frequency bands across the entire audible range. Any frequency-dependent variation of the harmonic content of a noise-like sound can give it a perceivable 'pitch', and it thus becomes playable.

All of these types of synthetic sounds can be used to make real sounds more interesting by combining the sounds into a hybrid composite (see Chapter 6).

Factory presets

Naming sounds is not as straightforward as it might appear at first. For example, if you have more than two or three piano sounds, then manufacturer's names or lots of adjectives tend to be used: 'Steinway piano' or 'Detuned pub piano' being simple examples. For sounds which are more synthetic in nature, the adjectives become more dense, or are abandoned altogether in favour of names which suggest the type of sound rather than try and describe it: 'crystal spheres' and 'thudblock' being two examples.

One final category of sound is found only in commercial sound synthesizers: the factory sounds which are intended to be used as demonstrations of the capabilities of the instrument when it is being auditioned by a potential purchaser. These sounds are typically produced rapidly at a late stage in the production process, and are not always a good guide to the true potential sound-making capabilities of the synthesizer. They also frequently suffer from a number of problems which are directly related to their design specification; they can be buried underneath excessive amounts of reverberation, they may use carefully timed cyclic variations and echo effects for special effects, and they are rarely organised in category groupings, favouring instead contrast and variation. Some techniques for making use of these sounds are described in Chapter 6.

1.1.5 Synthesis methods

Understanding how a synthesis technique works is essential for the adjustment (tweaking) of sounds to suit a musical context, and also knowing how the sound can be controlled in performance. This is just as much a part of the synthesists toolkit as playing ability.

There are many techniques which can be used to synthesise sound. Many of them use a 'source and modifier' model as a metaphor for the process which produces the sound: a raw sound source produces the basic tone, which is then modified by in some way to create the final sound. Another name for this model is the 'excitation and filter' model, as used in speech synthesis. The use of this model can be seen most clearly in analogue subtractive synthesizers, but it can also be applied to other methods of synthesis – sample and synthesis (S&S) or physical modelling, for example.

Some methods of synthesis are more complex: frequency modulation (FM), harmonic synthesis, Fonctions d'Onde Formantigue (FOF) (see Section 5.5) and granular synthesis. For these methods, the metaphors of a model can be more mathematical or abstracted, and thus may be harder to comprehend. This may be one of the reasons why the 'easier to understand' methods like subtractive synthesis and its derived variant called S&S have been so commercially successful.

1.1.6 Analogue synthesis

The word 'analogue' can also be spelt without the 'ue' ending. In this book, the longer version will be used.

'Analogue' refers to the use of audio signals, which can be produced using elements like oscillators, filters and amplifiers. Analogue synthesis methods split into three basic areas, although there are crossovers between them. The basic types are:

- subtractive
- additive
- wavetable.

Subtractive

Subtractive synthesis takes a 'raw' sound which is usually rich in harmonics, and filters it to remove some of the harmonic content. The raw sounds are traditionally simple mathematical waveshapes: square, sawtooth, triangle and sine, although modern subtractive synthesizers tend to provide longer 'samples' instead of single cycles of waveforms. The filtering tends to be a resonant low-pass filter, and changing the cut-off frequency of this filter produces the characteristic (and clichéd) 'filter sweep' sound, which is strongly associated with subtractive synthesis.

Additive

Additive synthesis adds together lots of sine waves with different frequencies to produce the final timbre. The main problem with this method is the complexity of controlling large numbers of sine waves, but see the digital section too.

Wavetable

Wavetable synthesis extends the ideas of subtractive synthesis by providing much more sophisticated waveshapes as the raw starting point for subsequent filtering and shaping. More than one cycle of a waveform can be stored, or many waveforms can be arranged so that they can be dynamically selected in real time – this produces a characteristic 'swept' sound which can be subtle, rough, metallic or even glassy in timbre.

1.1.7 Digital synthesis

Digital technology replaces signals with numerical representations, and uses computers to process those numbers. Digital methods of synthesising sounds are more varied than analogue methods, and research is still continuing to find new ways of making sounds. Some of the types that may be encountered include:

- FM
- wavetable
- sample replay

- additive
- S&S
- physical modelling
- software synthesis.

FM

FM is the technical term for the way that FM radio works, where the audio signal of music or speech is used to modulate a high frequency carrier signal which then broadcasts the audio as part of a radio signal. In audio FM, both signals are at audio frequencies, and complex frequency mirroring, phase inversions and cancellations happen that can produce a wide range of timbres. (In fact, most of the effects that audio FM uses are exactly the sort of distortions and problems that you try to avoid in radio FM!) The main problem with FM is that it is not possible to program it 'intuitively' without a lot of practice, but its major advantage in the 1970s was that it required very little memory to store a large number of sounds. With large falls in the cost of storage, this is no longer as crucially important in the 2000s. FM is used in some sound cards and portable keyboards, and like many synthesis techniques, its marketability seems to be influenced by the cycles of fashionability.

Wavetable

Wavetable synthesis uses the same idea as the analogue version, but extends the basic idea into more complex areas. The waveshapes are usually complete but short segments of real samples, and these can be looped to provide sustained sections of sound, or several segments of sound can be joined together to produce a composite 'sample'. Often this is used to 'splice' the start of one sound onto the sustain part of another. Because complete samples are not used, this method makes very efficient use of available memory space, but this results in a loss of quality. Wavetable synthesis is used in low-cost, mass-market sound cards and MIDI instruments.

Sample replay

Sample replay is the ultimate version of wavetable. Instead of looping short samples and splicing them together, sample replay does just that: it replays complete samples of sounds, with a loop for the sustained section of the sound. Sample replay uses lots of memory, and was thus initially used in more expensive sound cards and MIDI instruments only. Falling prices for memory (allegedly driven strongly downwards by huge sales of cartridges for video games consoles in the 1980s and 1990s) have led to sample replay becoming very widespread.

Additive

Digital techniques make the task of coping with lots of sine waves much easier, and digital additive synthesizers have been more successful than analogue versions, but they are still a very specialised field. There are very few sound cards and synthesizers that use only

9

additive synthesis, but additive is often an element within another type of synthesis, or can be part of a palette of techniques.

S&S

S&S is an acronym for 'samples and synthesis', and uses the techniques of wavetable and sample replay, but adds in the filtering and shaping of subtractive synthesis, but all in a digital form. This method is widely used in MIDI instruments, sound cards and professional electronic musical instruments.

Physical modelling

The term 'physical modelling' is still used where a mathematical model of an instrument is produced from the physics of that instrument, but the word 'modelling' has become a generic term for any mathematical modelling technique that can be applied to synthesis.

Physical modelling uses mathematical equations which attempt to describe how an instrument works. The results can be stunningly realistic, very synthetic or a mixture of both. The most important feature is the way that the model responds in much the same way as a real instrument; hence the playing techniques of the real instrument can often be employed by a performer.

Initially the 'real' instruments chosen were exactly that, and then plucked, hit and blown instruments were modelled to varying degrees of accuracy; but once these were established, then models of analogue synthesizers and even valve amplifiers and effects units began to be developed. The high processing demands of modelling meant that it was only found in professional equipment in the mid-1990s. But it rapidly became more widely adopted, and by the start of the 21st century it could be found, albeit in a simplified form, in low-cost keyboards intended for home usage, as well as computer sound cards, whilst in professional equipment, highly developed models are used to produce an increasingly wide range of 'modelled' sounds, instruments, amplifiers, effects, environments and loudspeakers.

Software synthesis

In the 2000s, the use of powerful general-purpose computers as audio processing and synthesis devices has given physical modelling a new role: *software synthesis*. Here, the computer replaces almost all of the 'traditional' equipment that might be expected by a time traveller from the 1970s. The computer can now integrate the functions of a sequencer for the notes, a synthesizer or sample-replay device to produce the sounds, a mixer to combine the sounds from several synthesizers or sample-replay devices, and process the mixed audio through effects-processing units, hard disk recording to capture the audio, and CD 'burning' software to produce finished CDs. The synthesizers and effects often use physical modelling techniques to emulate an analogue synthesizer, an analogue reverb line and more. All of these functions are carried out on digital signals, entirely within the computer – conversion to analogue audio is needed only for monitoring playback, and in the case of the CD, the audio signal output of the CD player is typically the first time that the audio signal has ever been in an analogue format. Chapter 9 explores this topic in more detail.

1.2 Beginnings

The beginnings of sound synthesis lie with the origins of the human, *Homo sapiens*, species. Many animals have an excellent hearing sense, and this serves a diverse variety of purposes: advance warning of danger, tracking prey and communication. In order to be effective, hearing needs to monitor the essential parts of the audio spectrum. This can involve very low frequencies in some underwater animals or ultrasonic frequencies for echo location purposes in bats; the dynamic range required can be very large.

Using distance as an analogy, a ratio of $10^{12}:1$ is equivalent to the ratio between one million kilometres and one millimetre.

Human hearing is more limited. Sounds from about 15 Hz to about 18 kHz can be heard, although this varies with the sound level, and the upper limit reduces with age. The dynamic range is more than 120 dB, which is a ratio of $10^{12}:1$. With two ears and the complex processing carried out by the brain, the average human being can accurately locate sounds and resolve frequency to fractions of a hertz, although this performance is dependent on the frequency, level and other factors.

Human beings also have a sophisticated means of producing sound: the vocal tract. The combination of vocal cords, throat, tongue, teeth, mouth cavity and lips provides a versatile way of making a wide variety of sounds: a biological sound synthesizer. The development of this particular instrument is long and still ongoing – it is probably the oldest and most important musical instrument (Figure 1.2.1).

The mixture of sophisticated hearing and an inbuilt sound synthesizer, plus the everyday usage via speech, singing or whistling, makes the human being a perceptive and interactive listener. The human voice is part of a feedback loop created by the ears and the brain. The brain not only controls the vocal tract to make the sounds, it also listens to the sounds created and adjusts the vocal tract dynamically. This analysis–synthesis approach is also used in resynthesis, as described in Section 5.6. The combination of sound production and analysis forms a powerful feedback mechanism; and it seems that knowing how to make sounds is an essential part of inferring the intended meaning when someone else makes the sounds.

Making sounds and listening to sounds is a fundamental part of human interactivity. Using the example of a human being from conception onwards, it is possible to see the range of possibilities:

- **Listening**: Pregnant mothers are often aware that sudden noises can startle a baby in the womb.
- **Mouth**: Most parents will confirm that babies are capable of making a vocal noise from just after birth!
- **Shaking**: Once control over hands and feet is possible, a baby will investigate objects by interacting with them. The rattle is specifically designed to provide an audible feedback when it is shaken.
- **Singing**: Part of the process of learning to speak involves long periods of experimentation by the infant, where the range of possible sounds is explored.

Figure 1.2.1 The human voice is a complex and sophisticated synthesizer capable of producing both speech and singing sounds. The main sound source is the vocal cords, although some sounds are produced by the interactions between the lips, tongue and teeth with air currents. The throat, nose, mouth, oesophagus and lungs form a set of resonant cavities that filter the sounds, and the mouth shape is dynamically variable.

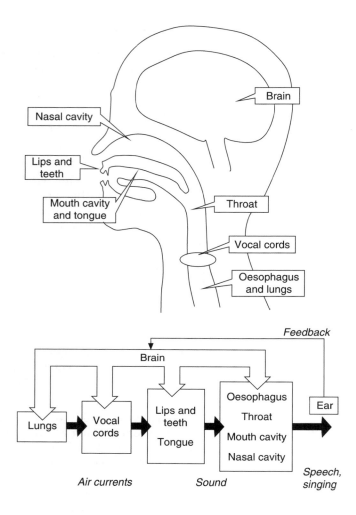

- **Speech**: The 'singing' sounds are then reduced down to the set of sounds which are heard from the parent's own speech.
- **Blowing**: Blowing (and spitting!) are part of the learning process for making speech sounds. Blowing into tubes and whistles may lead to playing real musical instruments.
- **Percussive pitching**: 'Open mouth' techniques for making sounds include slapping the cheek or top of the head, or tapping the teeth. The throat and mouth cavity are then altered to provide the pitching of the resulting sound.
- **Whistling**: Whistling requires the mastering of a musical instrument which is created by the lips.

These have been arranged in approximate chronological order, although the development of every human being is different. The important information here is the wide range of possible ways that sounds can be made, and the degree of control which is possible. Singing and whistling are both highly expressive musical instruments, and it is no accident that many musical instruments also use the foundations for the science of acoustics back in 1878, and Lee de Forest's triode amplifier of 1906 provided the electronics basis for controlling sound. E.C.Wente's condenser microphone in 1915 provided the first high quality audio microphone, and the tape recorder gave a way to store sounds.

Jacquard used holes punched in cards to control weaving machines in the early 1800s.

At the German Radio station NWDR in Cologne in 1951, Herbert Eimert began to use studio oscillators and tape recorders to produce electronically generated sounds. Rather than assemble test gear together, researchers at the Radio Corporation of America (RCA) Laboratories in the US produced a dedicated modular synthesizer in 1955, which was designed to simplify the tedious production process of creating sounds by using automation. A mark II model followed in 1957, and this was used at the Columbia-Princeton Electronic Music Centre for some years. Although the use of punched holes in paper tape to control the functions now appears primitive, the RCA synthesizer was one of the first integrated systems for producing electronic music.

Work at Bell Telephone Laboratories in the 1950s and 1960s led to the development of pulse code modulation (PCM), a technique for digitising or sampling sound and thus converting it into digital form. As is usual in telecommunications technology, the description and its acronym, pulse code modulation and PCM, are in a technical language that conveys little to the non-engineer. What PCM does is actually very straightforward. An audio signal is 'sampled' at regular intervals, and this gives a series of voltage values: one for each interval. These voltages represent the value of the audio signal at the instant that the sample was made. These voltages are converted into numbers, and these numbers are then converted into a series of electrical pulses, where the number and organisation of the pulses represent the size of the voltage. PCM thus refers to the coding scheme used to represent the numbers as pulses. (You may like to compare PCM with the pulse width modulation (PWM) as described in Chapter 2.)

PCM forms the basis of sampling. A great deal of work was done to formalise the theory and practice of converting audio signals into digital numbers. The concept of sampling at twice the highest wanted frequency is called the *Nyquist criterion* after work published by Nyquist and others. The filtering required to prevent unwanted frequencies being heard (they are a consequence of the sampling process) was also developed as a result of telecommunications research. Further work in the 1970s led to the invention of the digital signal processor (DSP), a specialised microprocessor chip which was produced in order to carry out the complex numerical calculations which were needed to enable audio coding algorithms to be developed. DSPs have since been used to produce many types of digital synthesizer, although the wide availability, powerful processing capability and low-cost of more general-purpose processors have meant that DSPs are no longer essential except in the most demanding applications.

Current telecommunications research continues to explore the outermost limits of acoustics, physics and electronics, although since telecommunications is now almost solely concerned with computers and the emphasis is increasingly on intercommunications between computers but not between people.

1.4 Tape techniques

1.4.1 *The analogue tape recorder*

The analogue tape recorder has been a major part of electronic music synthesis almost from the very beginning. It enables the user to splice together small sections of magnetic tape which represent audio, and then replay the results. This has the important elements of 'building up from small parts' that is the basis of the definition of synthesis.

The principle of the tape recorder is not new. The audio signal is converted into a changing magnetic field, which is then stored onto iron. Early examples recorded onto iron wire, then onto steel ribbon. There were also experiments with the use of paperbacked tape, but the most significant breakthrough was the use of plastic coated with magnetic material, which was developed in Germany in 1935. But it was not until after the end of World War II in 1945 that tape recording started to become widely available as a way of storing and replaying audio material. The tape that was used consisted of a thin acetate plastic tape coated with iron oxide, and polyester film still forms the backing of magnetic tape. More details of the technique of magnetic recording are given in Chapter 4.

Before recording, all music was live!

Tape recorders are very useful for synthesising sounds because they allow permanent records to be kept of a performance, or they allow a performance to be 'time-shifted': recorded for subsequent playback later. This may appear to be obvious to the modern reader, but before tape recording, the only way to record sound for later playback was literally to make a record! It is not feasible to break up and reassemble records (see Section 1.4.6), and so when it was first introduced, the tape recorder was a genuinely new and exciting musical tool.

Pitch and speed

A tape recording of a sound ties together two aspects of sound: the pitch and the duration. If you record the sound at 15 inches per second (ips), then playing back at twice the speed, 30 ips, will double the pitch and so it will be transposed up by one octave. But the duration will be halved, and so a 1-second sound will only last for half a second when played back at twice the recording speed. Breaking this interdependence is not at all easy using tape recorder technology, although it is relatively simple using digital techniques.

Being able to change the pitch of sounds once they are recorded can simplify the process of producing electronic sounds using oscillators. By changing the speed at which the sound is recorded, the same oscillators can be used to produce tape segments, which contain the same sound, but shifted in pitch by one or more octaves. This avoids some of the problems of continuously retuning oscillators, although it does depend on the tape speeds of the tape recorders being accurate.

Unfortunately, the tape speed of early tape recorders was not very accurate. Long-term drift of the speed affects the pitch of anything recorded, and so required careful monitoring. Short-term variations in tape speed are called *wow* and *flutter*. Wow implies a slow cyclic variation in pitch, whilst flutter implies a faster and more irregular variation in pitch. Depending on the type of sound, the ear can be very sensitive to pitch changes. Wow and flutter can be very obvious on solo piano playing, whilst some orchestral music can actually sound better!

Of course, changing the tape speed can be used as a creative tool: adjusting the tape speed whilst recording will permanently store the pitch changes in the recording, whilst changing the replay tape speed will only affect playback (but the speed changes are probably not as easy to reproduce on demand). Deliberately introducing wow and flutter can also be used to introduce vibrato and other pitch shifting effects.

Splicing

Once a sound has been recorded onto tape, it is then in a physical form which can be manipulated in ways which would be difficult or impossible for the actual sound itself. Cutting the tape into sections and then splicing them together allows the joining and juxtaposition of sounds. The main limits to this technique are the accuracy of finding the right place on the tape, and the length of the shortest section of tape which can be spliced together. Each join in a piece of tape produces a potential weak spot, and so an edited tape may need to be recorded onto another tape recorder. Each time that a tape is copied, the quality is degraded slightly, and so there is a need to compromise between the complexity of the editing and the fidelity of the final sound.

Reversing

Reversing the direction of playback of tape makes the sound play backwards. Unfortunately, because domestic tape recorders are designed to record in stereo on two sides (known as quarter-track format) merely turning the tape around does not work. Playing the back of the tape (the side which has the backing, rather than the oxide visible) does allow reversing of quarter-track tapes, but there is significant loss of audio quality. Professional tape recorders use the mono full-track and stereo half-track formats, where the tape direction is unidirectional, and these can be used to produce reversed audio (although the channels are swapped on a half-track tape!).

Playing sounds backwards has two main audible effects:

- Most naturally occurring musical instrument sounds have a sharp attack and a slow release or decay time, and this is reversed. This produces a characteristic 'rushing' sound or 'squashed' feel, since the main rhythmic information is on the beats and these are now at the end of the notes.

17

- Any reverberation becomes part of the start of the sound, whilst the end of the sound is 'dry'. Echoes precede the notes which produced them. Both of these serve to reinforce the crescendo effect of the notes.

Tape loops

Splicing the end of a section of tape back onto the start produces a loop of tape, and the sound will thus play back continuously. This can be used to produce repeated phrases, patterns and rhythms. Several tape loops of different lengths played back simultaneously can produce complex polyrhythmic sequences of sounds.

Sound on sound

Normally, a tape recorder will erase any pre-existing magnetic fields on the tape before recording onto it using its erase head. By turning this off, any new audio which is recorded will be mixed with anything already there. This is called 'sound on sound' because it literally allows sounds to be layered on top of each other. As with many tape manipulation techniques, there is a loss in the quality each time that this technique is used, specifically for the pre-existing audio in this case.

Delays, echoes and reverberation

By using two tape recorders, where one records audio onto the tape, and the second plays back the same tape, it is possible to produce time delays. The time delay can be controlled by altering the tape speed (which should be the same on both tape recorders) and the physical separation of the two recorders. Some tape recorders have additional playback heads, and these can be used to provide short time delays. Dedicated machines with one record head and several playback heads have been used to produce artificial echoes: the Watkins CopyCat being one example. The use of echo and time delays has become part of the performance technique of many performers, from guitarists to synthesists.

By taking the time delayed signal and mixing it into the recorded signal, it is possible to produce multiple echoes from only one playback head (or the playback tape recorder). With multiple playback heads spaced irregularly, it is possible to use this feedback to simulate reverberation. With too much feedback, the system may break into oscillation, and this can be used as an additional method of synthesising sounds, where a stimulus signal is used to initiate the oscillation.

Multi-tracking

Although early tape recorders had only one or two tracks of audio, experiments were carried out on producing tape recorders with more tracks. Linking two tape recorders together to give additional tracks was very awkward. Quarter-track tape recorders used four tracks, although usually only two of these could be played at once. Modified heads produced tape recorders with four separate tracks, and these 'multi-track' tape recorders were used to produce recordings where each of the tracks was recorded at a different time, with the complete performance only being heard when all four tracks were replayed simultaneously. This allowed the production of complete pieces of complex music using just one performer. Eight-track recorders followed, then 16-track machines, then 24, and additional tracks can be added by synchronising two or more machines together to produce 48- and even 96-track tape recorders.

1.4.2   Found sounds

Found sounds are ones which are not pre-prepared. They are literally recorded as they are 'found' *in situ*. Trains, cars, animals, factories and many other locations can be used as sources of found sounds.

The term *prepared sounds* is used for sounds which are specially set up, initiated and then recorded, rather than spontaneously occurring.

1.4.3   Collages

Just as with paper collages, multiple sounds can be combined to produce a composite sound. Loops can be very useful in providing a rhythmic basis, whilst found sounds, transposed sounds and reversed sounds can be used to add additional timbres and interest.

1.4.4   Musique concrète

Musique concrète is a French word which has come to be used as a description of music produced from ordinary sounds which are modified using the tape techniques described above. Pierre Schaeffer coined the term in 1948 as music made 'from … existing sonic fragments'.

1.4.5   Optical methods

Although magnetic tape provides a versatile method of recording and reproducing sounds using a physical medium, it is not the only way. The optical technique used for the soundtrack on film projectors has also been used. The sound is produced by controlling the amount of light which passes through the film to a detector. Conventional film uses a 'slot', which varies in width, although it is also possible to vary the transparency or opacity of the film. Optical systems suffer from problems of dynamic range, and physical degradation due to scratches, dust and other foreign objects.

Chapters 2 and 4 deal with optical techniques in more depth.

1.4.6   Disc manipulation

Before the wax cylinder, shellac disc, vinyl record or the tape recorder, all music happened live. Whilst the tape recorder was the obvious tool for manipulating music, it was not the only one. Records with more than one lead-in groove, leading to several different versions of the same audio being selected at random, have been used for diversions such as horse racing games, and in the 1970s, a 'Monty

Python' LP was deliberately crafted with a looped section which played: 'Sorry, squire: I've scratched your record!' continuously until the 'stylus' was lifted from the record.

Disc manipulation is often overlooked because vinyl discs are often perceived as playback-only devices. It is quite possible to produce many of the effects of tape using a turntable or disc-cutting lathe. For example:

- Large pitch changes can be produced by using turntables with large speed ranges.
- Some pickups can be used in reverse play to reverse the playback of sounds.
- Multiple pickups can be arranged on a disc so that echo effects can be produced.
- 'Scratching' involves using a turntable with a slip-mat under the disc, a bidirectional pickup cartridge, and considerable improvisational and cueing skill from the operator, who controls the playing of fragments of music from standard (or custom-cut) LP discs by playing them forwards and backwards, repeating phrases and mixing between two (or more) discs at once.

The live user manipulation of vinyl discs has become so successful that interfaces which attempt to emulate the same twin disc format have been produced for use with CDs. Software emulations of the technique are also available on a number of computer platforms.

1.4.7 Digital tape recorders

The basic tape recorder is an analogue device: the audio signal is converted directly into the magnetic field and stored on the tape. As with many analogue devices, the last 20 years of the 20th century have seen a gradual replacement of analogue techniques with digital technology, and this has also happened with the tape recorder. In the case of the tape recorder, much of the tape handling remains the same, although open reels of tape have typically been replaced with enclosed designs of cassettes, especially in a domestic context. In terms of tape manipulation techniques, the cassette makes access to the tape difficult, although for the ordinary home user of cassettes, it definitely makes tape recording more convenient.

The method of recording audio in a digital tape recorder involves converting the audio signal into a digital representation for subsequent storage on the tape. The data formats used to store the information on the tape are normally not designed to be edited physically, although there have been some attempts to produce formats which can be cut and spliced conventionally. In general, editing is done digitally rather than physically in a digital tape recorder, and so their creative uses are limited to storing the output of a performance or a session, rather than being a mechanism for manipulating sound.

One of the early formats for digital audio tape recording, the Sony F1 system, used Betamax video tape cassettes as the storage medium, and although intended for domestic and semi-pro usage, it was rapidly adopted by the professional music business in the 1980s. It was succeeded by DAT, which was not successful as a domestic format, where the compact cassette and latterly, the MiniDisc and CD-recordable (CD-R) optical disks have dominated, but DAT was widely accepted by the professional music industry.

The cost of hard disk capacity (and many other types of computer memory) appears to follow a permanently descending curve, and has rapid access time. Tape-based or optical storage is cheaper per byte, but has slower access time.

Hard disk recording replaces the tape storage with a hard disk drive, although these are normally backed up to a tape backup device or an optical drive like one of the variations on recordable digital versatile disk (DVD) technology.

1.5 Experimental versus popular musical uses of synthesis

There is a broad spectrum of possible applications for synthesis. At one extreme is the experimental research into the nature of sound, timbre and synthesis itself, whilst at the opposite extreme is the use of synthesizers in making popular music. In between these two, there is huge scope for using synthesis as a useful and creative tool.

1.5.1 Research

Research into music, sound and acoustics is a huge field. Ongoing research is being carried out into a wide range of topics. Just some of these include:

- alternative scalings
- alternative timbres
- processing of sounds
- rhythm, beats, timbre, scales, etc.
- understanding of how instruments work.

Much of this work involves multi-disciplinary research. For example, trying to work out how instruments work can require knowledge of physics, music, acoustics, electronics, computing and more. Some of the results of this research work can find application in commercial products: Yamaha's DX series of FM synthesizers and modelling-based software synthesis are just two of the many examples of the conversion of academic theory into practical reality. This is covered in more detail in Section 1.7 of this chapter.

1.5.2 Music

Music encompasses a huge variety of styles, sounds, rhythms and techniques. Some of the types of music in which a strong synthesised content may be found are:

- **Pop music**: Popular music has some marked preferences: it frequently uses a 4/4 time signature, and preferentially uses a strongly clichéd set of timbres and song forms (especially verse/chorus structures, and key changes to mark the end of a song). It often has a strong rhythmic element, which reflects one of its purposes: music to dance to. Pop music is also designed to be sung, with the vocal or instrumental hook often being a key part of the production effort.

- **Dance music**: Dance music has one purpose: music to dance to. Simplicity and repetition are therefore key elements. There are a large and evolving number of variants to describe the specific sub-genre: acid, melodic trance; house; drum and bass; jungle; garage; but the basic formula is one of a continuous 4/4 time signature with a solid bass and rhythm. Much dance music is remixed versions of pop and other types of music, or even remixed dance music.

- **New Age music**: New Age music mixes both natural and synthetic instruments into a form which concentrates on slower tempos than most popular music, and is more concerned with atmosphere.

- **Classical music**: Although much of classical music uses a standard palette of timbres, which can be readily produced by an orchestra, the augmentation by synthesizers is not unknown in some genres (particularly music intended for film, television and other media purposes).

- *Musique concrète*: Although *musique concrète* uses natural sounds as the source of its raw material, the techniques that it uses to modify those sounds are often the same as those used by synthesizers.

- **Electronic music**: Electronic music need not be produced by synthesizers, although this is often assumed to be the case. As with popular music, a number of clichés are commonly found: the 8- or 16-beat sequence and the resonant filter sweep are two examples.

Crossovers

There are some occasions when the boundaries between experimental uses of synthesizers crossover into more popular music areas, and vice versa. The use of synthesizers in orchestras typically occurs when conventional instrumentation is not suitable, or when a specific rare instrument cannot be hired. Music which is produced for use in many areas often requires to have elements of orchestral and non-orchestral instrumentation – adding synthesizer parts can enhance and extend the timbres available to the composer or arranger, and it avoids any need for the synthesizer to attempt to emulate a real orchestra. The use of orchestral scores for both movies and video games has produced a mass-market outlet for orchestration which is often augmented with synthesised instrumentation. Conversely, the use of orchestral instruments in experimental works also happens.

1.6 Electro-acoustic music

The study of the conversions between electrical energy and acoustic energy is called electro-acoustics. Unlike previous centuries, where the development of mechanical-based musical instruments has dominated the study of musical acoustics, in the 20th century, innovation has been concentrated on instruments that are electronic in nature. It is thus logical that the term electro-acoustics should also be used by musicians to describe music which is made using electronic musical instruments and other electronic techniques.

Unfortunately, the term 'electro-acoustic music' is not always used consistently, and it can also apply to music where acoustic instruments are amplified electronically. The term 'electronic music' implies a completely electronic method of generating the sound, and so represents a very different way of making music. In practice, both terms are now widely used to mean music which utilises electronics as an integral part of the creative process, and so it covers such diverse areas as amplified acoustic instruments (where the instruments are not merely made louder), music created by synthesizers and computers, and popular music from a wide range of genres (pop, dance, techno, etc.). Even classical music performed by an orchestra, but with additional electronic instrumentation, or even post-processing of the recorded orchestral sound, could be considered to be 'electronic'.

1.6.1   Electro-acoustics

Electro-acoustics is science tempered by human interaction and art. In fact, the close linking between the human being and most musical instruments, as well as the space in which they operate, can be a very emotional one. The electronic nature of many synthesizers does not fundamentally alter this relationship between human and instrument, although the details of the interface are still very clumsy. As synthesizers develop, they should gradually become performer-oriented, and less technological, which should make their electro-acoustic nature less and less important. Many conventional instruments have histories of many hundreds of years, whereas electro-acoustic music is less than a century old, and synthesizers are less than 50 years old. Electro-acoustics is comparatively new.

1.7   From academic research to commercial production ...

Synthesizers can be thought of as coming in two forms: academic and commercial. *Academic research* produces prototypes which are typically innovative, fragile and relatively extravagant in their use of resources. *Commercial synthesizers* are often cynically viewed as being almost the exact opposite: minor variations on existing technology; robust, perhaps even over-engineered; and very careful to maximise the use of available resources. Production development of research prototypes is often required to enable successful exploitation in the marketplace, although this is a difficult and exacting process, and there have been both successes and failures.

In order to be a success, there are a number of criteria that need to be met. Moving from a prototype to a product can require a complex exchange of information from the inventor to the manufacturer, and may often need additional development work. Custom chips or software may need to be produced, and this can introduce long delays into the timescales, as well as a difficult testing requirement.

Management tasks like organising contracts, temporary secondment of personnel, patents and licensing issues, all need to be monitored and controlled.

Robert Moog and Bob Oberheim have both used their names for companies and products, and then left that company ...

Even when the product has been produced, it needs to be promoted and marketed. This requires a different set of skills, and in fact, many successful companies split their operations into 'research and development' and a 'sales and marketing' parts. The synthesizer business has seen many companies with ability in one of these fields, but the failures have often been a result of a weakness in the other field. Success depends on talent in both areas, and the interchange of information between them. Very few of the companies who started out in the 1960s are still active; however, the creative driving force behind these companies, which is frequently only one person, is often still working in the field, albeit sometimes in a different company.

Apart from the development issues, the other main difference between academic research and commercial synthesizer products is the motivation behind them. Academic research is aimed at exploring and expanding of knowledge, whilst commercial manufacturers are more concerned with selling products. Unfortunately, this often means that products need to have wide appeal, simple user interfaces, and easy application in the popular music industry. The main end market for electronic musical instruments is where they are used to make the music that is heard on television, radio, films, DVDs and CDs, and the development process is aimed at just this area.

What follows are some brief notes on some of these 'developed' products.

1.7.1 Analogue modular

Analogue synthesizers were initially modular, and were probably aimed at academic and educational users. The market for 'popular' music users literally did not exist at the time. The design and approach used by early modular synthesizers was similar to those of analogue computers. Analogue computers were used in academic, military and commercial research institutions for much the same types of calculations that are now carried out by digital computers.

Pioneering work by electronic music composers using early modular synthesizers was more or less ignored by the media until Walter Carlos released some recordings of classical Bach produced using a Moog modular synthesizer. The subsequent release of this material as the 'Switched On Bach' album quickly became a major success with the public, and the album became one of the best selling classical music records ever. This success led to enquiries from the popular music business, with the Beatles and the Rolling Stones being early purchasers of Moog modular synthesizers.

By the beginning of the 21st century, software synthesis allowed the creation of emulated analogue modular synthesizers (and other electronic music instruments) on general-purpose computers. The comparatively low cost of the software means that musicians who could never afford a real modular synthesizer are able to explore sound-making, whilst the use of software means that patches can easily be stored and recalled: a huge advantage over analogue modular synthesizers.

1.7.2 FM

FM as a means of producing audio sounds was first comprehensively described by John Chowning, in a paper entitled: 'The synthesis of complex audio spectra by mean of frequency modulation', published in 1973. At the time, the only way that this type of FM could be realised was by using digital computers, which were expensive and not widely available to the general public.

As digital technology advanced, some synthesizer manufacturers began to look into ways of producing sounds digitally, and Yamaha bought the rights to use Chowning's 1977-patented FM ideas. Early prototypes used large numbers of simple transistor–transistor logic (TTL) chips, but these were quickly replaced by custom-designed chips which compressed these onto just a few more complex chips. The first functional all-digital FM synthesizer designed for consumer use was the Yamaha GS1, which was a pathfinder product designed to show expertise and competence, as well as test the market. Simple preset machines designed for the home market followed. Although the implementation of FM was very simple, the response from musicians and players was very favourable.

Perhaps as a consequence of the DX7, other synthesizers in the 1980s (and beyond) tended to concentrate on a similar price point and a two-model strategy: the basic 'mass-market' model, and a more expensive 'pro' model, often with more notes on the keyboard (or a weighted keyboard), and sometimes with extended functionality.

The DX1, DX7 and DX9 were released in late 1982, with the DX1 apparently intended as the professional player's instrument, the DX7 a mid-range, cut-down DX1, and the DX9 as the low-cost, large-volume 'best seller'. What actually happened is very interesting. The DX9 was so restricted in terms of functionality and sound that it did not sell at all well, whilst the DX7 was hugely in demand amongst both professional and semi-professional musicians, whilst the DX1 was interpreted as being a 'super' DX7 for a huge increase in price. Inevitably, it took Yamaha some time to increase the production of the DX7 to meet the demand, and this scarcity only served to make it all the more sought-after! By the time that the mark II DX7 was released, about a quarter of a million DX7s had been sold, which at the time was a record for a synthesizer.

The popularity of the DX7 was responsible for the mark II instrument, which was a major redesign, not a new instrument, a very rare approach, and one which shows how important the DX7, and FM, had become. For several years, between 1983 and 1986, Yamaha and FM enjoyed a popularity which ushered in the transition from analogue to digital technology. It also began the trend away from user programming, and towards the selling of pre-prepared sounds or patches. The complexity of programming FM meant that many users did not want to learn, and so purchased sounds from specialist companies which marketed the results of a small number of 'expert' FM programmers.

In the late 1990s, Yamaha released a new FM synthesizer module, the FS1R, which extended and enhanced the FM synthesis technique of the previous generation, and this was accompanied by a resurgence of interest in FM as a 'retro' method of making sounds. Fashion in synthesis is cyclic. In 2001, a software synthesizer version of the original DX series FM synthesizers was released, and DX-style FM joined the sonic palette of commercial software synthesis.

1.7.3 Sampling

Sampling is a musical reuse of technology which was originally developed for telephony applications. The principles behind the technique were worked out early in the 20th century, but it was not until the invention of the transistor in the 1950s that it became practical to convert continuous audio signals into discrete digital samples using PCM.

Commercial exploitation of sampling began with the Fairlight Computer Musical Instrument (CMI) in 1979, although this began as a wavetable synthesizer, as the size of the wavetables increased, it rapidly evolved into an expensive and fashionable professional sampling instrument, initially with only 8-bit sample resolution. Another 8-bit instrument, the Ensoniq Mirage, was the first instrument to make sampling affordable. E-mu released the Emulator in 1979, drum machines like the LinnDrum were released in 1979, and sampling even began to appear on low-cost 'fun' keyboards designed for consumer use in the home, during the mid-1980s.

The 8-bit resolution was replaced by 12 bits in the late 1980s, and 16 bits became widely adopted in the early 1990s. Before the end of the 20th century the CD standards of 16-bit resolution and 44.1 kHz had become widely adopted for samplers, with lower sampling rates only being used because of memory constraints. The 21st century has seen a wide adoption of software sample playback as an alternative to hardware: either as plug-ins to software MIDI and audio sequencers, or as stand-alone 'sample' sequencers.

The wide availability of pre-prepared sounds for samplers is analogous to the patches which are available for software-based analogue modular synthesizers (see Section 1.7.1). In both cases, most users make only minor changes to the sounds which have been created by a few highly skilled individuals.

CD-read-only memories (CD-ROMs) of pre-prepared samples have replaced *do it yourself* (DIY) sampling for the vast majority of users. Samplers have mostly become replay-only devices, with only a few creative individuals and companies producing samples on CD-ROMs, and many musicians using them. The use of music CDs as source material has become formalised, with royalty payments on this usage being 'business as usual' for many often-used artists of previous generations.

1.7.4 Modelling

Mathematical techniques like physical modelling seem to have made the transition from research to product in a number of parallel paths. There have been several speech coding schemes based on modelling the way that the human voice works, but these have been restricted to mainly telecommunications and military applications; only a few of these have found musical uses (see Chapter 5). Research results which have been reporting the gradual refinement of modelling techniques for musical instruments have been released, notably by Julius Smith, and commercial devices based on these began to appear in the mid-1990s. Yamaha's VL1 was the first major commercial synthesizer to use physical modelling based on blown or bowed tubes and strings, and many other manufacturers have followed, including many emulations of analogue synthesizers.

The development of electronic musical instruments is still continuing. The role of research is as strong as ever, although the pace of development is accelerating. Digital technology is driving synthesis towards general-purpose computing engines with customised audio output chips, and this means that the software is increasingly responsible for the operation and facilities that are offered, not the hardware. By the turn of the century, many companies had products which used general-purpose DSPs to synthesise their sounds, and had produced products which used mixtures of synthesis technologies to produce those sounds: FM, additive, emulations of analogue synthesis, physical modelling and more. But despite the flexibility and power of these systems, the popular choice has continued to be a combination of sample replay and synthesis; and perhaps the simplicity and familiarity of the metaphor used is a key part of this.

The 21st century has seen modelling technology become a standard tool to produce digital versions of both analogue electronic and natural instruments. Fast powerful processors have also greatly reduced dependence of the DSP as the processing engine, and opened up the desktop or laptop computer as a means of synthesising using modelling techniques. But the complex metaphor and interface demands of physical modelling and other advanced techniques have seen them pushed into niche roles, with only the analogue emulations enjoying wide commercial success.

1.8 Synthesised classics

One of the major forces, which popularised the use of synthesizers in popular music, was the use of synthesizers to produce recorded performances of classical music. Because these could be assembled onto tape with great precision, the timing control and pitch accuracy, which were used, were on a par with the best of human realisations, and so the results could be described as 'virtuoso' performances. In the 1950s, this suited the mood of the time, and so a large number of electronically produced versions of popular classical music were produced. This has continued to the present day, although it has become increasingly rare and uncommercial; perhaps the wide range of musical genres which co-exist in the 2000s means that it is no longer seen as relevant. Some of the relevant recordings in the author's collection are listed in Table 1.8.1.

There is no 'correct' way to use synthesizers to create music, although there are a number of distinct 'styles'. Individual synthesists have their own preferences, although some have less fixed boundaries than others, and can move from one style to another within a piece of music. I do not know of any formal means of categorising such styles, and so propose the following divisions:

- **Imitative**: Imitative synthesis attempts to use electronic means to realise a performance which is as close as possible to a recording of a conventional orchestra, band or group of musicians. The timbres and control techniques, which are used, are intended to mimic the real-world sounds and limitations of the instrumentation. Many film soundtracks fall into this category.
- **Suggestive**: This style does not necessarily use imitative instrumental sounds, but rather, aims to produce an overall end result, which is still suggestive of a conventional performance.
- **Sympathetic**: Although using instrumental sounds and timbres which may be well removed from those used in a normal performance, a 'sympathetic' realisation of a piece of music aims to choose sounds which are in keeping with some element of the conventional performance.
- **Synthetic**: Electronic music aims to free the performer from the constraints of conventional instrumentation, and so this category includes music where there is little that would be familiar in tone or rhythm to a casual listener.

1.8.1 Recordings

Many of the recordings in Table 1.8.1 have been 'deleted', and are thus unavailable for purchase. It is significant that the 'sounds' and synthesis techniques used in these recordings can still have an ongoing value, a longevity which can exceed that of the commercial recordings themselves.

There are a large number of recordings of popular classical music available which have been performed using electronic music equipment. These vary widely in style, sound quality, competence and 'collectability'. The examples listed in Table 1.8.1 from the author's collection have been chosen for their breadth of subjects and treatments, and include some non-classical 'electronic music'.

Table 1.8.1   Examples of synthesised music

Artist(s)	Title	Record company	Date	Catalogue number
Carlos, Walter	Switched on Bach	CBS	1973	MK 7194
Carlos, Wendy	Switched on Bach 2000	Telarc Records	1992	CD 80323
Ciani, Suzanne	Seven Waves	Private Music	1988	2046-2-P
Julian Colbeck and Jonathan Cohen	Back to Bach	Editions EG Records	1992	EEG 2104-2
Electrophon (Brian Hodgson and Dudley Simpson)	In a Covent Garden	Polydor Records	1973	2383 210
Team Metlay	Band of Fire	Atomic City	1991	ATOM CD-07
The Walter Murphy Band	A Fifth of Beethoven	Private Stock Records	1976	PVLP 1009
William Orbit	Pieces in a Modern Style	Warner Bros. Records	2000	9 47596-2
Xero Reynolds	D7Peacemaker:1 OST	Dementia 7 Studios llc	2002	:040102
Karlheinz Stockhausen	Kontakte	Wergo Records	1960	WER 60009
Isao Tomita	Snowflakes are Dancing	RCA	1974	ARL1-0488
Kit Watkins	A Different View	IC Records	1991	IC 720.142
Wuorinen, Charles	Time's Encomium	Nonesuch Records	1969	H-71225

1.9 Synthesis in context

Sound synthesis does not exist in isolation. But it is one method of producing sound and music. There are a large number of non-synthetic, non-electronic methods of producing musical sounds.

All musical instruments synthesise sounds, although most people would probably use the word 'make' rather than 'synthesise' in this context. In fact, the word synthesise has come to mean something which is un-natural; synthetic implies something which is similar to, but inferior to 'the real thing'. The ultimate example of this view is the sound synthesizer, which is often described as being capable of emulating any type of sound, but with the proviso that the emulation is usually not perfect.

1.9.1 'Synthetic' versus 'real'

As with anything new and/or different, there is a certain amount of prejudice against the use of electronic musical instruments in some contexts. This is often expressed in words like: 'What is wrong with *real* instruments?', and is frequently used to advocate the use of orchestral instruments rather than an electronic realisation using synthesizers.

There are two major elements to this prejudice: unfamiliarity and fear of technology. Many people are very much used to the sounds and timbres of conventional instrumentation, especially the orchestra. In contrast, the wider palette of synthetic sound is probably very unfamiliar to many casual listeners. So an unsympathetic rendering of a pseudo-classical piece of music, produced using sounds which are harsh, unsubtle, and obviously synthetic in origin, is almost certain to elicit an unfavourable response in many listeners. In contrast, careful use of synthesis can result in musical performances, which are acceptable even to a critical listener, especially if no clues are given as to the synthetic origins.

The technological aspect is more complex. Although the piano-forté was once considered too new and innovative to be considered for serious musical uses, it has now become accepted by familiarity. The concept of assembling together a large number of musicians into an orchestra was a more gradual process, but the same transition from 'new' to 'accepted' still occurred. It seems that there may be an inbuilt 'fear' of anything new or which is not understood. This extends far beyond the synthesizer: computers and most other technological inventions can suffer from the same aversion.

Attempting to draw the line between what is acceptable technologically and what is not can be very difficult. It also changes with time. Arguably, the only 'natural' musical instrument is the human voice, and anything which produces sounds by any other method is inherently 'synthetic'. This includes all musical instruments from simple tubes through to complex computer-based synthesizers. There seems to be a gradual acceptance of technological innovation over time, which results in the current wide acceptability of musical instruments which may well have been invented more than 500 years ago.

1.9.2 Film scoring

Film scoring is an excellent example of the way that sound synthesis has become integrated into conventional music production. The soundtracks of many films are a complex mixture of conventional orchestration combined with synthesis, but a large number of films have soundtracks which have been produced entirely electronically, with no actual orchestral content at all, although the result sounds like a performance by real performers. In some cases, although the music may sound 'realistic' to a casual listener, the performance techniques may be well beyond the capability of human performers, and some of the timbres used can be outside of the repertoire of an orchestra.

There are advantages and disadvantages to the 'all-synthetic' approach. The performer who creates the music synthetically has complete control over all aspects of the final music, which means that changes to the score can be made very rapidly, and this flexibility suits the restraints and demands which can result from film production schedules. But giving the music a human 'feel' can be more difficult, and arguably more time-consuming, than asking an orchestra to interpret the music in a slightly different way. The electronic equivalent of the conductor is still some way in the future, although there is considerable academic research into this aspect of controlling music synthesizers. One notable example which illustrates some of the possibilities are the violin bows which have been fitted with accelerometers by Todd Machover's team at the MIT Media Lab in Boston, USA, which can measure movement in three dimensions and which are not dissimilar to the sort of measurements which would be required for a conductor's baton.

Mixing conventional instrumentation with synthesizers is also used for a great deal of recorded music. This has the advantage that the orchestral instruments can be used to provide a basic sound, and additional timbres can be added into this to add atmosphere or evoke a specific feel. Many of the sounds used in this context are clichés of the particular time when the music was recorded. For example, soundtracks from the late 1970s often have a characteristic 'drum synthesizer' sound with a marked pitch sweep downwards – the height of fashion at the time, but quaint and hackneyed to people 10 or 20 years afterwards. But recycling of sounds (and tunes) does occur, and 'retro' fashionability is always ready to rediscover and reuse yesterday's clichés.

1.9.3 Sound effects

The real world is noisy, but the noises are often unwanted. For film and television work, background noise, wind noise and other extraneous sounds often mean that it is impossible to record the actual sound whilst recording the pictures, and so sound effects need to be added later. Everyday sounds like doors opening, shoes crunching on gravel paths, switches being turned on or off, cans of carbonated drink being opened and more are often required. Producing these sounds can be a complex and difficult process, especially since many sounds are very difficult to produce convincingly.

Years of exposure to film and television have produced a set of cliːchéd sounds which are often very different from reality. For example, does a real computer produce the typical busy whirring and bleeping sounds that are often used for anything in the context of computers or electronics? Sliding doors on spaceships always seem to open and close with a whoosh of air, which would seem to suggest a serious design fault. The guns in Western films suffer from a large number of ricochets and fight scenes often contain the noises of large numbers of bones being broken, whilst the combatants seem relatively uninjured.

Many of the sounds that you hear on film or television are dubbed on afterwards. Some of these are 'synthesised' live by humans using props on a 'Foley' stage, although often the prop is not what you might expect: rain can be emulated by dropping rice onto a piece of cardboard, for example. But many sounds are produced synthetically, especially when the real-world sound does not match expectations. One example is the noise made when a piece of electronic equipment fails catastrophically: often nothing is heard apart from a slight clunk or lack of hum, which is completely unsuitable for dramatic use. A loud and spectacular sound is needed to accompany the unrealistic shower of sparks and smoke which billow from the equipment.

This use of sounds to enhance the real world can also be used to extremes, especially in comedy. A commonly used set of 'cod' or comic sounds has become as much a part of the film or television medium as the 'fade to black'. Laboratory equipment blips and bloops, and elastic bands twang in an unrealistic but amusing way – the exaggeration is the key to making the sound effect funny. Many of these sounds are produced using synthesizers, or a combination of prop and subsequent processing in a synthesizer. Samplers are often used in order to reproduce these sound effects 'on cue', and the user interface to these can vary from a music keyboard to the large wooden pads connected to drum sensors that are used to add the fighting noises to Hong Kong 'Kung Fu' movies.

Given this mix of cliché, artificial recreation and exaggeration, it is not surprising that there is a wide variety of pre-prepared sound effect material in the form of sound effects libraries. As with all such 'canned' sample materials, the key to using it effectively is to become a 'chef' and to 'synthesise' something original using the library contents as raw materials. Discovering that the key 'weapon firing' sound is the same as the characteristic noise made by the lead robot in another television programme can have a serious effect on reputations of all concerned.

As the DVD (where video is often mistakenly taken to be the middle 'V' of the acronym) became the fastest-selling consumer item ever early in the 21st century, so surround sound has become more widely used, initially in films, but musicians are always keen to exploit any new technology. In film and television usage, the front channels are typically used for the speech and the music, with the rear channels being used for sound effects, atmosphere and special effects. In music, several systems which used variations of dummy-head recordings were experimented within the 1970s, 1980s and 1990s, and some commercial recordings were made using them and released on stereo CDs. But although these systems can enhance the stereo image by adding sounds which appear to be behind the listener, their limitations did mean that they tended to be used for special effects or for background ambience: a vocal performance that moved around your head, or raindrops surrounding you ... With no clear single contender for replacing stereo music on CD with a surround-based medium, exploiting the possibilities of surround music is not straightforward. Manufacturers are, of course, keen to see the replacement of recording equipment with surround-oriented new purchases.

1.10 Electronics and acoustics: fundamental principles

A knowledge of the electronic aspects of acoustics can be very useful when working with synthesizers, because synthesizers are just one of many tools which can be used to assist in the creation of music. So this section provides some background information on acoustics and electronics. Because some of the terminologies used in this section uses scientific unit symbols, some additional information on the use of units is also provided.

1.10.1 Acoustics

Acoustics is the science of sound. Sound is concerned with what happens when something vibrates. The vibration can be produced by vibrating vocal cords, wind whistling through a hole, a guitar string being plucked, a gong being struck, a loudspeaker being driven back and forth by an amplified signal, and more. Although most people think of sound as being carried only through the air, sound can also be transmitted through water, metals, wood, plastics and many other materials.

Whilst it is often easy to observe an object vibrating, sound waves are less tangible. The vibrations pass through the air, but the actual process of pressure changes is hard to visualise. One effective analogy is the stretched spring, which is vibrated at one end – the actual compression and rarefactions (the opposite of compression) of the 'waves' can then be seen travelling along the spring (Figure 1.10.1). Trying to amalgamate this idea of pressure waves on springs with the ripples spreading out on a pond is more difficult. As with light, the idea of spreading out from a source is hard to reconcile with what happens – people see

Figure 1.10.1 Pressure changes in the air can be thought of as being similar to a stretched spring which is vibrated at one end. The resulting pressure 'waves' can be seen travelling along the spring.

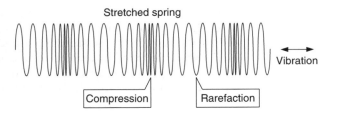

Stretched spring

Vibration

Compression

Rarefaction

and hear things, and waves and beams seem like very abstract notions. In real life, the only way to interact with sound waves is with your ears, or for very low frequencies, your body.

When an object vibrates, it moves between two limits; a vibrating string provides a good example, where the eye tends to see the limits (where the string is momentarily stopped whilst it changes direction) rather than where it is moving. This movement is coupled to the air (or another transmission medium) as pressure changes. The rate at which these pressure changes happen is called the *frequency*. The number of cycles of pressure change, which happen in 1 second is measured in a unit called hertz (cycles per second is an alternative unit). The time for one complete cycle of pressure change is called the *period*, and is measured in seconds.

Pitch and frequency

Frequency is also related to musical pitch. In many cases, the two are synonymous, but there are some circumstances in which they are different. Frequency can be measured, whilst pitch can sometimes be subjective to the listener. In this book frequency will be used in a technical context, whilst pitch will be used when the subject is musical in nature.

Apparently 'middle C' is so called because it is written in the middle: between the bass and treble staves.

The frequency usually used for the note A just above middle C is 440 Hz. There are local variations on this 'A-440 standard', but most electronic musical equipment can be tuned to compensate. Human hearing starts at about 20 Hz, although this depends on the loudness and listening conditions. Frequencies lower than this are called *subsonic* or, more rarely, *infrasonic*. For high frequencies human hearing varies with age and other physiological effects (like damage caused by over-exposure to very loud sounds or ear infections). For a normal teenager, frequencies of up to 18 kHz (18 000 Hz) can be heard; the 15 625-Hz line whistle from a 625-line Phase Alternation Line (PAL) television is one useful indicator. The aging process means that an average middle-aged person will probably only be able to hear frequencies of perhaps 12 or 13 kHz. The 'hi-fi' range of 20 Hz to 20 kHz is thus well in excess of most listeners ability to hear, although there is some debate about the ability of the ear to discern higher frequencies in the presence of other sounds (most hearing tests are made with isolated tones in quiet conditions).

Notes

The fundamentals of most musical notes are in the lower part of this range. The fundamental is the name given to the lowest major frequency which is present in a sound. The fundamental is the pitch, which most people would whistle when attempting to reproduce a given note. Harmonics, overtones or partials are the names for any additional frequencies which are present in a sound. Harmonics are those frequencies which are integer multiples of the fundamental – they form a series called the *harmonic series*. Overtones or partials are not related to the fundamental frequency. The upper part of the human frequency range contains these additional harmonics and partials. Table 1.10.1 shows the fundamental frequencies of the musical notes.

Table 1.10.1 Note frequencies in Hz

C	C#	D	D#	E	F
16.35	17.3239	18.354	19.4454	20.6017	21.8268
32.703196	34.6478	36.7081	38.8909	41.2034	43.6535
65.406391	69.2957	73.4162	77.7817	82.4069	87.3071
130.81278	138.591	146.832	155.563	164.814	174.614
261.62557	277.183	293.665	311.127	329.628	349.228
523.25113	554.365	587.33	622.254	659.255	698.456
1046.5023	1108.73	1174.66	1244.51	1318.51	1396.91
2093.0045	2217.46	2349.32	2489.02	2637.02	2793.83
4186.009	4434.92	4698.64	4978.03	5274.04	5587.65
8372.0181	8869.84	9397.27	9956.06	10548.1	11175.3
16744.036	17739.7	18794.5	19912.1	21096.2	22350.6

An 88-note piano keyboard will span A0 to C8, with the top C having a frequency of just over 4 kHz.

The study of musical scales is a complex subject. For further information see (Pierce, 1992).

Musical pitch is divided into octaves, and each octave represents a doubling of frequency. So A4 has a frequency of 440 Hz, whilst A5 has twice this frequency: 880 Hz. A3 has half the frequency: 220 Hz. Octaves are normally split into 12 parts, and the intervals are called *semitones*. The relationship between the individual semitones in an octave is called the *scale*. The table shows the equal tempered scale, where the intervals between the semitones are all the same: many other scalings are possible. Since there are 12 tones, and the frequency doubles in an octave interval, then semitone intervals in an equal tempered scale are each related by the 12th root of two, which is approximately 1.059463. Semitones are split up into 100 cents, but most human beings can only detect changes in pitch of 5 cents or more. Cent intervals are related by the 1200th root of 2, which is approximately 1.00057779. As an example of what this represents in terms of frequency: for a A5 note of 880 Hz, a cent is just under 0.51 Hz, and so 5 cents represent only 2.5 Hz!

Phase

When an object is vibrating, it repeatedly passes through the same position as it moves. A complete movement back and forth is called a *cycle* or an *oscillation*, which is why anything that produces a continuous vibration is called an *oscillator*. The particular point in a cycle at any instant is called the phase: the cycle is divided up into 360°, rather like a circle in geometry. Phase is thus measured in degrees, and zero is normally associated with either the start of the cycle, or where it crosses the resting position. The word 'zero crossing' is used to indicate when the position of the object passes through the rest position. A complete cycle conventionally starts at a zero crossing, passes through a second one, and then ends at the third zero crossing (Figure 1.10.2).

The change of position with time as an object vibrates is called the *waveform*. A simple oscillation will produce a sine wave, which looks

F#	G	G#	A	A#	B	Octave
23.1247	24.4997	25.9565	27.5	29.1352	30.8677	0
46.2493	48.9994	51.9131	55	58.2705	61.7354	1
92.4986	97.9989	103.826	110	116.541	123.471	2
184.997	195.998	207.652	220	233.082	246.942	3
369.994	391.995	415.305	440	466.164	493.883	4
739.989	783.991	830.609	880	932.328	987.767	5
1479.98	1567.98	1661.22	1760	1864.66	1975.53	6
2959.96	3135.96	3322.44	3520	3729.31	3951.07	7
5919.91	6271.93	6644.88	7040	7458.62	7902.13	8
11839.8	12543.9	13289.8	14080	14917.2	15804.3	9
23679.6	25087.7	26579.5	28160	29834.5	31608.5	10

Figure 1.10.2 A complete cycle crosses the zero axis three times.

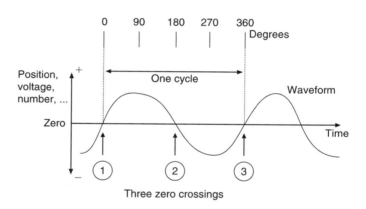

Three zero crossings

like a smooth curve. More complex vibrations will produce more complex waveforms – a guitar string has a complex waveform because it produces a number of harmonics at once. If two identical waveforms are mixed together, the phase can determine what happens to the resulting waveform. If the two are in phase, that is, they both have the same position in the cycle at the same instant, then they will be added together; this is called 'constructive interference', because the two waveforms add together as they 'interfere' with each other. Conversely, if the two waveforms are 180° out of phase, then the phases will be equal and opposite, and the two waveforms will tend to cancel each other out; this is called 'destructive interference'.

Beats

Slight differences of frequency between two waveforms can produce a different effect. Assuming that the two waveforms start at the same zero crossing, and with the same phase, then the waveform with the higher frequency will gradually move ahead of the slower waveform, and its phase will be ahead. This means that from an initial state of constructive interference, the waveforms will pass through destructive interference and then back to constructive interference repeatedly. The rate of passing through these adding and cancellation stages is determined by the difference in frequency. For a difference of one-tenth of a hertz, it will take 10 seconds for the cycle of constructive, destructive and constructive interference to occur. This cyclic variation in level of the mixed waveforms is called 'beating', and sounds like one sound which 'wobbles' in level. This beating is often used in analogue synthesizers to provide a 'lively' or 'interesting' sound.

If the difference in frequency between the two waveforms is increased, then the speed of the beats will increase. When the frequency of the beats is above 20 Hz, then the mixed sound begins to sound like two separate frequencies. As the difference increases, the two frequencies will pass through a series of ratios of frequency, which sound pleasant, and others which sound unpleasant. The ratios between the two frequencies is called an *interval*; the easiest and most 'pleasant' interval is a ratio of 2 : 1, an octave.

Timbre

Timbre is a description of the contents of a sound. The timbre of a sound is determined by the harmonic content: the relationship between the level of the fundamental, the levels of the harmonics or overtones, and their evolution in time (their envelopes, see below). Pure sounds tend to have only a few harmonics at low levels, whilst bright sounds tend to have many harmonics at high levels. Missing harmonics can also be important, and can produce 'hollow' sounding timbres. If the ratios of the frequencies between the fundamental and the other frequencies are not integers, then the timbre can sound bell-like or even like noise. The ability of the human ear to perceive timbre is related to the frequency. At low frequencies, the ear can detect phase differences and can follow changes in a large number of harmonics. As the frequency increases, then the phase discrimination ability of the ear diminishes above A4 (440 Hz), and the number of harmonics which can be heard decreases because of the response of the ear.

Timbre (tahm-brer) is derived from a French word. 'Tone colour' and 'tonal quality' are commonly used as synonyms for timbre.

For example, a sound which has a fundamental of 100 Hz has harmonics at 100-Hz intervals, and so the 150th harmonic is at 15 kHz. But a sound with a fundamental of 1 kHz has a 15th harmonic at 15 kHz. The number of audible harmonics are thus restricted as the fundamental frequency rises. Synthesizers provide comprehensive control over the frequency, phase and level of harmonics, and so give the user control of the timbre (Figure 1.10.3).

Loudness

When a string vibrates, the size of the string and the amount of movement determine how much energy is transferred to the surrounding medium (usually air). The larger the amount of energy that is turned into changes in air pressure, the louder the sound will be. This can be demonstrated by using a tuning fork: it becomes much louder when it is placed on a tabletop, because then it is moving a much larger amount of air. The amount of movement of a vibrating object is called

Figure 1.10.3 Timbre is set by the frequency content of a sound. In this example, the fundamental frequency of the sound is at frequency *f*, whilst there is a harmonic at twice this frequency, 2*f*. There is also an overtone or partial frequency at 3.75*f*. (Figure 2.3.7 provides an overview of spectrum plots like this one.)

Table 1.10.2 Decibels

Sound pressure level (dB)	Sound pressure (microbar)	Power (watts per square metre)		Equivalent	Musical dynamic
130	632	10 W	10	Threshold of pain	
120	200	1 W	1	Aircraft taking off	
110	63	100 mW	0.1	Loud amplified music	
100	20	10 mW	0.01	Circular saw	fff
90	6	1 mW	0.001	Train	ff
80	2	100 μW	0.0001	Motorway	f
70	0.6	10 μW	0.00001	Factory workshop	mf/mp
60	0.2	1 μW	0.000001	Street noise	p
50	0.06	100 nW	0.0000001	Noisy office	pp
40	0.02	10 nW	0.00000001	Conversation	ppp
30	0.06	1 nW	0.000000001	Quiet room	
20	0.002	100 pW	0.0000000001	Library	
10	0.006	10 pW	0.00000000001	Leaves rustling	
0	0.0002	1 pW	0.000000000001	Threshold of hearing	

the amplitude of the vibration, whilst the amount of energy in the sound, which is produced by the vibrating object is called the power or intensity of the sound. Power is measured in watts, but a relative logarithmic scale is commonly used to avoid large changes in units: dB or decibels. Named after Alexander Graham Bell, the pioneer of telephony, decibels are used to indicate the relative difference between sound intensities or sound pressure levels (Table 1.10.2). The perception of sound power or level by humans is subjective: a change in sound power of 1 dB is 'just audible', whilst for something to sound 'twice as loud', the change is about 10 dB. The entire scale of sound intensity, from silence to painful, is just 12 doublings of sound power!

Musicians use an alternative relative measure for sound level. The 'dynamics marks' used on musical scores provide guidance about the loudness of a specific note. These range from ppp (pianississimo, softest) to fff (fortississimo, loudest), although this tends to be a subjective measure, and is also dependent on the instrument producing the sound. On average, the range covered by dynamics marks is about 50 or 60 dB, which represents a ratio of about a million to one in sound intensity (Table 1.10.3).

Table 1.10.3   Dynamics

Musical dynamic	Name	Description	dB (approximate)
fff	Fortississimo	Loudest	100
ff	Fortissimo	Very loud	93
f	Forte	Loudly	85
mf	Mezzo-forte	Moderately loud	78
mv	Mezza-voce	Medium tone	70
mp	Mezzo-piano	Moderately soft	62
p	Piano	Softly	55
pp	Pianissimo	Very softly	47
ppp	Pianississimo	Softest	40

'Loudness' is a specific term, which means the subjective intensity of a sound, as opposed to intensity, which can be objectively measured by a sound intensity meter. The human hearing response to different frequencies is not flat: sounds between 3 and 5 kHz will sound louder than lower or higher pitched sounds, and a graph can be plotted showing this response, called an equal loudness contour. This topic is covered by the science of psycho-acoustics, which is the study of the inter-relationship between sound and its perception. Loudness is commonly used (incorrectly from a technical viewpoint) as a synonym for sound intensity.

Since sound is just pressure waves moving through a transmission medium like air, it can be measured in terms of the pressure changes which are caused. The unit for such pressure changes is the bar, although in common with many scientific units, smaller subdivisions like millibars or microbars are more likely to be encountered in normal acoustics measurements. Since sound loudness is dependent on the response of the ear, it is measured in phons, where the phon is based on a subjective measure of the apparent loudness of sounds at different frequencies and intensities.

Envelopes

The human sensory system seems to have a time resolution limit of about 10 ms, and so sounds which appear to start 'instantaneously' typically have attack times of less than about 10 ms.

Sounds do not start and stop instantaneously. It takes a finite time for a string to start vibrating, and time for it to reduce to a stationary state. The time from when an object is initiated into a vibrating state is called the *attack time*, whilst the time for the vibration to decay to a stationary state again is called the *decay time*. For instruments that can produce a continuous sound, like an organ, the decay time is defined as the time for the sound to decay to the steady-state *'sustain'* level, whilst the end of the vibration is called the *release time* (Figure 1.10.4).

Some instruments have long attack, decay and release times: bowed stringed instruments, for example. Plucked stringed instruments have shorter attack times. Some instruments have very fast attack times: pianos, percussion, for example. Very short times are often called *transients*. The combination of all the stages of a sound is called an envelope. It shows the change in volume of the sound plotted against time.

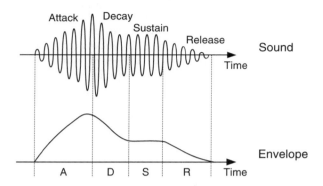

Figure 1.10.4 An envelope is the change in volume with time.

The word envelope can also be used in a more generic sense: it then refers to any complex time function. A typical example might be the envelope of one harmonic within a sound, which you would find in an additive synthesizer (see Chapter 2).

Gain and attenuation

The amplitude of a sound is a measurement of the extremes of its waveform: the most positive and most negative voltages. If the amplitude changes, then the ratio between the original and the changed amplitudes is called the *gain*. Gains can be positive or negative, and can refer to amplitude or power, and are usually measured in dB. Gains of less than one are called *attenuation*, so large attenuations mean that the audio signal can become very small, whilst large gains mean that the signal can become very large.

1.10.2 Electronics

Electronics is concerned with the study and design of devices which use electricity. Specifically it is concerned with the movement of electrons – tiny particles which carry a minute electrical charge and so produce electric currents when they move around circuits.

Voltage

Electrons flow through a conducting medium if there is a difference in the distribution of electrons, which means that there is an excess of electrons in one location, and too few electrons in another location. Such a difference is called a *potential difference*, or a *voltage*. Voltage is measured with a unit called the volt. The higher the voltage, the greater the potential difference, and the more electrons which want to move from one location to another. If the potential difference gets large enough, then the electrons will jump through air (which is what a spark is: electrons flowing through air). Normally electrons only flow through metals and other conducting materials in a more controlled manner.

Current

Current is the name given for the flow of electrons. Using water as an analogy, the current is the flow of water, whilst the potential difference is the height of the water tower above the tap. The higher the water tower, the greater the pressure and the larger the flow when the tap is opened. To put things into some sort of perspective: a current of 1 ampere (ampere is normally shortened to amp in common usage amongst electronics engineers) represents the movement of about 6000 million million electrons per second.

Resistors are materials which impede the progress of electrons. Most metals will allow electrons to pass with almost no resistance, although very few materials present no resistance to the flow of electrons. Materials which allow electrons to pass through with no resistance are called superconductors: conductors because they 'conduct' electrons along, and super because they have no resistance to the flow of electrons. The word 'resistance' is actually used in electronics, but with a refined meaning: the resistance of a material is a measure of how hard it is for electrons to flow through it. Materials which do not allow electrons to flow are called *insulators*, whilst materials which do allow the flow of electrons are called *conductors*.

Resistors

Electronic components are made which have specific resistances, and these are called resistors. Resistance is measured in ohms, and is the voltage divided by the current (Figure 1.10.5). If a current of 1 amp is flowing through a resistor, and there is a voltage of 1 volt across the resistor, then the resistance is 1 ohm:

$$R = V/I$$

where R is the resistance, V, the voltage and I, the current.

Conductance is the reciprocal of resistance, and it uses units called mhos, which shows that electronics engineers have a sense of humour!

Resistors can range in value from very low resistances (fractions of ohms) for short lengths of metal wire, through to very high resistances (millions of ohms) for some materials which are on the borders of being insulators. For very high resistances, an alternative measurement is used: conductance.

When the current flows through the resistor, it produces heat. The amount of heat is determined by the product of the voltage across the resistor and the current. This is called the power, which is given off by the resistor, and it is measured in watts.

$$\text{Power} = V \times I$$

Figure 1.10.5 Another way of looking at the relationships between voltage, current and resistance is by considering the voltage across a resistor. If current *I* is flowing through a resistance of *R* ohms, then the voltage which will be present across the resistor will be *V* volts, where *V = IR*.

Current flowing through the resistor is *I* amps

Resistor value is *R* ohms

Voltage across the resistor is *V* volts

Ohm's law

$$V = IR$$

If 1 amp flows through a resistor with a resistance of 1 ohm, then the voltage across that resistor will be 1 volt, and 1 watt of power will be dissipated as heat by the resistor. The small resistors that are found in most domestic electronic equipment, like radios and hi-fi, will be 1/4 or 1/8 watt, and will be just under a centimetre long and a couple of millimetres across. A 'typical' value would be 10 000 ohms.

Capacitors

Having said that electrons carry a charge, and that the flow of charge is called a current, what happens if no current flows? Charge can be stored by having a device, which stores electrons, and this is called a capacitor; since it has a 'capacity' for holding charge. You 'charge up' a capacitor by applying a voltage to it. Once it has stored a charge you can remove the voltage and the charge will stay in the capacitor (although it will gradually decay away in time). The size of a capacitor is measured in farads (named after Michael Faraday, a major pioneer in early electricity and magnetism experiments) and has the symbol F. This is a very large unit, so large that 1 F capacitors are very rare. Capacitors are normally measured in smaller subunits of farads: μF, nF or pF being the most common units. Large capacitors are often quoted in tens of thousands of μF, which represents a few hundredths of a farad.

Inductors

Inductors do not like change. If you pass a current through an inductor, it initally tries to prevent the current flowing as the magnetic field is produced. When you try to stop the current flowing, then the magnetic field is converted back into current to try and maintain the current flow. This is why you sometimes get arcing at the contacts of devices that have lots of coils inside: like electric motors. The current is trying to keep flowing, even across gaps in the circuit!

Inductors are almost the opposite of capacitors – instead of storing charge, they temporarily store current. An inductor is often made from a coil of wire, and the action of current flowing causes a magnetic field to be produced. The energy from the current flow is thus stored as a magnetic field. If the current is removed, then the magnetic field will collapse, and produce a current as it does so. The energy is thus converted from current to magnetic field and back again. The 'size' of an inductor is measured in henrys, and again, this is such a large unit that hundredths and thousandths of henrys are much more likely to be found in common use.

Transistors

The transistor is a device which uses special materials called semiconductors (Figure 1.10.6). Silicon and germanium are two examples. A semiconductor is a material whose resistance is normally very high, but to which the addition of tiny amounts of other elements can alter the resistance. By controlling exactly how these other elements are placed in the semiconductor material it is possible to produce devices, which can control the flow of currents. A transistor is one such device. It has three terminals: current flows between two of these only when the third has a small current flow too. The control current is much smaller than the main current flow, and so the device can be used as an amplifier. If the control current turns on and off, then the main current turns on and off too, so the transistor can also be used as a switching device. Transistors are the basis of almost all electronics.

Figure 1.10.6 A transistor uses one current to control another. It can be used as an amplifier, a voltage-to-current converter or a switch.

When a current is applied to this terminal of the transistor ...

... then a current flows through these terminals

Transistors which use current as the control are called bipolar or junction transistors, whilst there are other types which use electric fields to control the main current, and these are called field effect transistors, abbreviated to FETs. FETs use very small currents indeed, and are widely used in electronics, particularly in making chips (see below).

Diodes

Diodes are simple semiconducting devices which allow a current to flow only one way. Inside there is a barrier which prevents the flow of electrons in one direction, but which breaks down and lets the electrons flow past in the other direction. When the current flows, the diode then behaves like a low value resistor, and so some heat is produced. If the barrier is made from a special material, then the effective resistance is higher, but instead of heat, light is produced, and these are called light-emitting diodes, or LEDs (Figure 1.10.7).

Current only flows in this direction

Figure 1.10.7 Current only flows through a diode in one direction.

The functions of both diodes and transistors used to be produced by using valves. Valves were small evacuated glass tubes which had a small heating element which was used to excite a special material so that it emitted electrons which travelled across the valve to a collecting plate. Current could only flow from the emitter (called a cathode) to the collecting plate (called the anode), and so a diode was produced. By putting a grid in between the cathode and anode, the flow of current could be controlled in much the same way as a transistor.

Integrated circuits

Integrated circuits, or ICs for short, are an extension of the process which is used to make transistors. Instead of putting just one transistor onto a piece of silicon, the first ICs 'integrated' a complete two-transistor circuit onto one piece of silicon. As the technology developed, resistors and capacitors were added, and the number of transistors increased rapidly. By the mid-1990s, ICs made from hundreds of thousands of transistors had become common.

The development of sophisticated stand-alone computer ICs from very humble cash register origins has produced the microprocessor. Commonly known as a 'chip', microprocessors carry out a vast range of processing tasks: a typical item of consumer electronic equipment will contain several: a video cassette recorder (VCR) could have a 'chip' dedicated to dealing with the front panel and IR remote control commands. Another would keep track of the programming and time functions, whilst another might handle the tape transport mechanism.

The use of microprocessor chips has had a major effect on the evolution of synthesizers: most notably in the change from analogue to digital methods of producing sounds. In a wider context, chips have become ubiquitous in most items of electronic equipment, but their function is completely unknown to all but a very small number of users of the equipment. One illustration of this change is hobby electronics. In the 1970s it was possible to buy kits of parts to build your own analogue synthesizer, and large numbers of people actually built or adapted these kits and constructed synthesizers, the author included. In the process of building the synthesizer, the constructor would learn a lot about how it worked, and so would be able to repair it if it went wrong. In the 2000s, such kits are considerably rarer, and fewer people have the time or skill to build them – few items of electronics fail, and those that do are often replaced rather than repaired. The 21st century equivalent of the synthesizer kit is the personal computer (PC), and although it is certainly possible to program your own synthesizer, the complexities are such that few people do. However, many people utilise software written by those few to make music.

Analogue electronics

Analogue electronics is concerned with signals: audio, video, instrumentation or control signals. Usually direct representations of the real-world value, but converted to an electrical signal by some sort of transducer or converter, analogue signals indicate the value by a voltage or current. For example, a device for measuring the level of a liquid in a tank might produce a voltage and by connecting this voltage to a calibrated indicator or meter, the level can be monitored remotely.

Signals in analogue electronics are often shown as plots of the value against time. These waveforms are often interpreted and drawn as if they were centred on a value of zero. So the use of the term 'zero crossing' does not necessarily mean that the waveform actually passes through a zero position, but merely that it passes an arbitrary line, which is approximately mid-way between the highest and lowest points of the oscillation.

Operational amplifiers

Operational amplifiers, or op-amps, are one of the basic building blocks of analogue electronics (Figure 1.10.8). Whilst transistors can be used as amplifiers, op-amps provide idealised, near-perfect gain blocks which are easy to control and use in circuits. Op-amps have a very large gain, and in normal use this is deliberately reduced by feeding some of the output back into the input, rather like the way that people tend to shout if they can not hear themselves.

The integrator is one example of an analogue processing element which can be created using an op-amp. By connecting the output of an op-amp to the input through a capacitor, the resulting circuit can only change its output slowly – at a rate set by the time that the capacitor takes to charge. An integrator can be used to convert a sudden

Figure 1.10.8 An op-amp has a very large gain and so it needs feedback to be applied in order to reduce the gain to a known amount.

change into a smooth transition, and is a simple filter circuit which 'filters' out rapid changes.

The oscillator is a variation on the integrator. If a capacitor is arranged so that it acts as a timing element for an op-amp circuit, then the circuit will repeat the cycle of charging and discharging the capacitor continuously. This produces a repetitive output at a frequency set by the time it takes for the capacitor to charge and discharge.

Filters are sophisticated versions of integrators. They come in many forms, and most have a gain, which is dependent on frequency, exactly the opposite of a hi-fi amplifier that aims to produce a 'flat' or consistent gain for any input frequency. Filters have a wide range of uses in synthesizers.

Digital electronics

Digital electronics uses signals which represent real-world values as numbers. The numbers are held in binary form: voltages or current which can have only two values or states, on and off or one and zero. By using groups of these two-valued voltages, any number can be stored in digital form. One familiar digital circuit is a light switch: assuming that there is no dimmer, then a light is either on or off.

Gates are simple electronic circuits which take one or more of these digital inputs and produce an output which is a logical function of them. For example, an output might only occur if both inputs are the same, or an output might be the opposite of the input. The rules for determining how these interactions take place is called Boolean algebra, and this is the branch of mathematics which is used to solve problems of the form: 'John does not cycle to work at the weekend. Bill travels to work, but only at the weekend. Simon has a car and he gives a lift to a colleague on Saturday. Who does Simon give the lift to?' (It could be John or Bill: more information is needed to provide a more definitive answer).

Registers are simple circuits which can store a binary value. Sets of registers can be used to hold whole numbers, and are known as memory. Real-world values which can change are represented as sequences of numbers, and this can occupy large amounts of memory. Audio signals require high precision and the frequent measuring of the value, and so synthesizers, and especially samplers, may need to contain lots of memory chips in order to store audio signals.

Memory comes in two forms. Permanent storage is called *read-only memory* (ROM) and is used to store the instructions, which control how a piece of equipment works called the operating system. Digital information (data) is stored in ROM memory by physically breaking links inside the ROM with short bursts of high current. Temporary storage is called *random access memory* (RAM), since any of the data it contains can be directly accessed as it is required; in contrast to a serial memory like a tape, where you need to wind through the tape to get to the required data. Some variants of ROM memory can be erased and rewritten: instead of using a permanent break in a wire, they store the data as charges on capacitors. These reprogrammable ROMs are called erasable programmable ROMs (EPROMs) or flash EPROMs, although this is commonly becoming abbreviated to just flash memory.

Microprocessors are stand-alone general-purpose computers which are designed to carry out lots of logical operations very quickly and efficiently. They do this by having memory stores and registers for values, a special arithmetic section which can carry out logic functions on the values, and a way to control the movement and processing of the data: usually a sequence of instructions called a *program*.

DSPs are microprocessors which have been optimised to deal with manipulating signals: often audio signals, although video and other types of signal are also possible. DSPs have a streamlined architecture and special circuitry to carry out functions rapidly and efficiently.

Form and function

The end product that uses digital and analogue electronics is often defined by its functions. The user does not need to know what type of storage is used in a dictation machine as long as it captures and plays back speech. In simple products the functionality is expressed in the 'form' of the device. A user could guess what a dictation machine was by observing the microphone and the tiny cassette or flash memory card slot, plus the control labelling. And a few investigatory presses of buttons would quickly reveal how to use it.

But PCs are intended to be generic devices. The function or operation of a dictation program on a computer might not be obvious at all, and for complex programs then training might be required to be able to do anything! In many ways, a synthesizer is a generic instrument in much the same way. You do not need to know how it works inside in detail, but you do need to have a usable model for how it makes the sounds, and how to control them. Playing a synthesizer might not be obvious, and training might be required to make any noises at all!

The preceding sections are not intended to be complete guides to either electronics or acoustics. Instead they aim to give an overview of some of the major concepts and terms which are used in these

Table 1.10.4 Units

Name	Symbol	Ratio	
Peta	P	1 000 000 000 000 000	One thousand million million times
Tera	T	1 000 000 000 000	One million million times
Giga	G	1 000 000 000	One billion times (one thousand million times)
Mega	M	1 048 576	1 048 576 times
Mega	M	1 000 000	One million times
K	K	1 024	One thousand and twenty-four times
Kilo	k	1 000	One thousand times
Milli	m	1/1 000	One thousandth
Micro	μ	1/1 000 000	One millionth
Nano	n	1/1 000 000 000	One billionth (One thousand millionth)
Pico	p	1/1 000 000 000 000	One million millionth

subjects. Further information can be found by following the references in the bibliography.

1.10.3 Units

Technical literature is full of units, and these units are often prefixed with any of a number of symbols which show the relative size of the unit. A familiar example is the use of the metre for measuring the dimensions of a room, but kilometres are used for measuring the dimensions of a country. A kilometre is 1000 metres, and this is shown by the 'kilo' prefix to the basic unit: the metre.

Table 1.10.4 gives some conversions between units and prefixes.

So one microsecond (μs) is one millionth of a second, whilst one megahertz (MHz) is one million hertz.

When the size of computer memory is described, two prefixes are used which refer to powers of two instead of powers of ten. A Kbyte of memory does not mean 1000 bytes of memory, instead it refers to 1024 bytes of memory. Kbytes are sometimes mistakenly called kilobytes. A similar confusion can arise over the use of the prefix M. One Mbyte of memory is 1 048 576 bytes of memory, and not a million! In this case, although the use of the word megabyte in this context is technically wrong, it has entered into common usage and become widely accepted.

1.11 Digital and sampling

This section brings together the background principles behind the two major technologies used in digital musical instruments: digital and sampling.

1.11.1 Digital

The word 'digital' can be applied to any technology where sound is created and manipulated in a discrete or quantised way, as samples (numbers which represent the sounds) rather than continuous values. This tends to imply the use of computers and sophisticated electronics, although an emphasis on the technology is often a marketing ploy rather than a result of using digital methods to make sounds. Physical modelling synthesizers are an excellent counter example where much of the complexity of the digital processing is deliberately hidden from the user, and as a result, the synthesizer is perceived by the performer as merely a very flexible and responsive 'instrument'. Perhaps in the future we will see digital 'instruments' where the synthetic method of sound production is not apparent from the external appearance. (Although a power-supply cable might be a useful clue!)

An understanding of binary numbers would have been essential for an understanding of computers in the 1970s or 1980s. In the 21st century, the underlying electronics is much less important, and an understanding of the operating system (Windows, MacOS, Linux, etc.) is essential.

In order for digital techniques to work with sound, there needs to be a way to represent sounds and values as numbers. Digital systems use binary digits, or bits, as their basic way of storing and manipulating numbers. Bits tend to be organised into groups of eight, for various historical and mathematical reasons. A single bit can have one of two values: on or off, usually given the values 1 and 0, respectively. Eight bits can represent any of 256 values, from 0 to 255, or %0000 0000 to %1111 1111 in binary notation. The '%' is often used to indicate a binary number, and the binary digits (bits) are grouped into blocks of four to aid reading. A collection of 8 bits is known as a byte, and the blocks of 4 are called nibbles(!). Sixteen bits can be used to represent numbers from 0 to 65 535, and more bits can provide larger ranges of numbers. The 2 bytes which make up a 16-bit 'word' are called the most significant and the least significant bytes, which is normally abbreviated to MSB and LSB, respectively. Notice that these simple numbers are integers – only whole numbers can be represented with this method.

Larger numbers, and especially decimal numbers, require a different method of representing them. Floating point numbers split the number into two parts: a *decimal number* part from 0 to 9.9, and a *multiplier or exponent part*, which is a power of ten. The value 2312 could thus be regarded as being 2.312×1000, and this would be stored as 2.312×10^3 in floating point representation. For binary numbers, a power of two is used instead of a power of ten, but the principle of splitting the number into a decimal number and a multiplier is the same. Floating point numbers can be processed using either a microprocessor or with special-purpose arithmetic chips called DSPs, which are optimised for the carrying out of complicated mathematical operations on numbers (some are designed for integers, whilst others are intended for floating point numbers); these are typically used for filtering, equalisation and 'effects' like echo, reverb, phasing and flanging.

1.11.2 *Sampling*

Sampling always produces numbers which are an incomplete representation of the analogue original. But the amount of incompleteness can be made insignificant and unimportant with careful design.

Sampling is the process of conversion from an analogue to a digital representation. Whereas an audio signal is a continuous series of values, which can be displayed on an oscilloscope as a waveform, a digital 'signal' is a series of numbers. The numbers represent the value of the audio signal at specific points in time and these are called samples. The sampling process has three stages:

1. The audio signal is 'sampled'.
2. The sample value is converted to a number.
3. The number is presented at an output port.

Samples are thus just numbers which represent the value of an audio waveform at a specific instant of time.

The opposite of sampling is the conversion from digital to analogue. This is called 'sample replay'. Replaying samples has three stages:

1. The number is presented to an input port.
2. The number is converted to an analogue value.
3. The analogue value forms part of an audio signal.

Sample replay is the basis of almost all digital synthesizers. Regardless of how the digital sample is produced, the conversion from digital to audio is what produces the sound that is heard. Chapter 3 shows how sample replay has progressed from single cycle waveforms to complex looped sample replay.

1.11.3 Conversion

The conversion process from analogue to digital and back again is at the heart of sampling technology. A complete digital audio conversion system, as used in a sampler, a direct-to-disk recording system, or a digital effect processor, typically consists of two sections. An 'analogue-to-digital' section converts the audio signal into digital form and temporarily stores it in the sample RAM memory:

- audio signal
- anti-aliasing filters
- sample and hold
- analogue-to-digital conversion (ADC) chip
- sample RAM containing digital sample values.

On the other hand, the 'digital-to-analogue' section reverses the process and converts the digital representation of the audio back into an analogue audio signal:

- sample RAM containing digital sample values
- digital-to-analogue conversion (DAC) chip
- deglitcher
- reconstruction filter
- audio signal.

The majority of the actual conversion is achieved by two chips: ADC is carried out by an ADC chip, whilst the reverse process of DAC is done by a DAC chip. DACs are also commonly used inside ADCs (see Figure 1.11.3).

Although the names are different, the circuits which make up the two parts on either side of the sample RAM memory have very similar functions. The anti-aliasing filters prevent any unwanted audio

Figure 1.11.1 An overview of a complete sampling system. Note that the filter at the input permanently removes the higher frequencies, and that the filter at the output reconstructs just the filtered version of the original audio signal.

frequencies from being converted by the ADC, whilst the reconstruction filter prevents any additional frequencies produced by the DAC's stepped waveform output from being heard at the audio output. The sample and hold circuit improves the quality of the conversion by presenting a constant level whilst the conversion is taking place, and the deglitcher prevents any momentary unwanted outputs from the DAC from being converted back into audio clicks (Figure 1.11.1).

Digital to analogue

A typical DAC has three parts:

- A latch to hold the digital number.
- A network of resistors to convert the number to a voltage.
- An output buffer amplifier.

The latch holds the digital number which represents the sample value. The network of resistors is arranged so that the bits in the digital number produce voltage, which are proportional to their position in the number. Large value bits produce big voltages, and small value bits produce small voltages. These voltages are added together by the output amplifier, whose output is thus an analogue voltage, which represents the value of the digital number (Figure 1.11.2).

Analogue to digital

In a typical ADC, the audio waveform is examined at regular intervals of time and the value is held in an analogue memory circuit called a *sample and hold*. The sample and hold circuit is the point in the conversion circuitry where the audio signal is actually 'sampled', and it is designed to capture the instantaneous value of the voltage at that point in the waveform and hold it whilst the conversion process proceeds. If the sample and hold circuit takes too long to capture the audio waveform, or if the held value changes whilst the conversion is taking place, then this too can degrade the quality of the conversion.

49

Figure 1.11.2 A DAC converts digital numbers to analogue voltages by using a network of resistors. The network is arranged so that the bits in the latch change the output voltage depending on their value, so the most significant bits have the largest effect.

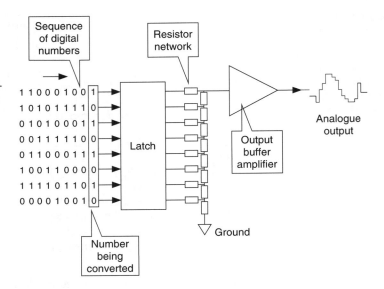

Once the sample and hold circuit has captured the value of the audio waveform, the held value is compared with a value which is produced by a counter and a DAC – in much the same type of circuit as that found in a wavetable synthesizer producing a sawtooth waveform. The counter counts up from zero, and as it does so, the ascending count of numbers is converted into a rising voltage by the DAC. A comparator circuit looks at the held value from the sample and hold circuit and the output of the DAC, and when they are the same, the comparator indicates that the two are equal, and the counter output is conveyed to the output of the ADC and held in a latch. The output of the ADC now holds a number which represents the value of the sample. The counter then resets and the ADC can begin to process the next value from the sample and hold circuit (Figure 1.11.3).

The detailed operation of some ADCs may differ from this, but the principle is the same – the audio signal is sampled, the sample value is converted to a numerical representation of the value, and the number appears at the output port of the ADC. Some ADCs achieve the conversion process in different ways, and the output format of the number can be either serial (one stream of bits) or parallel (several streams carrying a complete sample).

1.11.4 Sampling theory

In order for the process of taking a sample of the audio signal and transforming it into a number to work correctly, a number of criteria have to be met. Firstly, the rate at which the samples are taken must be at least twice the highest frequency, which is required to be converted, this in practice means that the input is normally filtered so that the highest frequency which can be present is known. Secondly, the samples must be taken at regular intervals – any jitter or uncertainty in the timing can significantly degrade the conversion quality.

Figure 1.11.3 An ADC takes an audio signal, samples it and then compares the value with the output of a DAC driven by a counter. When the two values are the same, the comparator latches the output of the counter into the ADC output latch. The analogue signal sample has then been converted into an equivalent digital value. This process repeats at the sampling frequency.

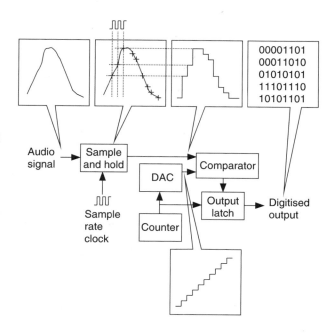

Finally, the numbers used to represent the signal must have enough resolution to adequately represent its dynamic range.

1.11.5 Sample rate

The simplest representation of a waveform at a given frequency is two sample values: ideally the top and bottom peaks. The time between these two peaks represents half of the period of the waveform drawn through them, which assumes that it is a sine wave and if this is the highest frequency, then it must be a sine wave. So because two points are needed, the sampling needs to be at least twice as fast as the frequency which is being sampled. This requirement that the sampling frequency is at least twice the highest frequency in the signal is called the *Nyquist criterion*. Note that if the sampling rate was exactly twice the frequency of the audio signal, then the two points would always be at exactly the same values on the waveform, which could include zero, and so there might be no output at all. Sampling is not normally done synchronously, and so a sampling rate, which is at least twice that of the highest frequency in the audio signal, will enable the same waveform to be reconstructed at the output of a subsequent DAC section.

If the audio signal is sampled at a rate, which is higher than twice the highest frequency which is present in the audio signal, then no additional information is provided by using a higher sampling rate. The Nyquist criteria thus represent the most efficient rate at which to sample a given audio signal with a specific highest frequency component. However, sampling at higher frequencies can simplify the design and implementation of the filtering and some other parts of the circuitry. In the limiting case, some ADCs sample at several

51

Figure 1.11.4 The Nyquist rate is twice the highest frequency which is present in the signal to be sampled. At least two samples are needed in order to provide a single cycle of a waveform.

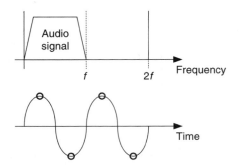

hundred times the Nyquist rate and then process the resulting 1-bit representation to produce the equivalent of more bits sampled at a lower rate. But the basic amount of information, which is required in order to be able to reconstruct the audio signal, still remains constant (Figure 1.11.4).

1.11.6 Filtering and aliasing

If any frequencies are present in the audio signal which are above the 'half-sampling' frequency, then they will still be sampled, but the effect will be to make them appear to be lower in frequency; this process is called aliasing. It can be likened to a security camera, which looks at a room for a few seconds every 5 min. If the room has some-one inside the first time that the camera is active, but not the second time, then there are several scenarios for what has happened. The obvious case is that the person was present for the first 5 min and so was observed when the camera was active, but who then left before the camera was active the second time. Alternatively, the person could have been in and out of the room several times, and just happened to be present first time, but not the second. The important point is that the two cases appear the same from the viewpoint of the camera. Aliasing behaves in the same way – an aliased high frequency appears as a lower frequency in the digital representation, and it is not possible to reconstruct the original higher frequency. It is a 'one-way' process where information is lost or becomes ambiguous.

To prevent information being lost by the sampling process, anti-aliasing filters are used to constrain the audio signal to below the half-sampling frequency. This ensures that only frequencies which can be reproduced are sampled, and so guarantees that the DAC will be able to output the same audio signal. The design of these filters affects the quality of the conversion, since they need to pass all frequencies below the half-sampling rate, but completely reject all frequencies above the half-sampling rate. 'Brick wall' filters with flat pass-bands and high stop-band rejection are difficult to design and fabricate, and in practice, the cut-off frequency of the filter is set to slightly lower than the half-sampling rate, and the stop-band rejection is chosen so that any frequencies which pass through to the

Figure 1.11.5 (i) If an audio signal contains some frequencies which are higher than half of the sampling frequency, then aliasing can occur. (ii) An anti-aliasing filter prevents this by having a pass-band which is set so that frequencies above the half-sampling frequency are in the stop-band of the filter. Theoretically this filter should pass everything below the half-sampling frequency and stop everything above. (iii) In practice, filters with a realisable cut-off slope and sufficient stop-band attenuation to prevent audible aliasing are used.

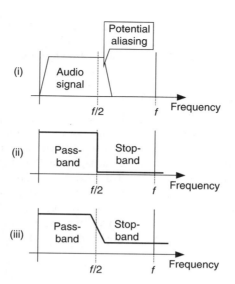

conversion process will be so small that they will be lost in the inherent noise of the converter (Figure 1.11.5).

The half-sampling frequency sets the highest frequency which was present in the original audio signal before it was sampled, and this is the highest frequency which will be reproduced when the sample is replayed. So for a sample rate of 44 kHz, the highest frequency which can be reproduced by the replay circuitry will be just under 22 kHz. But reproducing sample values from a memory device also produces unwanted additional frequencies. Consider again the limiting case of two adjacent sample values, which represent a sine wave at just under the half-sampling frequency: when these are read out from a memory device, they will form the equivalent of a square-shaped waveform. Additional filtering is required to remove these extra frequencies, and a sharp low-pass filter with a cut-off frequency set to near the half-sampling rate is normally used. This filter is often called a reconstruction filter, and it limits the output spectrum of the sample replay to those frequencies which are below the half-sampling frequency. Any unwanted frequencies which are not removed by this filter are called aliasing frequencies.

Domestic CD players with a 44.1 kHz sample rate, and thus a half-sampling rate is 22.05 kHz, normally quote the upper limit of the audio signal frequency response as being 20 kHz. Professional DAT recorders typically use 48 kHz sampling, and also quote a 20 kHz upper frequency response. This 44.1/48 kHz sample rate/20 kHz bandwidth has become a 'de facto' standard for samplers and digital synthesizers.

Sample rates of 96 and 192 kHz began to appear at the close of the 20th century, and have increasingly been used for converting audio at the start of an otherwise all-digital processing chain based on computers and hard disk storage. Since CDs and DATs are designed around the 44.1/48 kHz, a number of alternative enhanced CD formats, plus DVD audio, have been produced that use higher sampling rates, but these have not seen wide public acceptance.

1.11.7 Resolution

The size of the numbers which are used to represent the sample values determine the fidelity with which the audio signal can be reproduced. In digital circuitry, the number of bits, which are used to represent the sample value, limits the range of available numbers. In the simplest case, a 1-bit number can have only two values: 1 and 0. For each additional bit which is used, the number of available values doubles: so for 2 bits, four values are available. Three bits provides eight values, 4 bits 16 values and so on. In general, the number of available sample value numbers is given by:

$$D = 2^n$$

where D is the number of available values and n is the number of bits used. The number of available numbers to represent the sample values affects the precision of the digital version of the original audio signal. If only one bit is available, then only a very crude version of the audio signal is possible. As more bits are used to represent the sample values, the ratio between the largest and the smallest number which can be represented increases and it is the size of the smallest change which determines how good the resolution is. As the number of bits increases, the detail which can be represented by the numbers improves. This reduces the distortion, and for a typical 16-bit conversion system the distortion will be more than 90 dB below the maximum output signal.

The number of bits, which are used to represent a sample is important because it sets the limiting value on the output quality of the signal. The relationship can be approximated by the simple formula:

$$S = 6n \, \text{dB}$$

where S is the signal-to-noise (and distortion) ratio (SNR): the ratio between the loudest audio signal and the inherent noise and distortion of the system, often called the dynamic range, measured in dB, and n is the number of bits. This is the performance of a perfect system, and represents the 'ideal' case: real-world digital audio systems will only approach these figures (Table 1.11.1).

Table 1.11.1 shows the number of bits versus the 'ideal' dynamic range.

As the table shows, a 'CD quality' output should have a dynamic range of nearly 96 dB: 'better than 90 dB' is frequently quoted in manufacturer's specifications. Note that the entire audible range, from silence to painful, can be covered by 20 bits. It thus appears that using between 16 and 20 bits should be adequate for almost all purposes. Unfortunately, this is not the case, and the simple example of volume control illustrates the problem.

Table 1.11.1 Bits and SNR

Number of bits	Dynamic range
8	48
10	60
12	72
14	84
16	96
18	108
20	120
22	132
24	144

It should be noted that shifting digital numbers to the right is not a very useful way of making changes to the volume of a signal, since the 6-dB steps are very coarse. In practice, the numbers are reduced or increased by using a multiplication device: often a special-purpose signal processing chip.

Suppose that a digital synthesizer design uses 16-bit numbers to represent the audio samples, and that the volume control is implemented by manipulating the digital audio signal. So, for maximum output volume (0 dB), all of the 16 bits in the audio samples will be used in the replay of the signal. A crude method of reducing the volume could be achieved by using less bits: shifting the digital numbers to the right. Each bit which is removed reduces the volume by 6 dB, so a coarse volume control might work by shifting the digital words to the right so that less bits are used, with zeroes added from the left. So as bits are removed the volume decreases, but there is a corresponding decrease in the dynamic range of the signal. For an audio signal which is at −48 dB, only 8 of the original 16 bits are being used to produce the audio signal, which means that the output signal effectively has only an 8-bit resolution – the remainder of the signal has been filled with eight zeroes.

Using only half of the available bits for an audio signal has two major effects on the audio. The reduction in dynamic range means that there is a corresponding increase in the background noise level, whilst the release of notes can become distorted, especially if reverb is used. This characteristic 'grainy' distortion is called 'quantisation noise', and is caused by the transition between silence and audio represented by just changing 1 bit: effectively the audio waveform has been converted into a pulse wave.

In audio signal terms, with 8-bit integer numbers at a rate of 8 kHz, the sound quality is comparable to a low quality cassette or telephone. Sixteen-bit numbers at a rate of 44.1 kHz are often quoted as being of 'CD audio quality', since this is the basic storage used by a CD player for audio.

Reducing the volume by having less bits in the output signal is thus very different from always using all the available bits and changing the volume with an analogue volume control. With the analogue control, the full bit resolution is always available, and so a −48 dB signal would still have the same dynamic range as the original sample; even if some of this is buried in the background electrical noise of the system. Some types of DAC chips allow exactly this type of output. Multiplying DACs and floating point DACs can be used with two inputs: one of which represents the audio signal at the full bit resolution, whilst the other input represents the volume control bits. This type of 'fixed with scaling' conversion system is in widespread use. For example, telephones do not use linear coding but their basic performance is approximately 8 bit for SNR, with about the equivalent of 12 bits for the dynamic performance; although the restricted bandwidth significantly affects the perceived quality. In synthesizers, the sample resolution is normally 16 bits, whilst the volume control can be 6 bits or more, which is sometimes translated as '24-bit DACs' in manufacturers' literature.

1.12 MIDI (Musical instrument digital interface)

MIDI (Rumsey, 1994) has played a major role in the development of electronic music since 1983. Wherever possible, this book has deliberately avoided making explicit references to MIDI in order to prevent it becoming 'Yet another book on MIDI'.

For example, the envelopes described in Chapter 2 are mostly dealt with in terms of control voltages, gate pulses and trigger signals because these are likely to be the native interfacing for many analogue synthesizer, even though many users will also use a MIDI-to-control voltage converter box to enable the use of MIDI control.

Since some readers may not be familiar with MIDI, the remainder of this section provides some background information.

1.12.1 Overview

MIDI provides an interface for the exchange of information between electronic musical instruments and computers. It is based around musical events, except in rare circumstances, musical sounds are not conveyed via MIDI. Instead, MIDI carries information about what is happening and occurrences such as: when a note has been pressed, when a drum is hit and when the sequencer has stopped. A MIDI-equipped keyboard will thus output information about what is happening on its own keyboard; so if some notes are played, it will output MIDI information as 'messages' which describe which notes are being played, as they are being played.

1.12.2 Ports

The MIDI interface is called a MIDI port. There are three types, although only one or two of the types may be present on a given piece of equipment.

- The in-port accepts MIDI data.
- The out-port transmits MIDI data.
- The 'thru' (American spelling: MIDI was originally specified in the USA) port merely transmits a copy of the MIDI data which arrives at the in-port.

All MIDI ports look alike: they consist of 180° 5-pin Deutsche Industrie Norm (DIN) sockets, although each port will normally be marked with its function (in, out or thru).

1.12.3 Connections

Connecting MIDI ports together requires just one simple rule:

Always connect an out or a thru to an in.

In a MIDI 'network', information flows from a controller source to an information sink. A keyboard is often used as a source of control information, whilst a synthesizer module is usually an information sink. So the out-port of the keyboard would be connected to the in-port of the synthesizer module. MIDI messages would then flow from the keyboard to the synthesizer module, and the synthesizer could then be 'played' from the keyboard.

1.12.4 Channels

MIDI provides 16 separate channels, which can be thought of as television channels. A piece of MIDI equipment can be 'tuned' so that it receives only one channel, and it will then only respond to MIDI messages which are on that channel. Alternatively, it is possible to set a piece of MIDI equipment so that it will respond to messages on any channel called 'omni'. Some important MIDI messages are sent to all 16 channels.

If more than 16 channels are required, then multiple MIDI ports and cables are used. Each additional MIDI port or cable provides another 16 separate channels.

1.12.5   Modes

MIDI has several modes of operation. The important ones are as follows:

- Monophonic (one instrument: one note at once).
- Polyphonic (one instrument: several notes at once).
- Multi-timbral (several different instruments at once: several notes at once).

Modes are normally only important to users of guitar controllers or other specialised uses.

1.12.6   Program changes

Continuing with the television analogy, MIDI calls sounds or patches 'programs'. The message which indicates that a program should change is called a 'program change' message. Any of 128 programs can be selected. If more programs are required, then a bank change message allows the selection of banks of 128 programs. For most applications, a program change number does not indicate a specific sound, but a specialised mapping called 'general MIDI' or GM does specify which program change number calls up what sort of sound from a sound module.

1.12.7   Notes

One of the commonest MIDI messages is the note on message. This indicates that a note has been played on a keyboard, although it could also mean that a sequencer is replaying a stored performance. The note on message contains information on the MIDI channel that is being used, what key has been played and how quickly the key was pressed: this is called the 'velocity'. As a shorthand method of sending messages, a velocity of zero is taken to mean a note off message, even though a separate note off message exists. The MIDI note on message does not contain any timing information about when the key was pressed – the message itself is used to indicate that the key has just been pressed.

Other common note specific messages include:

- **Pitch-bend message**, which transmits any changes in the position of the pitch-bend wheel.
- **After-touch messages** (polyphonic and monophonic) which transmit information about how hard the keys are being pressed once they have reached the end of their travel. This is intended

57

as an additional control source for introducing vibrato or other modulation by increasing the finger pressure on a key which is being held down.

1.12.8   MIDI controllers

A MIDI controller is something which is used to control part of a performance like the modulation wheel which is often found on the left-hand side of the keyboard on many synthesizers, and which can be used to introduce vibrato or other modulation effects into the sound. Another example might be a foot volume pedal which plugs into a synthesizer – it controls the volume of the synthesizer directly, but it may also cause the synthesizer to transmit MIDI volume messages which indicate the position of the foot pedal.

There are a large number of possible controllers, with functions ranging from volume or portamento, through to one which can control the timbre of a sound or set an effect parameter. Only a few of the controllers are defined; many are deliberately left undefined so that manufacturers can allocate them for their own purposes.

1.12.9   System exclusive

Although there are lots of MIDI controller messages, there is an alternative way to provide control over a remote MIDI device. System exclusive (sysex) messages are designed to allow manufacturers to make their own MIDI messages. The sysex messages can be used to edit synthesizer parameters, to store sound data and to transmit samples.

1.12.10   MIDI files

MIDI files are a way to move MIDI sequencer file information between different sequencers. The PC standard disk format allows MIDI files to be moved freely between any computers that can read 3.5-inch IBM PC compatible floppy disks. In the 2000s, USB flash memory devices are increasingly replacing the floppy disk.

1.12.11   Reference

The book by Francis Rumsey (1994) gives excellent detailed information on MIDI and is recommended reading. The official MIDI documentation (The MIDI Specification, published by the International MIDI Association) is very formal, rather technical and not intended for the general reader.

1.13   After MIDI

MIDI has proved to be a major unifying force within sound synthesis and sampling for more than 20 years. But the leading edge of technology in the early 1980s was the transmission of 31 250 sets of 7 bits per second using affordable mass-market chips. A designer of a replacement for MIDI in the 20th anniversary year, 2003, would be looking at general-purpose rapid data transfer technologies like IEEE-1394, known as FireWire, with a rate of 50 million sets of 7 bits per second. FireWire has seen wide acceptance as a method of connecting digital video camcorders to computers.

One emerging standard protocol based on IEEE-1394 (FireWire), called mLAN, and developed by Yamaha, is designed for moving digital audio and other music-related data like MIDI and synchronisation signals around recording studios, although the number of pieces of equipment which supported mLAN was very small by the middle of 2003. The integration of several separate pieces of conventional cabling into one mLAN cable carrying them all within a single digital link has the advantage of neatness, but has less of the immediacy of MIDI or audio cabling, where connecting a keyboard to a sound module, or that module to a mixer, has a simple directness.

ZIPI (*The Computer Music Journal*, Vol. 18, No. 4) was a proposal made in 1996 for an alternative interface for connecting a performance controller to a synthesizer. It provided much more detail and flexibility in the way that pitch and performance information are conveyed than MIDI offers. In particular, it avoided making any assumptions about the type of controller that would be used, so there were no keyboard-specific limitations, unlike some of the assumptions built into MIDI. But ZIPI was not intended as a replacement for MIDI, instead it was more a means of transporting performance control from a player's instrumental controller to a remote synthesizer. The exact format of the player's controller was not defined: it could be a keyboard, stringed instrument or wind instrument. ZIPI did not see wide commercial adoption, but many of its aims of increased performance controllability could conceivably be implemented with future mLAN-based instruments.

In one way, the success of MIDI has led to it becoming a ubiquitous lower limit of connectivity; but this wide acceptance has meant that the conventions of MIDI (and its limitations) have become part of the technology of music making, and are unchallenged or questioned. This may have restricted the development of extensions for specific purposes like ZIPI, and may also mean that the MIDI socket is going to be an essential element of rear-panel design long after the mLAN socket also becomes a 'must-have' item of a specification.

1.14 Questions

This section is designed to act as a brief review of the subject covered in the preceding chapter. The answers are in the text.

1. What is sound synthesis?
2. What is the difference between a modular and a performance synthesizer?
3. Outline the major methods of sound synthesis.
4. What is acoustics?
5. What is electronics?
6. Outline the processes that are required to take a product from laboratory prototype to commercial production.

7. Describe some ways in which synthesizers can be used to make music.
8. Categorise 10 different sounds under the following categories: realistic, synthetic, imitative, suggestive or sympathetic.

Time line

Date	Name	Event	Notes
1500s	Barrel Organ	The barrel organ. Pipe organ driven by barrel covered with metal spikes	The forerunner of the synthesizer, sequencer and expander module!
1582	Galileo	Galileo conceives the idea of using a pendulum as a means of keeping time	
1600s	Gottfried Leibnitz	Developed the mathematical theories of logic and binary numbers	
1600	William Gilbert	Electricity is named after the Greek word for Amber	William Gilbert was the court physician to Elizabeth the first
1612	Francis Bacon	Publishes 'New Atlantis, which describes all sorts of now current sound 'wonders in a passage starting': 'We also have sound houses ...'	An essential quote in most books on electronic music
1642	Blaise Pascal	First mechanical calculator	Addition or subtraction only
1657	Christian Huygens	Christian Huygens uses the pendulum to regulate the time-keeping of a clock	
1676	Thomas Mace	Thomas Mace uses a thread and a heavy round object to mark musical time	Also designed a lute with 50 strings in 1672
1694	Gottfried Leibnitz	Devised a mechanical calculator which could multiply and divide	
1696	Etienne Loulie	Etienne Loulie invents the 'Chronometre', an improvement on Mace's idea, but with a variable length thread	
1700	J.C. Denner	Invented the Clarinet	Single-reed woodwind instrument
1752	Benjamin Franklin	Flew a kite in a thunderstorm to prove that lighting is electrical	
1756–1827	Ernst Chladni	Worked out the basis for the mathematics governing the transmission of sound	The 'Father of Acoustics'
1768–1830	Jean Baptiste Fourier	French mathematician who showed that any waveform could be expressed as a sum of sine waves	Basis of Fourier (additive) synthesis and fast Fourier transform (FFT)
1801	Valve Trumpet	The modern valve trumpet was invented	Not all musical instruments are old!
1804	Jacquard	Jacquard punched cards invented	Basis of stored program control, as used in computers, pianolas, etc.
1807	Jean Baptiste Joseph Fourier	Fourier published details of his theorem, which described how any periodic waveform can be produced by using a series of sine waves	The basis of additive synthesis

Date	Name	Event	Notes
1812	D.N. Winkel	Winkel invented a clockwork-driven double-pendulum timer, which is very much like a metronome	
1815	J.N. Maelzel, brother of Leonard Maelzel	Invented the metronome and patented it	Some dispute about Maelzel versus Winkel as who actually invented the metronome
1818	Beethoven	Beethoven started to use metronome marks in scores	
1820	Oersted	Discovery of Electromagnetism	The basis of electronics
1821	Michael Faraday	Discovered the dynamo, and formalised link between magnetism, electricity, force and motion	Used in motors, microphones, solenoids
1833	Charles Babbage	Invented the difference engine – mechanical calculator intended for producing log tables	The electronic calculator eventually made log tables obsolete!
1837	Samuel Morse	Invented Morse code	
1844	Samuel Morse	Invented the electric telegraph	The first telegraph message was 'What hath God wrought?'
1846	Adolphe Sax	Invented the saxophone	
1849	Heinrich Steinweg	Steinway pianos founded by Heinrich Steinweg	
1862	Helmholtz	Published 'On the Sensations of Tone'	Laid the foundations of musical acoustics
1866–1941	Dayton Miller	Worked on photographing sound waves and turned musicology into a science	
1868–1919	Wallace Sabine	Founded the science of architectural acoustics as the result of a study of reverberation in a lecture room at Harvard where he was Professor of Physics	
1876	Alexander Graham Bell	Invented the telephone	Start of the marriage between electronics and audio
1877	Thomas Edison	Thomas Alva Edison invented the cylinder audio recorder called the 'Phonograph'. Playing time was a couple of minutes!	Cylinder was brass with a tin foil surface, replaced with metal cylinder coated with wax for commercial release
1878	David Hughes	Invented moving coil microphone	
1878	Lord Rayleigh	Published 'The Theory of Sound'	Laid the foundations of acoustics
1887	Heinrich Hertz	Produced radio waves	
1888	Emile Berliner	First demonstration of a disc-based recording system, the 'Gramophone'	Disk was made of zinc, and the groove was recorded by removing fat from the surface, and then acid etching the zinc
1895	Marconi	Invented radio telegraphy	

Date	Name	Event	Notes
1896	Thomas Edison	Invented motion picture	
1897	Yamaha	Nippon Gakki (Yamaha) founded	
1898	Valdemar Poulsen	Invented the Telegraphone, which recorded telephone audio onto iron piano wire (also known as the Dynamophone)	30 second recording time, and poor audio quality
1899	William Duddell	Turned the noise emitted by a carbon arc lamp into a novelty musical instrument	Known as 'The Singing Arc'
1901	Harry Partch	Experimented with 13th tones and other microtonal scales	Mostly self-taught
1901	Guglielmo Marconi	Marconi sends a radio signal across the Atlantic	
1903	Double-sided LP	The Odeon label release the first double-sided LP	Two single-sided LPs stuck together?
1904–1915	Valve	Development of the valve	The first amplifying device, the beginnings of electronics
1906	Lee de Forest	Invented the triode amplifier	The beginning of electronics
1908	Oliver Messiaen	Serialism, Eastern Rhythms and exotic sonorities	Some of his music used up to six Ondes Martenot
1910–1920	Futurists	Futurists	Category of music
1912	John Cage	Pioneer in experimental and electronic music	Famous for 'prepared' pianos and '4 minutes 33 seconds'; a silent work
1914	Hornbostel and Sachs	Published a classification of musical instruments based on their method of producing sound	Idiophones, Membranophones, Chordohones, Aerophones, …
1915	E.C. Wente	Produced the first 'Condenser' microphone using a metal plated insulating diaphragm over a metal plate	Now known as a 'Capacitor' microphone
1915	Lee de Forest	The first valve-based oscillator	
1916	Luigi Russolo	Categorises sounds into six types of noise	Also invented the Russolophone, which could make seven different noises
1920–1950	Musique Concrète	Musique Concrète	Tape manipulation
1920s	Cinema organs	Cinema organs, using electrical connection between the console keyboard and the sound generation	Also started to use real percussion and more: car horns, etc.; mainly to provide effects for silent movie accompaniment
1920s	Harry Nyquist	Developed the theoretical basis behind sampling theory	Nyquist frequency named after him
1920s	Microphone recordings	First major electrical recordings made using microphones	Previously, many recordings were 'acoustic' using large horns to capture the sound of the performers
1920	Lev Theremin	The Theremin, patented in 1928 in the US. Originally called the 'Aetherophone'	Based on interfering radio waves

Date	Name	Event	Notes
1920	Louis Blattner	The first magnetic tape recorder	Blattner was a US film producer
1923	John Logie Baird	Began experiments with light sources and disks with holes in them for scanning images	The beginning of television and computer monitors
1925	Pierre Boulez	Pioneer of serialims and avante-garde music	
1925	John Logie Baird	First television transmission	… across an attic workshop!
1928	Maurice Martinot and Ondes	Invented the Ondes Martenot, an early synthesizer	Controlled by a ring on a wire, finger operated
1929	Couplet and Givelet	Four voice, paper tape-driven 'automatically operating oscillation type'	Control was provided for pitch, amplitude, modulation, articulation and timbre
1930s	Baldwin, Welte, Kimball and others	Opto-electric organ tone generators	
1930s	Bell Telephone Laboratories	Invented the Vocoder, a device for splitting sound into frequency bands for processing	More musical uses than telephone uses!
1930s	LP groove direction	Some dictation machines recorded LPs from the centre out instead of edge in	This pre-empted the CD 'centre out' philosophy
1930s	Ondes	Ondioline, an early synthesizer	Uses a relaxation oscillator as a sound source
1930s	Run-in Grooves	Invented run-in grooves on records	Previously, you put the needle into the 'silence' at the beginning of the track …
1934	John Compton	UK patent for rotating loudspeaker	
1934	Laurens Hammond	Hammond 'Tone Wheel' Organ uses rotating iron gears and electromagnetic pickups	Additive sine waves
1935	AEG, Berlin	AEG in Germany used iron oxide backed plastic tapes produced by BASF to record and replay audio	Previously, wire recorders had used wire instead of tape
1937	Tape recorder	Magnetophon magnetic tape recorder developed in Germany	The first true tape recorder
1940s	Arnold Schoenberg	12-tone technique and atonality	
1940s	Wire and Ribbon recorders	Major audio recording technology uses either steel wire or ribbon	High speed, heavy and bulky, and dangerous if the wire or ribbon breaks!
1943	Colossus	The world's first electronic calculator	Built to crack codes and ciphers
1945	Metronome	First pocket metronome produced in Switzerland	
1945	Ronald Leslie	Patents rotating speaker system	
1947	Conn	Independent electromechanical generators used in organ	

Date	Name	Event	Notes
1948	Baldwin	Blocking divider system used in organ	
1948	Pierre Schaeffer	Musique concrète	
1948	Pierre Schaeffer	'Concert of Noises' futurist movement. Invented music concrete	Music concrete is made up of pre-existing elements
1949	Allen	Organs using independent oscillators	
1949	C.E. Shannon	Published book *The Mathematical Theory of Communications*, which is the basis for subject of information theory	Shannon's sampling theorem is the basis of sampling theory
1950s	Charles Wuorinen	Quarter tones	
1950s	Tape recorder	Magnetic tape recorders gradually replace wire and ribbon recorders	There were even domestic wire recorders in the 1950s!
1950	John Leslie	Reintroduction of Leslie speakers	This time they are a success
1951	Hammond	Melochord	
1951	Herbert Eimert	Northwest German Radio NWDR in Cologne started experimenting with sound using studio test gear	Used oscillators and tape recorders to make electronic sounds
1954	Milton Babbitt, H.F. Olsen and H. Belar	RCA Music Synthesizer mark I	Only monophonic
1955–1956	Stockhausen	'Gesang der Junglinge' mixed natural sounds with purely synthetic sounds	
1955	E.L. Kent	Kent Music Box in Chicago. Inspired RCA mark II synthesizer	
1955	Louis and Bebe Barron	Soundtrack to 'Forbidden Planet' is a 'tour de force' of music concrete using synthetic sounds	
1957	RCA	RCA Music Synthesizer mark II	Used punched paper tape to provide automation
1958	Charlie Watkins	Charlie Watkins produced the CopyCat tape echo device	
1958	Edgard Varese	Produced some 'electronic poems' for the Brussels Expo	
1958	RCA	RCA announced the first 'cassette' tape, a reel of tape in an enclosure	Not a success
1960s	Mellotron	The Mellotron, which used tape to reproduce real sounds	Tape-based sample playback machine
1960s	Wurlitzer, Korg	Mechanical rhythm units built into home organs by Wurlitzer and Korg	
1962	Ligetti	Ligetti used the metronome as a musical instrument	Actually he used 100 of them in concert
1962	Telstar	The first telecommunications satellite to transmit telephone and television signals	

Date	Name	Event	Notes
1963	Don Buchla	Simple voltage controlled oscillator (VCO), filter (VCF) and amplifier (VCA) based modular synthesizer: 'The Black Box'	Not well publicised
1963	Herb Deutsch	First meeting with Robert Moog. Initial discussions about voltage controlled synthesizers	
1963	Philips	Philips in Holland announced the 'Compact Cassette', two reels plus tape in a single case	A success well beyond the original expectations!
1965	Early Bird	First geo-stationary satellite	
1965	Paul Ketoff	Built the 'Synket', a live performance analogue synthesizer for composer John Eaton	Commercial examples like the Mini Moog and Arp Odyssey, soon followed
1966	Don Buchla	Launched the Buchla Modular Electronic Music System, a solid-state, modular, analogue synthesizer	Result of collaboration with Morton Subotnick and Ramon Sender
1966	Rhythm machine	Rhythm machines appeared on electronic organs	Non-programmable and very simple rhythms
1968	Walter Carlos	'Switched On Bach', an album of 'electronic realisations' of classical music, became the best seller	Moog synthesizers suddenly change from obscurity to stardom
1969	Philips	Digital master oscillator and divider system	
1970s	Ralph Deutsch	Digital generators followed by tone-forming circuits	The popularisation of the electronic organ and piano
1970	ARP Instruments	ARP 2600 'Blue Meanie' modular-in-a-box released	
1970	Tom Oberheim	Oberheim Electronics founded	US company
1972	E-mu	E-mu founded by Dave Rossum. Initial products are custom modular synthesizers	
1972	Hot Butter	'Popcorn' became a hit single	
1972	Roland	Ikutaro Kakehashi founded Roland in Japan, designed for R&D into electronic musical instruments	First products are drum machines
1972	Timmy Thomas	Timmy Thomas' 'Why Can't We Live Together' is probably the first record to replace the drummer with a drum machine	
1973	John Chowning	Published paper: 'The synthesis of complex audio spectra by means of frequency modulation', the definitive work on FM	FM introduced by Yamaha in the DX series of synthesizers 10 years later
1973	Oberheim	First digital sequencer	The first of many
1974	Kraftwerk	'Autobahn' album was a huge success. A mix of music concrete technique and synthetic sounds	

Date	Name	Event	Notes
1974	Sequential Circuits	Sequential Circuits was founded by Dave Smith	US company
1975	Fairlight	Fairlight was founded by Kim Ryrie and Peter Vogel	Australian company
1977	Roland	MC8 microcomposer launched: the first 'computer music composer', essentially a sophisticated digital sequencer	Cassette storage, this was 1977!
1978	Electronic Dream Plant	Wasp Synthesizer launched. Monophonic, all-plastic casing, very low-cost, touch keyboard, but it sounded much more expensive	Designed by Chris Hugget and Adrian Wagner
1978	Philips	Philips announced the CD	This was the announcement, getting the technology right took a little longer …
1979	First Digital LPs	First LPs produced from digital recordings made in Vienna	A mix of analogue playback and digital recording technology
1980	Electronic Dream Plant	Spider Sequencer for Wasp Synthesizer. One of the first low-cost digital sequencers	252-note memory, and used the Wasp DIN plug interface
1981	Moog	Robert Moog was presented with the last Mini Moog at NAMM in Chicago	The end of an era
1981	Roland	Roland Jupiter-8. Analogue eight-note polyphonic synthesizer	
1981	Yamaha	Yamaha R&D Studio opened in Glendale, California, USA	
1982	Moog	Memory Moog, six-note polyphonic synthesizer with 100 user memories	Cassette storage! Six Mini Moogs in a box!
1982	Philips/Sony	Sony launched CDs in Japan	First domestic digital audio playback device
1982	PPG	Wave 2.2, polyphonic hybrid synthesizer, was launched	German hybrid of digital wavetables with analogue filtering
1982	Robert Moog/MIDI	First MIDI specification announced by Robert Moog in his column in *Keyboard* magazine	
1982	Roland	Jupiter 6 launched: the first Japanese MIDI synthesizer	Very limited MIDI specification. Six-note polyphonic analogue synth
1982	Sequential	Prophet 600 launched: the first US MIDI synthesizer	Six-note polyphonic analogue synth, marred by a membrane numeric keypad
1983	Oxford Synthesiser Company	Chris Huggett launched the Oscar, a sophisticated programmable monophonic synthesizer	One of the few mono-synths to have MIDI as standard
1983	Philips/Sony	Philips launched CDs in Europe	Limited catalogue of CDs rapidly expanded
1983	Roland	Roland launched the TR-909, the first MIDI-equipped drum machine	

Date	Name	Event	Notes
1983	Sequential Circuit	Sequential Circuit's Prophet 600 was the first synthesizer to implement MIDI	Prophet 600 was marred by awful membrane switch keypad
1983	Yamaha	'Clavinova' electronic piano launched	
1983	Yamaha	MSX Music Computer: CX-5 launched	The MSX standard failed to make any real impression in a market already full of 8-bit microprocessors
1983	Yamaha	Yamaha DX7 was released. First all-digital synthesizer to enjoy huge commercial success. Based on FM synthesis work of John Chowning	First public test of MIDI was Prophet 600 connected to DX7 at the NAMM show, and it works (partially!)
1984	Yamaha	Marketing of custom FM LSI begins	Yamaha began to market their in-house FM expertise to the world
1985	Akai	The S612 was the first affordable rackmount sampler, and the first in Akai's range	12-bit, quick-disk storage and only six-note polyphonic
1985	Ensoniq	Introduced the 'Mirage', an affordable 8-bit sample recording and replay instrument	
1985	Korg	Korg announced the DDM-110, the first low-cost digital drum machine	The beginning of a large number of digital drum machines …
1985	Yamaha	Yamaha R&D Studio opened in Tokyo, Japan	
1986	Sequential	Sequential launched the Prophet VS, a 'Vector' synth which used a joystick to mix sounds in real-time	One of the last sequential products before the demise of the company
1986	Steinberg	Steinberg's Pro 16 software for the Commodore C64	The start of the explosion of MIDI-based music software
1986	Yamaha	Clavinova CLP series electronic pianos launched	CLP pianos were pianos; the CVP series added on auto-accompaniment features
1986	Yamaha	DX7II was revised DX7 (a mark II)	Optional floppy disk drive
1987	Casio	Introduced the Casio CZ-101, probably the first low-cost multi-timbral digital synthesizer	Used Phase Distortion, a variant of waveshaping
1987	DAT	DAT was launched. The first digital audio recording system intended for domestic use	Worries over piracy severely prevented its mass marketing
1987	Roland	MT-32 brought multi-timbral S&S in a module	The start of the 'keyboard' and 'module' duality
1987	Roland	Roland D-50 combined sample technology with synthesis in a low-cost mass-produced instrument	S&S
1987	Yamaha	Yamaha DX7II centennial model, second-generation DX7, but with extended keyboard (88 notes) and gold plating everywhere	Limited edition

Date	Name	Event	Notes
1987	Yamaha	Yamaha R&D Studio opened in London, England	After a move out of the inner city, still active in 2003
1988	Korg	Korg M1 was launched. Uses digital S&S techniques with an excellent set of ROM sounds	A runaway best seller. Filter has no resonance
1989	Breakaway	The Breakaway Vocaliser 1000 is a pitch-to-MIDI device that translates singing into MIDI messages and sounds via its on-board sampled sounds	Somewhat marred by a disastrous live demonstration on the BBC's 'Tomorrow's World' programme
1990	Technos	French–Canadian company Technos announced the Axcel, first resynthesizer	Not a commercial success
1991	General MIDI (GM)	First formalisation of synthesizer sounds and drums	Specifies sounds, program change tables and drum note allocation
1992	MiniDisc	Recordable digital audio disc format was released by Sony	
1995	Yamaha	VL1, world's first physical modelling instrument was launched	Duophonic and very expensive
1996	SMF	Standard MIDI File specification announced by MIDI Manufacturer's Association	
1997	Korg	Z1 polyphonic physical modelling synthesizer	
1997	DVD	First DVD video players are released. DVD Audio standard does not appear until 1999 …	
1998	Yamaha	DJ-X, a dance performance keyboard disguised as a 'fun' keyboard	Followed by a keyboardless DJ version, the DJXIIB
1999	MP3	First MP3 audio players for computers appeared	Internet music downloading began …
1999	General MIDI 2 (GM2)	Enhanced superset of GM, now called GM1	Backwards compatible. Increased the tightness of the specification
2000	Yamaha	mLAN, a FireWire-based, single cable for digital audio and MIDI	Slow acceptance for a brilliant concept
2001	Korg	Karma, a combination of a synthesizer with a powerful set of algorithmic time and timbre processing	
2001	General MIDI Lite	Reduced GM1 specification designed for use in devices like mobile phones	
2002	Hartmann Music	Neuron Resynthesizer	Arguably the first commercially produced resynthesizer
2002	SP-MIDI	Scalable polyphony was layered, note priority approach to making GM files to drive polyphonic ring tones in mobile phones	Some mobile phones are GM1 compatible
2003	Yamaha	Vocaloid, mass-market singing synthesis software	Backing vocals will never be the same again!

Techniques

2   Analogue synthesis

This chapter describes analogue synthesis: from voltage control to musical instrument digital interface (MIDI); from monophonic to polyphonic; from modular to performance-oriented; from subtractive synthesis to formant synthesis and beyond.

The word 'analogue' can also be spelt without the 'ue' ending. In this book, the longer version will be used.

2.1   Analogue and digital

The word 'analogue' means that a range of values are presented in a continuous rather than a discrete way. Continuous implies making measurements all the time, and also infinite resolution – although inherent physical limitations like the grain size on photographic flim, or the noise level in an electronic circuit will prevent any real-world system from being truly continuous. Discrete means that you use individual finite sample values taken at regular intervals rather than measure all the time, with the assumption that the samples are a good representation of the original signal. Digital synthesis uses these discrete values.

An analogue synthesizer is thus usually defined as one that uses voltages and currents to directly represent both audio signals and any control signals which are used to manipulate those audio signals. In fact, 'analogue' can also refer to any technology in which sound is created and manipulated in any way where the representation is continuous rather than discrete. Analogue computers were used before low-cost digital circuitry became widely available, and they used voltages and currents to represent numbers. They were used to solve complex problems in navigation, dynamics and mathematics.

Analogue electronics happens to be a convenient way of producing sound signals – but there are many other ways: mechanical, hydraulic, electrostatic, chemical, etc. For example, vinyl discs use analogue technology where the mechanical movement of the stylus is converted into sound. Tape recorders reproduce sound from analogue signals stored on magnetic tape.

In synthesisers, the use of the word 'analogue' often implies voltage controlled oscillators (VCOs) and filters (VCFs). These have a characteristic set of audio characteristics: VCOs can have tuning stability or modulation linearity problems, for example; and analogue filters can break into self-oscillation or may distort the signal passing through them. These features of the analogue electronics which are used in the design can contribute to the overall 'tone quality' of the instruments.

Analogue synthesizers are commonly regarded as being very useful for producing bass, brass and the synthesizer 'cliché' sounds, but not a very good choice for simulating 'real' sounds. The typical clichéd sound is usually a 'synthy' sound consisting of slightly detuned oscillators beating against each other, with a resonant filter swept by a decaying envelope.

Digital synthesizers can deliberately introduce randomness, and these attempts to emulate reality are known as modelling; see Section 5.3.

In contrast, digital synthesizers use discrete numerical representations of the audio and control signals. They are thus capable of reproducing pre-recorded samples of real instruments with a very high fidelity. They also tend to be very precise and predictable, with none of the inherent uncertainty of analogue instruments. Some of the many digital synthesis techniques are described in Chapter 5.

The difference between analogue and digital representations can be likened to an experiment to measure the traffic flow through a road junction. The actual passage of cars can be observed and the number of cars passing a specific point in a given time interval are noted down. The movement of the cars is analogue in nature, since it is continuous, whereas the numbers are digital, since they only provide numbers at specific times (Figure 2.1.1).

This link between a physical experiment and the numbers, which can be used to describe it, is also significant because the first analogue synthesizers, and in fact, the first computers, were analogue, not digital. An *analogue computer* is a device which is used to solve mathematical problems by providing an electrical circuit which behaves in the same way as a real system, and then observing that happens when some of the parameters are changed. A simple example is what happens when two containers filled with water are connected together. This can be modelled by using an integrator circuit: a capacitor in a feedback loop (Figure 2.1.2). A step voltage applied to the integrator input simulates pouring water into one container – the voltage at the output of the integrator will rise steadily until the voltage is the same as the applied voltage, and then stops. If the integrator time constant is made larger, which is equivalent to reducing the flow of water between the containers (or making the second

Figure 2.1.1 The movement of the cars is continuous or analogue, whereas the number of cars is discrete or digital.

Figure 2.1.2 Two connected buckets can model an integrator circuit.

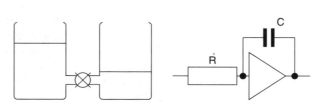

container larger), then the integrator will take longer to reach a steady state after a step voltage has been applied.

More sophisticated situations require more complex models, but the basic idea of using linear electronic circuits to simulate the behaviour of real-world mechanical systems can be very successful. For more information on modelling techniques, see Section 5.3.

2.1.1 Voltage control

One of the major innovations in the development of the synthesiser was voltage control. Instead of providing *mechanical control* over the many parameters which are used to set the operation of a synthesiser, voltages are used. Since the component parts of the synthesizer produce audio signals which are also voltages, the same signals which are used for audio can also be used for control purposes. One example is an oscillator used for tremolo or vibrato modulation when used at a frequency of a few tens of hertz, but the same oscillator becomes a sound source itself if the frequency is a few hundred hertz.

'Mechanical control' here means human-operated switches and knobs ...

Controlling a synthesizer with voltages requires some way of manipulating the voltages themselves, and for this voltage controlled amplifiers (VCAs) are used. These use a control voltage (also known as CV) to alter the gain of the amplifier, and can be used to control the gain of audio signals or control voltages. Using VCAs means that a synthesizer can provide a single common gain control element.

Although not all analogue synthesizers contain the same elements, many of the parts are common, and the method of control is the same throughout. Voltage control requires two main parts: sources and destinations.

Voltage control sources include the following:

- **Low frequency oscillators (LFOs)**: These are required for vibrato, tremolo and other cyclic effects.
- **Envelope generators (EGs)**: These produce multi-segment control voltages, where the time and slope of each segment can be controlled independently.
- **Pitch control**: Typically provided by a pitch wheel or lever, which provides a control voltage where the amount of pitch bend is proportional to the voltage.
- **Keyboard control**: The output from a music keyboard – provides a control voltage where the pitch is proportional to the voltage.
- **VCFs**: These can self-oscillate and so provide control signals.
- **VCOs**: These can be used as part of FM or ring modulation sounds.

Voltage controlled destinations include:

- **LFOs**, where the voltage is used to control the frequency or the waveshape.
- **EGs**, where the voltages can be used to control the times or slopes of each of the segments.

- **VCFs**, where the voltage is used to control the cut-off frequency of the filter, and perhaps the Q or resonance of the filter.
- **VCOs**, where the voltage is used to control the frequency of the oscillator, or sometimes the shape or pulse width of the output waveform.
- **Voltage controlled pan**, where the voltage is used to control the stereo positioning of the sound.
- **VCAs**, where the voltage is used to control the gain of the amplifier.

Each of these modules will be explained in more depth in this chapter.

2.1.2 Tape and models

Not all analogue synthesizers have to be voltage controlled. The use of tape manipulation and real physical instruments to synthesise sounds might be regarded as the ultimate in 'analogue' synthesis, since it is actually possible to interact with the actual sounds directly and continuously. Despite this, the word 'analogue' usually implies the use of electronic synthesizers.

The 'source and modifier' model is often applied to analogue synthesizers, where the VCOs are the source of the raw audio, and the VCF, VCA and attack decay sustain release (ADSR) envelopes form the modifiers. But the same model can be applied to sample and synthesis (S&S) synthesizers, or even to physical modelling. Even real-world musical instruments tend to have a source (for a violin, you vibrate

Figure 2.1.3 Performance controls are altered during the playing of the instrument, whilst fixed parameter controls normally remain unchanged.

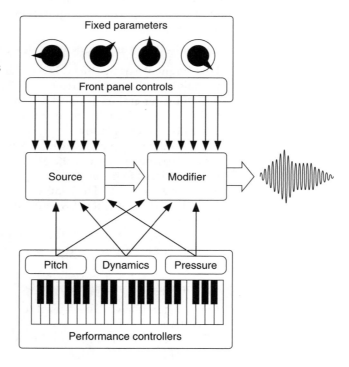

the string using the bow) and modifier structure (for a violin it is the resonance of the body that gives the final 'tone' of the sound).

The controls of the sound source and the modifier can be split into two parts: *performance controls* which are altered during the playing of the instrument and *fixed parameter controls* which tend to remain unchanged whilst the instrument is played (Figure 2.1.3).

Because it came first, many of the terminology, models, and metaphors of analogue synthesis are re-used in the more recent digital methods. Although this serves to improve the familiarity for anyone who has used an analogue synthesizer, it does not help a more conventional musician who has never used anything other than a real instrument.

2.2   Subtractive synthesis

Subtractive synthesis is often mistakenly regarded as the *only* method of analogue sound synthesis. Although there are other methods of synthesis, the majority of commercial analogue synthesizers use subtractive synthesis. Because it is often presented with a user interface consisting of a large number of knobs and switches, it can be intimidating to the beginner. Because there is often a one-to-one relationship between the available controls and the knobs and switches, it is well suited to educational purposes. It can also be used to illustrate a number of important principles and models which are used in acoustics and sound theory.

Theory: source and modifier

Subtractive synthesis is based around the idea that real instruments can be broken down into three major parts: a source of sound, a modifier (which processes the output of the source) and some controllers (which act as the interface between the performer and the instrument). This is most obviously apparent in many wind instruments, where the individual parts can be examined in isolation (Figure 2.2.1).

For example: a clarinet, where a vibrating reed is coupled to a tube, can be taken apart and the two parts can be investigated independently. On its own, the reed produces a harsh, strident tone, whilst the body of the instrument is merely a tube which can be shown to have a series of acoustic resonances related to its length, the diameter of the

Figure 2.2.1 The performer uses the instrument controllers to alter the source and modifier parameters.

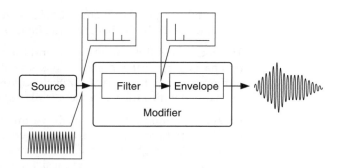

Figure 2.2.2 The source produces a constant raw waveform. The filter changes the harmonic structure, whilst the envelope shapes the sound.

longitudinal hole and other physical characteristic, in other words, it behaves like a series of resonant filters. Put together, the reed produces a sound which is then modified by the resonances of the body of the instrument to produce the final characteristic sound of the clarinet.

Although this model is a powerful metaphor for helping to understand how some musical instruments work, it is by no means a complete or unique answer. Attempting to apply the same concept to an instrument like a guitar is more difficult, since the source of the sound appears to be the plucked string, and the body of the guitar must therefore be the modifier of the sound produced by the string. Unfortunately, in a guitar, the source and modifier are much more closely coupled, and it is much harder to split them into separate parts. For example, the string cannot be played in isolation in quite the same way as the reed of a clarinet can, nor can all of the resonances of the guitar body be determined without the strings being present.

Despite this, the idea of modifying the output of a sound source is easy to grasp, and it can be used to produce a wide range of synthetic and imitative timbres. In fact, the underlying idea of source and modifier is a common theme in most types of sound synthesis.

Subtractive synthesis

Subtractive synthesis uses a subset of this generalised idea of source and modifier, where the source produces a sound which contains all the required harmonic content for the final sound, whilst the modifier is used to filter out any unwanted harmonics and shape the sound's volume envelope. The filter thus 'subtracts' the harmonics which are not required; hence the name of the synthesis method (Figure 2.2.2).

2.2.1 Sources

The waveshapes in analogue synthesizers are merely approximations to the mathematical shapes, and the differences give part of the appeal of analogue sounds.

The sound sources used in analogue subtractive synthesizers tend to be based on mathematics. There are two basic types: waveforms and random. The waveforms are typically simple *waveshapes*: sawtooth, square, pulse, sine and triangle are the most common. The shapes are the ones which are easy to describe mathematically, and also to produce electronically. Random waveshapes produce noise, which contains a constantly changing mixture of all frequencies.

Figure 2.2.3 A block diagram of a typical VCO.

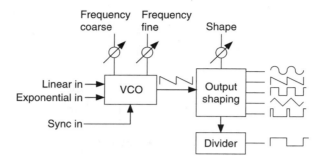

Oscillators are related to one of the component parts of analogue synthesizers: function generators. A function generator produces an output waveform, and this can be of arbitrary shape, and can be continuous or triggered. An oscillator which is intended to be used in a basic analogue subtractive synthesizer normally produces just a few continuous waveshapes, and the frequency needs to be controlled by a voltage.

VCOs

The VCOs provide voltage control of the frequency or pitch of their output. Some VCOs also provide voltage control inputs for modulation (usually FM) and for varying the shape of the output waveforms (usually the pulse width of the rectangular waveshape, although some VCOs allow the shape of other waveforms to be altered as well). Many VCOs have an additional input for another VCO audio signal, to which the VCO can be synchronised. Hard synchronisation forces the VCO to reset its output to keep in sync with the incoming signal, which means that the VCO can only operate at the same or multiple frequencies of the input frequency. This produces a characteristic harsh sound. Other 'softer' synchronisation schemes can be used to produce timbral changes in the output rather than locking of the VCO frequency.

A typical VCO has controls for the coarse (semitones) and fine (cents) tuning of its pitch, some sort of waveform selector (usually one of sine, triangle, square, sawtooth and pulse), a pulse width control for the shape of the pulse waveform, and an output level control (Figure 2.2.3). Sometimes multiple simultaneous output waveforms are available, and some VCOs also provide 'sub-octave' outputs which are one or two octaves lower in pitch. A control voltage for the pulse width allows the shape of the pulse waveform (and sometimes other waveforms as well), to be altered. This is called pulse width modulation (PWM), or shape modulation.

Harmonic content of waveforms

The Mini Moog waveforms are arranged in the order of increasing harmonic content.

The ordering of waveforms on some early analogue synthesizers was not random. The waveforms are deliberately arranged so that the harmonic content increases as the rotary control is twisted.

Figure 2.2.4 A sine waveform and harmonic spectrum, and the same diagrams with actual frequencies shown.

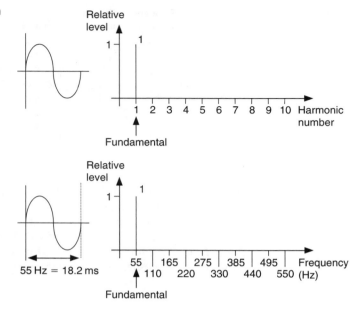

Figure 2.2.5 A triangle waveform and spectrum.

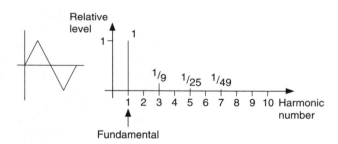

The simplest waveshape is the sine wave (Figure 2.2.4). This is a smooth, rounded waveform based on the mathematical sine function. A sine wave contains just one 'harmonic', the first or fundamental. This makes it somewhat unsuitable for subtractive synthesis since it has no harmonics to be filtered.

A triangle waveshape has two linear slopes (Figure 2.2.5). It has small amounts of odd-numbered harmonics, which give it enough harmonic content for a filter to work on.

A square wave contains only odd harmonics (Figure 2.2.6). It has a distinctive 'hollow' sound and a very synthetic feel.

A sawtooth wave contains both odd and even harmonics (Figure 2.2.7). It sounds bright, although many pulse waves can actually have more harmonic content. 'Super-sawtooth' waveshapes replace the linear slope with exponential slopes, as well as gapped sawtooths: these can contain greater levels of the upper harmonics than the basic sawtooth.

Figure 2.2.6 A square waveform and spectrum, with a typical clarinet spectrum for comparison.

Figure 2.2.7 A sawtooth waveform and spectrum, with the spectrum also shown on a vertical decibel scale.

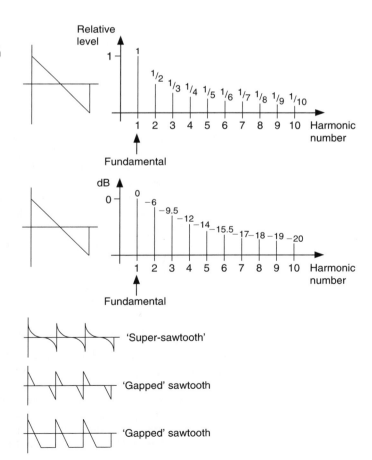

Figure 2.2.8 A pulse wave and spectrum. The relative levels of the harmonics depend on the width of the pulse.

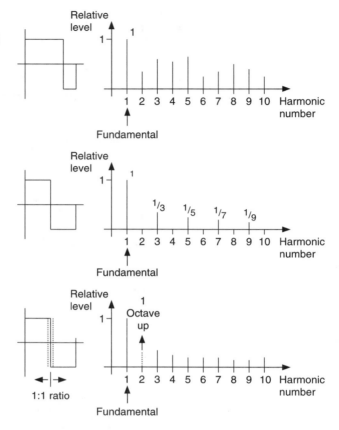

It is possible to adjust the pulse width to give a square by ear: listening to the fundamental, the pulse width is adjusted until the note one octave up fades away. This note is the second harmonic, and is thus not present in a square waveform. See also Figure 2.2.38.

Depending on the ratio between the two parts (known as the mark–space ratio, shape, duty cycle or symmetry), pulse waveforms (Figure 2.2.8) can contain both odd and even harmonics, although not all of the harmonics are always present. The overall harmonic content of pulse waves increases as the pulse width narrows, although if a pulse gets too narrow it can completely disappear (the depth of PWM needs to be carefully adjusted to prevent this). A special case of a pulse waveshape is the square wave, where the even harmonics are not present. Pulse width modulated pulse waveforms are known as PWM waveforms, and their harmonic content changes as the width of the pulse varies. PWM waveforms are normally controlled with LFO or an envelope, so that the pulse width changes with time. The audible effect when a PWM waveform is cyclically changed by an LFO is similar to two oscillators beating together.

All of the waveshapes and harmonic contents shown above are idealised. In the real world the edges are not as sharp, the shapes are not so linear, and the spectra are not as mathematically precise. Figure 2.2.6 shows a more realistic spectrum with dotted lines. For example, the 'sine' wave output on many VCOs is usually produced by shaping a triangle wave through a non-linear amplifier which rounds off the top of the triangle so that it looks like a true sine wave.

Figure 2.2.9 Analogue waveshaping allows the conversion of one waveform shape into others. In this example the sawtooth is the source waveform, although others are possible.

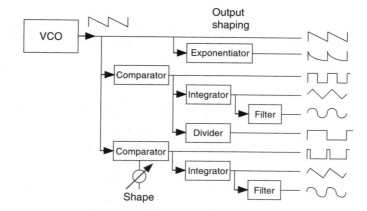

The resulting waveform resembles a sine wave, although it will have additional harmonics – but for the purposes of subtractive synthesis, it is perfectly adequate. Section 2.3 on additive synthesis shows what real-world waveforms look like when they are constructed from simpler waveforms, rather than the perfect cases shown above (Figure 2.2.9).

2.2.2   Modifiers

Effects are not normally included as 'modifiers' in analogue synthesizers.

There are two major modifiers for audio signals in analogue synthesizers: filters and amplifiers. Filtering is used to change the harmonic content or timbre of the sound, whilst amplification is used to change the volume or 'shape' of the sound. Both types of modifiers are typically controlled by EGs, which produce complex control voltages which change with time.

2.2.3   Filters

A filter is an amplifier whose gain changes with frequency. It is usually the convention to have filters whose maximum gain is one, and so it is more correct to say that for a filter, the attenuation changes with frequency. A VCF is one where one or more parameters can be altered using a control voltage. Filters are powerful modifiers of timbre, because they can change the relative proportions of harmonics in a sound.

Filters come in many different forms. One classification method is based on the shape of the attenuation curve. If a sine wave test signal is passed through a filter, then the output represents the attenuation of the filter at that frequency; this is called the frequency response of the filter. An alternative method injects a noise signal into the filter and then monitors the output spectrum, but the sine wave method is easier to carry out. The major types of frequency response curve are:

* low-pass
* band-pass
* high-pass
* notch.

In general, analogue synthesizer filters have two or four poles, whilst digital filters can have up to eight or more.

Low-pass

A low-pass filter has more attenuation as the frequency increases. The point at which the attenuation is 3 dB is called the cut-off frequency, since this is the frequency at which the attenuation first becomes apparent. It is also the point at which half of the power in the audio signal has been lost, and so it is sometimes called the half-power point. Below the cut-off frequency, a low-pass filter has no effect on the audio signal, and it is said to have a flat response (the attenuation does not change with frequency). Above the cut-off frequency, the attenuation increases at a rate which is called a slope. The slope of the attenuation varies with the design of the filter. Simple filters with one resistor and capacitor (RC) will have slopes of 6 dB/octave, which means that for each doubling of frequency, the attenuation increases by 6 dB. Each pair of RC elements is called a pole, and the slope increases as the number of poles increases. A two-pole filter will have an attenuation of 12 dB/octave, whilst a four-pole filter will have 24 dB/octave. Audibly, a four-pole filter has a more 'synthetic' tone, and makes much larger changes to the timbre of the sound as the cut-off frequency is changed. A two-pole filter is usually associated with a more 'natural' sound and more subtle changes to the timbre (Figure 2.2.10).

Low-pass VCFs usually have the cut-off frequency as the main controlled parameter. A sweep of cut-off frequency from high to low frequencies makes any audio signal progressively 'darker', with the lower frequencies emphasised and less high frequencies present. A filter sweeping from high frequency to low frequency of cut-off is often referred to as changing from 'open' to 'closed'. When the cut-off frequency is set to maximum, and the filter is 'open', then all frequencies can pass through the filter.

As the cut-off frequency of a low-pass filter is raised from zero, the first frequency which is heard is usually the fundamental. As the frequency rises, each of the successive harmonics (if any) of the sound will be heard. The audible effect of this is an initial sine wave (the fundamental), followed by a gradual increase in the 'brightness' of the sound as any additional frequencies are allowed through the filter. If the cut-off frequency of a low-pass filter is set to allow just the fundamental to pass through the filter, then the resulting sine wave will be identical for any input signal waveform. It is only when the cut-off frequency is increased and additional harmonics are heard, then the differences between the different waveforms will become apparent. For example, a sawtooth will have a second harmonic, whilst a square wave will not.

High-pass

A high-pass filter has the opposite filtering action to a low-pass filter: it attenuates all frequencies which are below the cut-off frequency. As with the low-pass VCF, the parameter which is voltage controlled is the cut-off frequency. High-pass filters remove harmonics from a signal waveform, but as the frequency is raised from zero, then it is

Figure 2.2.10 Filter responses are normally shown on a log frequency scale since a dB/octave cut-off slope then appears as a straight line. But harmonics are based on linear frequency scales, and on these graphs the filter appears as a curve. Low-pass filtering a sawtooth waveform with the cut-off frequency set to four different values:
(i) At 100 Hz the filter cut-off frequency is the same as the fundamental frequency of the sawtooth waveform. The second harmonic is 30 dB below the fundamental and so the ear will hear an impure sine wave at 100 Hz.
(ii) At 300 Hz the first three harmonics are in the pass-band of the filter, and the output will sound considerably brighter.
(iii) At 500 Hz, the first five harmonics are in the filter pass-band, and so the output will sound like a slightly dull sawtooth waveform.
(iv) At 1 kHz, the first ten harmonics are all in the pass-band of the filter and the output will sound like a sawtooth waveform.

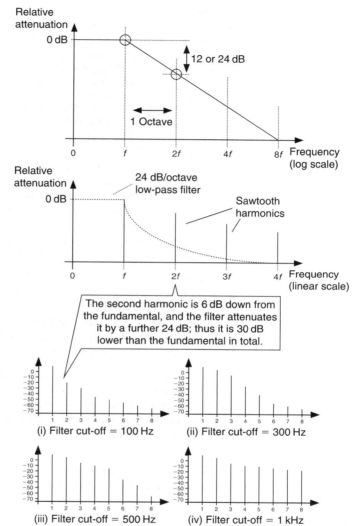

the fundamental which is removed first. As additional harmonics are removed, the timbre becomes 'thinner' and brighter, with less low frequency content and more high frequency content, and the perceived pitch of the sound may change because the fundamental is missing.

Some subtractive synthesizers have a high-pass non-VCF connected either before or after the low-pass VCF in the signal path. This allows limited additional control over the low frequencies which are passed by the low-pass filter. It is usually used to remove or change the level of the fundamental, which is useful for imitating the timbre of instruments where the fundamental is not the largest frequency component.

Band-pass (and notch) filters are the equivalent of the resonances that happen in the real world. A wine-glass can be stimulated to oscillate at its resonant frequency by running a wet finger around the rim.

Band-pass

A band-pass filter only allows a set range of frequencies to pass through it unchanged – all other frequencies are attenuated. The range of frequencies which are passed is called the bandwidth, or more usually, the pass-band, of the filter. Band-pass VCFs usually have control over the cut-off frequency and bandwidth.

A band-pass filter can be thought of as a combination of a high-pass and a low-pass filter, connected in series, one after the other in the signal path. By using the same control voltage to the cut-off frequency inputs of two VCFs (one high-pass and the other low-pass), then the cut-off frequencies will 'track' each other, and the effective bandwidth of the band-pass filter will stay constant as the cut-off frequencies are changed. The width of the band-pass filter's pass-band can be controlled by adding an extra control voltage offset to one of the filters. If the cut-off frequency of the low-pass filter is set below that of the high-pass filter, then the pass-band does not exist, and no frequencies will pass through the filter (Figure 2.2.11).

Band-pass filters are often described in terms of the shape of their pass-band response. Narrow pass-bands are referred to as 'narrow' or 'sharp', and they produce marked changes in the frequency content of an audio signal. Wider pass-bands have less effect on the timbre, since they merely emphasise a range of frequencies. The middle frequency of the pass-band is called the centre frequency.

Very narrow band-pass filters can be used to examine a waveform and determine its frequency content. By sweeping through the frequency range, each harmonic frequency will be heard as a sine wave when the centre frequency of the band-pass filter is the same as the frequency of the harmonic (Figure 2.2.12).

Notch

A notch filter is the opposite of a band-pass filter. Instead of passing a band of frequencies, it attenuates just those frequencies and allows

Figure 2.2.11 A band-pass filter only passes frequencies in a specific range. This is normally the two points at which the filter attenuates by 3 dB. It can be thought of as a low-pass and a high-pass filter connected in series (one after the other). In the example shown, the lower cut-off frequency is about 0.6*f* (for the high-pass filter), whilst the upper cut-off frequency is about 1.6*f* (for the low-pass filter). The bandwidth of the filter is the difference between these two cut-off frequencies. Small differences are referred to as 'narrow', whilst large differences are known as 'wide'.

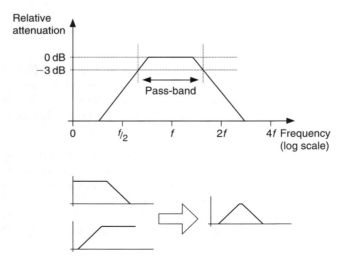

Figure 2.2.12 If a narrow band-pass filter is used to process a sound which has a rich harmonic content, then the harmonics which are in the pass-band of the filter will be emphasised, whilst the remainder will be attenuated. This produces a characteristic resonant sound. If the band-pass filter is moved up and down the frequency axis, then a characteristic 'wah-wah' sound will be heard – this is sometimes used on electric guitar sounds.

Figure 2.2.13 A notch filter is the opposite of a band-pass filter, which it attenuates a band of frequencies. It can also be formed from a series combination of a low-pass and a high-pass filter, provided that the low-pass cut-off frequency is lower than the high-pass cut-off frequency. If not, then no notch will be present.

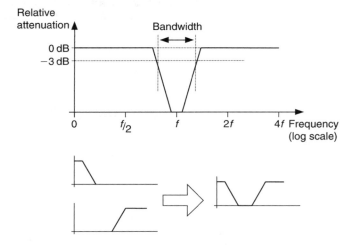

all others to pass through unaffected. Notch filters are used to remove or attenuate specific ranges of frequencies, and narrow 'notches' can be used to remove single harmonic frequencies from a sound. Notch VCFs usually provide control over both the cut-off and bandwidth (or 'stop-band') of the filter (Figure 2.2.13).

Scaling

If the keyboard pitch voltage is connected to the cut-off frequency control voltage input of a VCF, then the cut-off frequency can be made to track the pitch being played on the keyboard. This means that any note played on the keyboard is subjected to the same relative filtering, since the cut-off frequency will follow the pitch being played. This is called pitch tracking or keyboard scaling (Figure 2.2.14).

Resonance

Low-pass and high-pass filters can have different response curves depending on a parameter called resonance or Q (short for 'quality', but rarely referred to as such). Resonance is a peaking or accentuation

Figure 2.2.14 Filter scaling, tracking or following is the term used to describe changing the filter cut-off so that it follows changes in the pitch of a sound. This allows the spectrum of the sound produced to stay the same. In the example shown, the filter peak tracks the changes in the pitch of the sound when two notes two octaves apart are played – the peak coincides with the fundamental frequency in each case. With no filter scaling then the note with a fundamental of $4f$ two octaves up would be strongly attenuated if the filter cut-off frequency did not change from the peak at a frequency of f.

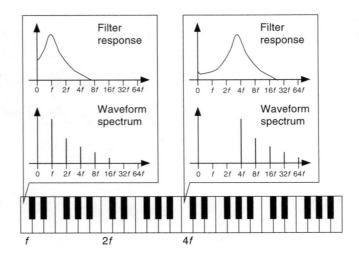

Figure 2.2.15 Resonance changes the shape of a low-pass filter response most markedly at the cut-off frequency. The result is a smooth and continuous transition from a low-pass to something like a narrow band-pass filter.

of the frequency response of the filter at a specific frequency. For band-pass filters, the Q figure is given by the formula:

$$Q = \frac{\text{Centre frequency}}{\text{Bandwidth (or pass-band)}}$$

This formula is often also used for the resonance in the low-pass and high-pass filters used in synthesizers. For these low-pass and high-pass filters the resonance is usually at the cut-off frequency, and it forms a 'peak' in the frequency response (Figure 2.2.15).

In many VCFs, internal feedback is used to produce resonance. By taking some of the output signal and adding it back into the input of the filter, the response of the filter can be emphasised at the cut-off frequency. See Section 2.6 for more information on the implementation of filters.

Most subtractive synthesizers implement only low-pass and band-pass filtering, where the band-pass is often produced by increasing the Q of the low-pass filter so that it is a 'peaky' low-pass rather than a true band-pass filter. This phenomenon of a peak of gain in an otherwise low-pass (or high-pass) response is called 'corner peaking'. Some models of analogue synthesizer also have an additional simple high-pass filter, whilst notch filters or band-rejects are very uncommon.

There are two types of filters: constant-Q and constant bandwidth. Constant-Q filters do not change their Q as the frequency of the filter is changed. This means that they are good for applications where the filter is used to produce a sense of pitch from an unpitched source like noise. Since the Q is constant, the bandwidth varies with the filter frequency and so sounds 'musical'. Constant-bandwidth filters have the same bandwidth regardless of the filter frequency. This means that a relatively narrow bandwidth of 100 Hz for a filter frequency of 4 kHz, is very wide for a 400-Hz frequency: the Q of a constant-bandwidth filter changes with the filter frequency. Most analogue synthesizer filters are constant-Q.

The effect of changing the cut-off frequency of a highly resonant low-pass filter in 'real time' is quite distinctive, and can be approximated by singing 'eee-yah-oh-ooh' as a continuous sweep of vowel sounds.

Filter oscillation

If the resonance of a peaky low-pass or a band-pass VCF is increased to the point at which the filter plus its feedback has a cumulative gain of more than one at the cut-off frequency, then it will break into self-oscillation. In fact, this is one method of producing an oscillator – you put a circuit with a narrow band-pass frequency response into the feedback loop of an amplifier or operational amplifier (op-amp) (Figure 2.2.16). The oscillation produces a sine wave, sometimes much purer than the 'sine' waves produced by the VCOs!

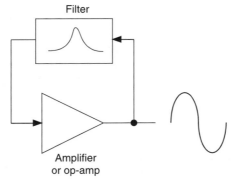

Figure 2.2.16 If a filter with a strong resonant peak in its response is connected around an amplifier, then the circuit will tend to oscillate at the frequency with the highest gain – at the peak of the filter response. This can be easily demonstrated (perhaps too easily) with a microphone and a public address system.

2.2.4 Envelopes

An envelope is the overall 'shape' of the volume of a sound, plotted against time (Figure 2.2.17). In an analogue synthesizer, the volume of the sound output at any time is controlled by a voltage controlled amplifier (see VCA), and the voltage which is used is called an envelope. Envelopes are produced by 'EGs', and have many variants. EGs are categorised by the number of controls which they provide over the shape of the envelope. The simplest provide control only over the start and end of a sound, whilst the most complex may have a very large number of parameters.

Envelopes are split into segments or parts (Figure 2.2.18). The time from silence to the initial loudest point is called the *attack time*, whilst the time for the envelope to decrease or decay to a steady value is

Figure 2.2.17 The 'envelope' of a sound is the overall shape – the change in volume with time. The shape of an envelope often forms a distinctive part of a sound.

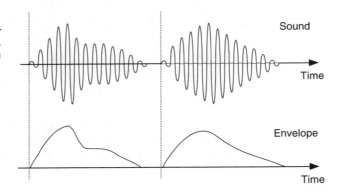

Figure 2.2.18 Envelopes are divided into segments depending on their position. The start of the sound is called the 'attack segment'. After the loudest part of the sound, the fall to a steady 'sustain' segment is called the 'decay' segment. When the sound ends, the fall from the sustain segment is called the 'release' segment.

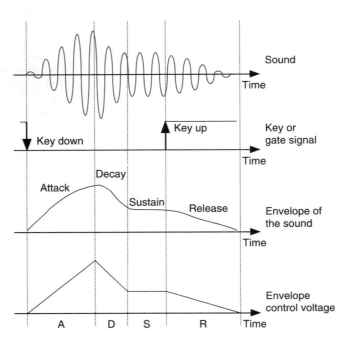

called the *decay time*. For instruments that can produce a continuous sound, like an organ, the decay time is defined as the time for the sound to decay to the steady-state 'sustain' level, whilst the time that it takes for the sound to decay to silence when it ends is called the *release time*.

Bowed stringed instruments can have long attack, decay and release times, whilst plucked stringed instruments have shorter attack times and no sustain time. Pianos and percussion instruments can have very fast attack times and complex decay/sustain segments. There is an almost standardised set of names for the segments of envelopes in analogue synthesizers, which is in contrast to the more diverse naming schemes used in digital synthesizers.

Envelopes are usually referred to in terms of the control voltage that they produce, and it is normally assumed that they are started by a key being pressed on a keyboard. Envelopes can be considered to be sophisticated time-based function generators with manual key triggering.

The following are some of the common types of EGs.

Attack release

Attack release (AR) envelopes only provide control over the start and end of a sound (Figure 2.2.19). The two-segment envelope control

Figure 2.2.19 In an AR envelope the pressing down of a key (or a similar gating device on a synthesizer that does not use keys) starts the attack segment. When the peak level has been reached, then the envelope stays at this level until the key is released (of the gating signal is removed) and the envelope falls in the release segment. If the key is released whilst the envelope is in the attack segment, then the envelope normally moves to the release segment, and need not reach the peak level (see also Figure 2.2.27). Some synthesizers provide a control which forces the whole of the attack segment to be completed.

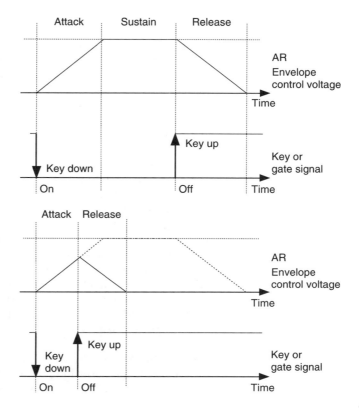

Figure 2.2.20 An AD envelope is similar to an AR envelope, except that there is no sustain segment. When the peak level is reached, the envelope decays, even if the key is held down.

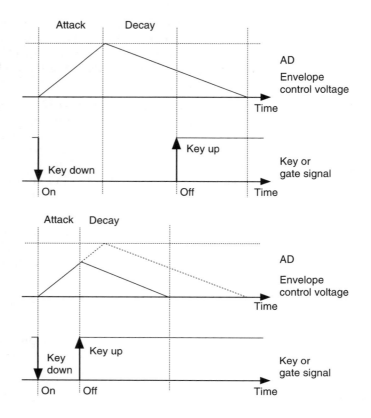

voltage, which is produced, rises up to the maximum level, and then falls back to the quiescent level, which is usually 0 volts. AR envelopes are often found on 1970s vintage string machines: simple polyphonic keyboards which used organ 'master oscillator and divider' technology with simple filtering and chorus effects processing to give an emulation of an orchestral string sound (see Section 3.3 on Digitally controlled oscillators for more information).

Attack decay

If the envelope moves into the decay segment as soon as the attack segment has reached its maximum level, then the decay time sets how long it takes for the envelope to drop to zero. This means that only percussive (non-sustaining) envelopes can be produced (unless the decay time is set to be very long, as in some attack decay release (ADR) envelopes). These two-segment attack delay (AD) envelopes (Figure 2.2.20) are often found connected to the frequency control input of VCOs, where the envelope then produces a rapid change in pitch at the start of the note, known as a 'chirp'. This can be effective for vocal and brass sounds. Inverting the envelope can produce changes downwards in pitch instead of upwards.

Attack decay release

The ADR envelope uses long decay times to simulate a high sustain level, in which case the resulting envelope is very much like an AR

Figure 2.2.21 The ADR envelope provides control over separate decay and release segments. This allows more complex envelope shapes to be produced than is possible with AR or AD EGs. If the key or gate is released during the attack segment, then the envelope moves to the release segment and ignores the decay segment.

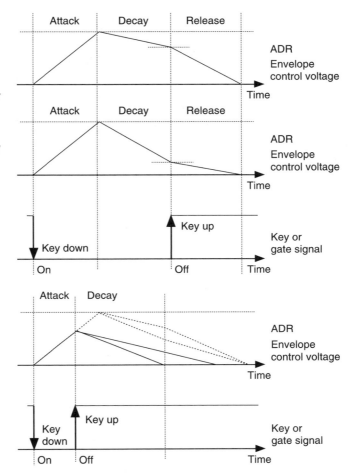

envelope, or else a percussive AD envelope by using shorter decay times (Figure 2.2.21).

Attack decay sustain

If a sustain level is added to an AD envelope, then the attack decay sustain (ADS) EG is the result (Figure 2.2.22). The attack segment reaches a maximum value, and the decay time then sets how long it takes for the envelope to reach the sustain level. Some ADS EGs have switches that make the release time the same as the decay time, or else have a very short release time. The type of envelope that is produced depends on the sustain level. If the sustain level is set to the maximum level (the same as the attack reaches) then two-segment AR-type envelopes are produced. If the sustain level is set to zero, then only two-segment AD envelopes are produced. With the sustain level set mid-way, then four-segment ADSR-type envelopes can be produced. These have an initial attack and decay portion, then the sustain portion whilst the key is held down, and then a release portion when the key is released.

Figure 2.2.22 An ADS envelope adds a sustain segment at the end of the decay segment. The 'release' time is normally set to the same as the decay time, although some synthesizers provide a switch which forces a fast release time regardless of the setting of the decay time. An ADS EG can be used to produce a wide variety of envelopes, including ones which have many of the characteristics of ADSR (see below), AR and AD envelopes.

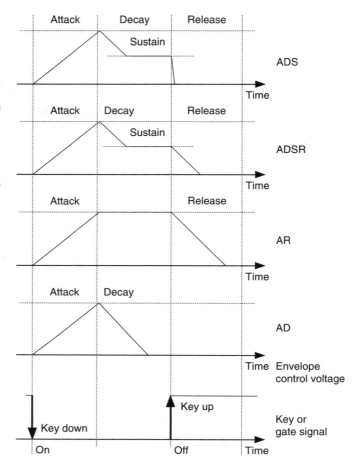

Attack decay sustain release

The most widely adopted EG is probably the ADSR (Figure 2.2.23). With just four controls, it is capable of producing a wide variety of envelope shapes; with only the attack decay 1 break decay 2 release (ADBDR) dual-decay variant offering superior flexibility at the cost of one extra control. The ADSR EG's main weakness is that the sustain segment is static, it is a fixed level. For this reason, ADSR-type envelopes are not particularly well suited in producing percussive piano-type envelopes, where the 'sustain' portion of the sound gradually decays to zero.

Attack hold decay sustain release

Some envelopes force the envelope to stay at the maximum or peak level for a fixed time when the attack segment has finished, and before the decay segment can start (Figure 2.2.24). These are called attack hold decay sustain release (AHDSR) envelopes. This is useful when a percussive envelope is set with very rapid attack and decay times, and the minimum length of the envelope needs to be controlled. For some sounds, an AD envelope with fast times (less than 10 ms) can be too short to be audible.

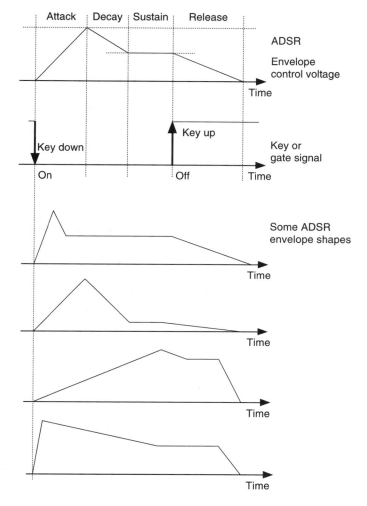

Figure 2.2.23 The ADSR envelope adds a separate control for the release time. This provides enough flexibility to produce a large number of envelopes with a small number of controls, and the ADSR envelope is widely used in synthesizers.

Figure 2.2.24 An AHDSR envelope adds a 'hold' segment at the end of the attack segment, rather like the sustain segment, but the length is set by a time rather than when the key or gate is released.
As with other envelope shapes, if the key is released before the sustain segment, then the envelope moves to the release segment.

A variation on the hold segment being after the 'attack' segment of the envelope is the attack decay hold release (ADHR) envelope, where the 'sustain' segment is only held up to a specific time, after which it begins to decay. This is arguably better suited to percussive and piano sounds than the ADSR.

Attack decay 1 break decay 2 release

By splitting the decay segment into two portions, with a 'break-point' level controlling when one decay portion finishes and the other starts, then a wide range of envelope shapes can be produced (Figure 2.2.25). By setting the second decay to very long times, then it can be used in much the same way as a sustain segment, although it has the advantage that it can still decay away slowly. This is arguably a better emulation of real-world envelopes for instruments like pianos, where the sustain segment is actually a long decay time. In some implementations of ADBDR envelopes, this second decay is called the 'slope' segment to distinguish it from the decay segment.

Advanced EGs

There are many sophisticated enhancements of the basic analogue ADSR EG (Figure 2.2.26) . Most of these are ADSRs with the addition of initial time delay, break-points in the attack or decay segments, and times for the peak and sustain levels. Although the extra controls provide more possibilities for envelope shapes, they also greatly increase the complexity of the user interface. Delayed envelopes (denoted by an initial 'D' in the abbreviation: DADSR for delayed ADSR) are used when the start of the envelope needs to be delayed in time without the need for using a long attack time, or where the attack needs to be rapid after the delay time.

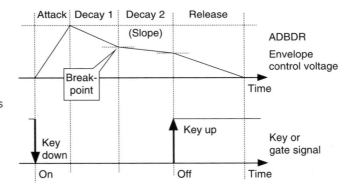

Figure 2.2.25 The ADBDR envelope has two decay segments and the transition from one decay is set by a variable level control, rather like a sustain level control. By setting the decay time to a long value, they can be used as pseudo-sustain segments, and so an ADBDR envelope can produce similar envelopes to an ADSR type.

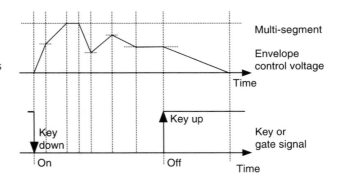

Figure 2.2.26 Multi-segment envelopes can have several attack, decay and release segments, as well as hold and sustain segments. Break-points can also be used to split a segment into smaller segments.

Some of these EGs provide a break-point in the attack segment, so that two different attack times can be controlled. This is especially useful for long attack times, where the start of the audio signal is too quiet to be heard, and the initial portion of the attack segment is heard as a delay. By having a rapid rise to a level where the audio signal is audible, followed by a slower second attack portion, this unwanted apparent delay can be avoided. This extra break-point is also useful for simulating more complicated attack curves.

Break-points are not always explicitly named as such. The interaction between the gate signal and the envelope often has implied break-points at the transitions between attack, decay, sustain and release. These are frequently not documented in the manufacturer's product information. The usual method of operation is shown in Figure 2.2.27. If the key is only held down for a short time, and the envelope is still in the attack segment when the key is released, then the envelope will go into the release segment. In this case the envelope may not reach the maximum level, although some EGs always rise to the maximum level. If there is a hold time associated with the maximum level, then this is usually not affected by the key being released. If the envelope has reached the decay segment, then when the key is released, the envelope will go into the release segment.

Figure 2.2.27 The transition from the attack segment to the release segment when the key or gate is released can be thought of as adding in a break-point to the attack segment.

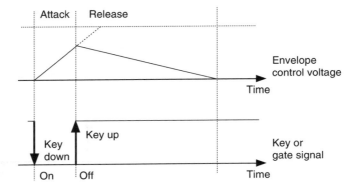

If the initial, final, peak and sustain levels are all controllable, then the envelope flexibility can become approximately equivalent to the multi-segment envelopes often found in digital synthesizers, although the terminology is normally very different. See Chapter 5 for more details on digital envelopes.

Some analogue synthesizers only have one EG, which is then used to control both the VCF and VCA. If two envelopes are available, then patching one to the filter and the other to the amplifier provides independent control over the volume and timbre. A third envelope could be used to control the pitch of the VCOs, or perhaps the stereo position of the sound using two VCAs arranged as a pan control.

Linear or exponential?

Many real-world quantities change in a non-linear way. This can be because of the process involved, or because of the way that the change is perceived. For example, the theoretical population growth curve of many animal species shows an exponential or power-law growth because the initial two animals produce two new individuals, who then eventually join the breeding population, and then these four individuals produce four new offsprings. The doubling of the population in each successive generation produces a rapidly increasing population curve. Conversely, because human ears perceive sound in a non-linear way, each doubling of the apparent volume level requires about ten times the energy in the sound. Again, the relationship connecting the two variables is a non-linear one.

Many natural sound envelopes have non-linear curves. Changes are usually rapid at first, and gradually slow down (Figure 2.2.28). This is particularly apparent with the attack segment of envelopes, where a linear rise in volume sounds too slow at first, whereas an exponential rise in volume sounds 'correct' – in fact, it sounds 'linear' to the human ear! Some EGs enable a switched selection between linear and exponential curves. EGs with break points in the attack, decay and release segments can produce similar effects to exponential curves, albeit with a crude approximation.

Figure 2.2.28 An exponential envelope does not use linear slopes, and often provides more realistic sounding envelopes.

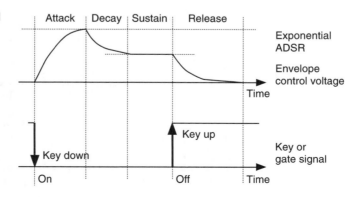

Triggering

The initiation of an EG is often assumed to be caused by a key being pressed on a music keyboard. Although this is the way that many synthesizers are set up, it is not the only way that envelopes can be started – an LFO or VCO could provide a trigger which will start the EG. In this case, the envelope is not tied to the keyboard, and can be used when a complex repeated control voltage is required (Figure 2.2.29).

When the keyboard is used to start an envelope, two separate signals are produced. The 'gate' signal indicates when the key is up or down, whilst the start of the key depression is shown by a 'trigger' pulse (Figure 2.2.30). The response of an EG to these two signals depends on how the EG is configured .

'Single trigger' EGs start when they receive a gate and a trigger, and progress through the envelope, entering the release segment when the gate signal ends to indicate that the key is no longer being held down.

'Multi trigger' EGs start when they receive a gate signal and a trigger pulse, but additional trigger pulses will restart part of the attack segment and the decay segment. These extra trigger pulses are normally produced by monophonic synthesizers (one note at once) only when a key is held down and another key is pressed.

'LFO trigger' or 'external trigger' EGs normally ignore the trigger pulse, and treat the input signal as a gate. The width of the LFO waveform or the length of the external signal set the length of the gate signal.

Whereas sources of audio signals or control voltages can be routed to almost any destination in a synthesizer, the routing of trigger and gate signals is often much more restricted – usually they are hard-wired from the keyboard.

Figure 2.2.29 The retriggering of an EG can sometimes be used to add in a break-point and start a new attack, normally from the level which had been reached by the envelope. The overall length of the envelope is controlled by the key being pressed down, or a similar gate control in synthesizer which are not controlled by a keyboard. The retriggering of the envelope is controlled by a trigger signal which is generated by the start of each new note. This is normally found on monophonic synthesizers, where the gate is produced globally from any keys which are being held down, whilst the triggers are produced individually by each key.

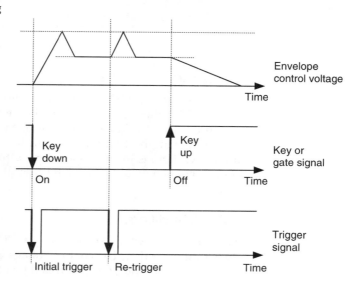

Figure 2.2.30 The gate and trigger routing from a keyboard to the EG is normally fixed, whilst the keyboard control voltage can be routed to a number of destinations.

Table 2.2.1 Envelope segments

Symbol	Segment	Description	Type
D	Delay	The time from the start of the envelope to the start of the attack segment	Time
I	Initial	The first level of the envelope. The quiescent level	Level
A	Attack	The time taken for the envelope to rise from the initial level to the maximum (peak) level	Time
H	Hold	The time that the envelope stays at the maximum (peak) level	Time
P	Peak	The level to which the envelope rises at the end of the attack time	Level
D	Decay	The time for the envelope to fall from the maximum (peak) level, to the sustain or final level	Time
B	Break-point	The level at which one decay segment changes to another	Level
D	Decay	The time for the second decay segment to fall from the break-point level, to the sustain or final level	Time
S	Sustain	The level at which the envelope stays whilst the key is held down. (Gate signal On)	Level
S	Sustain	The time for which the sustain segment lasts. (Often the minimum time)	Time
R	Release	The time for the envelope to fall, from the sustain level, to the final level	Time
F	Final	The final level of the envelope (usually the same as the initial)	Level

Voltage controlled parameters

Some EGs provide voltage control of the segment times and levels. This enables the shape of the envelope to be changed with one or more control voltages. One use of this facility is for 'scaling', where the length of all the times in the envelope are changed to imitate variations in envelope shape with pitch, in which case the control voltage would be derived from the keyboard pitch control voltage.

2.2.5 Amplifiers

Most analogue synthesizers have a VCA as the final stage of the modifier section. The control voltage is used to change the gain of an amplifier. The VCA controls the volume of the audio signal, and is sometimes connected directly to the output of an EG. An offset voltage can also be used to provide a volume control; so even the output volume of a synthesizer can be voltage controlled.

There are two types of input to VCAs:

- **Linear inputs** are used for tremolo and AM. They are also used with exponential curve envelopes.
- **Exponential inputs** are used for volume changes and linear curve envelopes.

The combination of linear and exponential envelopes with linear and exponential VCAs provides much scope for confusion. Using an exponential curve envelope with an exponential VCA produces a result which has sudden or abrupt changes rather than steady transitions.

Tremolo is a cyclic variation in the volume of a sound. It is produced by using an LFO control voltage to alter the gain of a VCA. Tremolo normally uses a sine or triangle waveform at frequencies between 5 and 20 Hz. Higher frequencies from an LFO or a VCO produce AM, where the output of the VCA is a combination of the audio signal and the LFO or VCO frequency. See Section 2.4.1 for more details on AM.

Apart from their normal use as volume-controlling devices, VCAs can also be used to provide 'filtering' effects. By connecting the keyboard pitch voltage to the control voltage input of a VCA, the gain of the VCA is then dependent on the pitch control voltage from the keyboard. Since the keyboard pitch voltage normally rises as the keyboard note position rises, then the VCA will act much as in a high-pass filter, since low notes will be at a lower volume than higher notes. By inverting

Figure 2.2.31 A VCA can be used to produce control of volume which follows the keyboard by routing the keyboard control voltage to the VCA gain control. This is similar to the tracking of a filter, and produces a coarse high-pass filtering effect, where higher notes are attenuated less than lower notes.

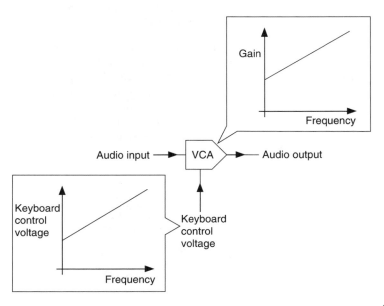

the keyboard pitch voltage, a low-pass 'filter' effect can be produced. This coupling of the VCA to the keyboard pitch voltage is called 'scaling', since the output of the VCA is scaled according to the pitch (Figure 2.2.31).

2.2.6 Other modifiers

LFOs

LFOs are used to produce low frequency control voltages. They are in two forms: VCOs and special-purpose oscillators. VCO-based LFOs can have their frequency controlled with an external control voltage, whilst special-purpose oscillators cannot.

Unlike audio frequency VCOs, LFOs need to produce waveforms where the shape is normally more important than the harmonic content. So, in addition to the sine, square, pulse and sawtooth waveforms, additional shapes like an inverted sawtooth are also provided. These might be used when the LFO is connected to a source like a VCO, and is controlling the pitch of the VCO. The basic sawtooth, or ramp-up waveform, would then produce a pitch that rose slowly and dropped quickly. The inverted shape, although still called a sawtooth, would now be a ramp-down waveform, and would give a pitch that rose quickly and dropped slowly.

Two specialised LFO waveform outputs are often found on LFOs: (i) sample and hold and (ii) arbitrary.

Sample and hold is the name given to a random or repetitive sequence of control voltages, which are produced by using the LFO to repeatedly take the value of another voltage source, and then keeping that value until the next time that it measures the value again (Figure 2.2.32). This process is called 'sampling' the value, and that value is then 'held' until the next sample is taken. The technique is thus called sample and hold.

If the voltage source that is sampled is noise, then the sample values will be random in level. This produces a series of values which do not repeat, and are not regular or predictable. The regular timing from the periodic sampling is the only known quantity. If another LFO or VCO is sampled, then one of two results is possible. If the second LFO or VCO is not synchronised to the sampling LFO, then the output of the sample and hold will be a series of values which are partly random

Figure 2.2.32 Sample and hold circuits take regular 'samples' of a noise (or other waveform) and then maintain that level until the next sample is taken. The rate of the samples is normally controlled by an LFO. The output consists of a series of steady voltages with rapid transitions, but whose level is not predictable. If the noise source is replaced with a repetitive waveform, then the output levels depend on the timing relationships between the sample LFO and the waveform being sampled.

and partly repetitive – the exact pattern depends on the relative frequencies and the LFO/VCO waveform. If the sampling LFO and the second LFO/VCO are synchronised so that they are locked together with the LFO/VCO being a multiple or fraction of the sampling LFO, then the output pattern will repeat.

Sample and hold is often used to control the cut-off frequency of a resonant low-pass filter. This is an effective way of providing 'interest' and 'movement' in a sound when it is in the sustain segment of an ADSR envelope. Unfortunately, the rhythmic random changing timbres that this produces have become an over-used cliché.

Arbitrary waveforms are the ones which are constructed from a series of simpler waveform segments (Figure 2.2.33). There are many variations possible:

- two or more levels (rather like a simple sequencer)
- two or more straight-line slopes (much like an envelope)
- two or more curves (exponential, linear, sine, power law, etc.).

Arbitrary waveform generators are also called function generators. They can be used to replace EGs, control panning and effects settings, and even act as simple sequencers to produce a series of pitched notes.

LFO output waveforms are frequently available simultaneously, so that a sine wave can be used at the same time as a square waveform (Figure 2.2.34). The common outputs are:

- sine
- triangle
- square
- sawtooth/ramp-up
- inverted sawtooth/ramp-down
- pulse
- inverted pulse (100% pulse width)
- sample and hold
- arbitrary.

Envelope follower

An envelope follower takes an audio signal or control voltage, converts it to just positive values, and then low-pass filters it with a filter which has a very low cut-off frequency – a few hertz (Figure 2.2.35). This removes any high frequencies from the input, and leaves just a control voltage which represents the envelope of the input audio or control voltage. It is thus almost the opposite of a VCA: a VCA uses

Figure 2.2.33 Arbitrary wave-shape generators extend the concept of the multi-segment EG by providing additional shapes for the transition from one break-point to the next.

Figure 2.2.34 LFO outputs are normally provided in a variety of shapes to give additional control possibilities; although in practice, the sine wave is almost always used for vibrato or tremolo, and the square wave is almost exclusively used for trills. The other shapes are often presented in normal and inverted forms, and are often used for special effects sounds.

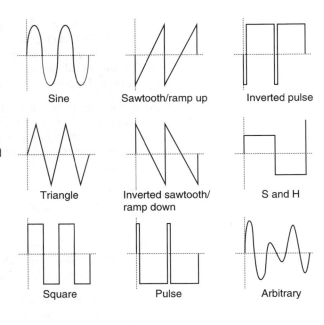

Sine

Sawtooth/ramp up

Inverted pulse

Triangle

Inverted sawtooth/ramp down

S and H

Square

Pulse

Arbitrary

Figure 2.2.35 An envelope follower is used to 'extract' the envelope from an audio signal. This can be used to process external signals in a synthesizer. The audio signal is low-pass filtered, and then a diode-pump circuit is used to provide the final output voltage.

Low-pass filter

Diode

C   R

Diode-pump

a control voltage to change the envelope of an audio signal, whilst an envelope follower takes an audio signal and produces a control voltage.

Some envelope followers also produce gate and trigger outputs which are suitable for controlling EGs – the envelope follower is then a complete module for interfacing an external audio signal with an analogue synthesizer. If the envelope follower is used to process source control voltages, then it can be used to 'smooth' rapidly changing waveforms which have sharp transitions, or even produce portamento effects if the keyboard pitch control voltage is processed.

Externally triggered sample and hold

If a sample and hold circuit has an external sample clock input, then it can be used to sample voltage sources at non-periodic intervals. One suitable sample clock source is the keyboard gate or trigger signals. Using the keyboard to control the sample and hold, an output is produced which changes only when a new key is pressed on the keyboard. By using an envelope follower to produce gate or trigger signals from an external audio input, then the sample and hold can be

Figure 2.2.36 A waveshaper uses a non-linear transfer function to change the shape of a waveform. This is often used to convert a triangle waveform into an approximation of a sine wave, and is adequate for shaping LFO and VCO outputs.

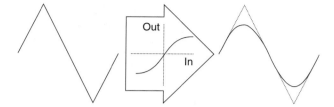

driven from an external audio signal. In this way, any audio signal can be used as a source of control voltages.

Waveshaper

Although rarely implemented on analogue synthesizers, the waveshaper is a non-linear amplifier which allows control over the relationship between the input and output signals. Any non-linearity in this relationship changes the shape of the waveform passing through the waveshaper, and this changes the harmonic content of the signal. Chapter 5 contains more information on the use of waveshaping in digital synthesizers.

Another interpretation of an analogue waveshaper is that it adds distortion to the signal, and so it is best used for monophonic signals. A more familiar waveshaper is the 'fuzz box' used by guitarists, where the passing of polyphonic audio signals through a clipping circuit produces large amounts of distortion (Figure 2.2.36).

Modulation

Modulation is another type of modifier. Any parameter which can be voltage controlled is a potential means of modulation. Although VCAs are available from the front panels of many analogue synthesizers, they are also used inside to allow control voltages to act as modulators – anywhere where a control voltage is used to change the amplitude or level of a signal or control voltage.

Some of the many possible ways that sources can be modified using modulation are:

- **LFO (LFO/envelope/keyboard)**: LFO modulation changes the rate or frequency of the LFO. This can be used to produce vibrato or tremolo whose rate is not fixed.
- **VCO mod (LFO/envelope/keyboard)**: LFO modulation of a VCO produces vibrato. Envelope modulation produces pitch sweeps. Keyboard modulation changes the scaling of the VCO: it can change the keyboard so that an octave on the keyboard represents any pitch interval to the VCO.
- **Filter mod (LFO/envelope/keyboard)**: LFO modulation of a filter produces cyclic timbre changes. Envelope modulation produces dynamic timbral changes during the course of a single

note. Keyboard modulation controls how the filter 'tracks' the note on the keyboard.

- **PWM (LFO/envelope/keyboard)**: PWM changes the timbre of the source waveform.
- **AM (LFO/VCO)**: AM with low-frequencies produces tremolo. At higher frequencies it adds extra frequencies to the audio signal (see Section 2.4).
- **FM (LFO/VCO)**: FM uses the linear frequency control voltage input of the VCOs. It produces additional frequencies in the output signal (see Section 2.4 and Chapter 5).
- **Cross-modulation (VCO)**: Cross-modulation connects the outputs of two VCOs to their opposite's frequency control voltage input, and so each frequency modulates the other. This produces complex FM-like timbres, but it can be difficult to control and keep in tune.
- **Pan (LFO/VCO/envelope/keyboard)**: LFO modulation of the stereo pan position produces 'auto-pan', where the audio signal moves cyclically from one side of the stereo image to the other. VCO modulation can spread individual harmonics across the stereo image. Envelope modulation moves the image with the note envelope. Keyboard modulation places notes in the stereo image dependent on their position on the keyboard.
- **Other sources**: Many other sources and modifiers can be modulated. The effects section of many analogue synthesizers allows parameters like the reverberation time, flange speed and others to be controlled.

Controllers

In conventional instruments, the control of the sound production is often a mechanical linkage between the performer and the instrument. A saxophone player uses a number of levers to control the opening and closing of the holes which determine the effective length of the saxophone. Control over the timbre can be accomplished by how the lips grasp the mouthpiece and the reed, as well as the use of the tongue. Further expression comes from the lungs with control over air pressure.

The interfacing between the performer and the synthesizer sound generation circuitry is accomplished by one or more controller devices. The main note-pitch controller is usually a modified organ-type keyboard, although sometimes weighted action piano-type keyboards are used. Changes in pitch are normally produced with a rotary control called a pitch-bend wheel, and a similar control is used to add in modulation effects like vibrato or tremolo. Control over volume and timbre can be accomplished by using a foot pedal – as used in organs for volume.

Keyboard

The familiar music keyboard with its patterned combination of black and white keys is widely used as the main discrete pitch control for note selection, as well as initiating envelopes. Although normally connected together, the pitch selection and envelope triggering functions can be separated.

Pitch bend

Continuous control over the pitch is achieved by using a 'pitch-bend' controller. These are normally rotating wheels or levers, and usually change the pitch of the entire instrument over a specified range (often a semitone or a fifth). They produce a control voltage whose value is proportional to the angle of the control. Pitch-bend controls normally have a spring arrangement which always returns the control to the centre 'zero' position (no pitch change) when it is released. This central position is often also mechanically detented, so that it can be felt by the operator, since it will require force to move it away from the centre position.

Modulation

Modulation is controlled using rotary wheels or lever, where the control voltage is proportional to the angle of the control. Modulation controllers are not normally sprung so that they return to the centre position. Some instruments allow pressure on the keyboard to be used as a modulation controller.

There have been some attempts to combine the functions of pitch bend and modulation into a single 'joystick' controller, but the most popular arrangement remains the two wheels: pitch bend and modulation.

Foot controllers

Foot controllers are pedals which provide a control voltage which is proportional to the angle of the pedal. Although associated with volume control, they can be used as modulation controls, or even as pitch-bend controls.

Foot switches

Foot switches are foot-operated switches which normally only have two values (some multi-valued variants are produced, but these are rare). They are used to control parameters such as sustain, portamento, etc.

See Chapters 7 and 8 for more details on controllers.

2.2.7 Using analogue synthesis

Learning how to make the best use of the available facilities provided by an analogue synthesizer requires time and effort. Although there are a number of 'standard' configurations of VCO, VCF, VCA and envelopes, the key to making the most of an analogue synthesizer is understanding how the separate parts work: both in isolation and in combination. If copies can be located, then Roland (1978, 1979) and De Furia (1986) are excellent references for further reading on this subject.

As a brief introduction to some of the techniques of using an analogue synthesizer, the remainder of this section shows how a subtractive analogue synthesizer can be an excellent learning tool for exploring

Figure 2.2.37 By varying the cut-off frequency of a resonant low-pass filter, the harmonic content of a waveform can be heard. As each of the harmonics which are present in the spectrum pass through the peak of the filter, they will be clearly heard. The frequency of the harmonic can be determined by noting the frequency of the filter when the harmonic is heard.

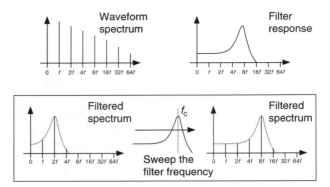

some of the principles of audio and acoustics. Here are some of the demonstrations which can be carried out using a subtractive synthesizer.

Harmonic content of waveforms

The harmonic content of different waveshapes can be audibly demonstrated by using a low-pass VCF with high resonance (set just below self-oscillation), or a narrow band-pass filter. Each VCO waveform is connected to the filter input, and the filter cut-off frequency is slowly increased from zero to maximum (Figure 2.2.37). As the resonant peak passes the fundamental, the filter output will be a sine wave at that frequency. As the cut-off frequency is increased further, the fundamental sine wave will disappear, and the next harmonic will be heard as the cut-off frequency matches the frequency of the harmonic. The audible result is a series of sine waves, whose frequency matches the frequencies of the harmonics.

If noise is passed through the filter, then the output will be sine waves whose frequencies will be within the pass-band of the resonant peak, and whose levels will change randomly. The audible result is rather like whistling.

Harmonic content of pulses

The harmonic content of different pulse widths of pulse waveforms can be demonstrated by listening to the pulse waveform and changing the pulse width manually (Figure 2.2.38). At a pulse width of 50%, the sound will be noticeably hollow in timbre: this is a square wave. The square wave position can be heard because the second harmonic, which is one octave above the fundamental, will disappear. Using the resonant filter technique described in the previous example, individual harmonics can be examined – tuning the filter to the harmonic which disappears for a square wave can be used to emphasise this effect.

As the pulse width is reduced, then the timbre will become brighter and brighter, and with very small pulse widths, the sound may disappear entirely. (This is a consequence of the design of the VCO circuitry, and not an acoustic effect!) Conversely, increasing the pulse

Figure 2.2.38 The harmonic content of a square wave and a rectangular wave are different, especially the even harmonics. The second harmonic is not present in a square wave, and yet can be clearly heard in a rectangular waveform. This can be used to produce square waves from a VCO which provides control over the width of the pulse. By adjusting the pulse width control and listening for the disappearance of the second harmonic, a square wave can be produced.

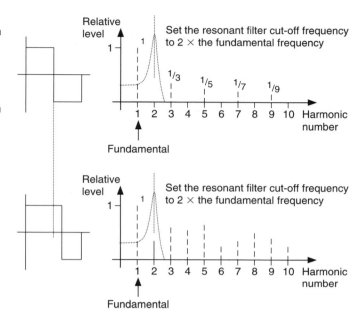

Figure 2.2.39 If a strongly resonant filter is 'triggered' by a brief pulse of noise or an envelope pulse, then it can 'ring' producing a decaying oscillation at the cut-off or peak frequency.

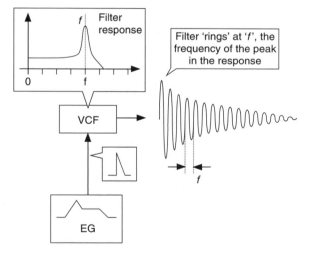

width from 50% produces the same changes in the timbre, and, again at very large pulse widths, may result in the loss of the sound.

Filtering

Many resonance and ringing filter effects can be demonstrated by connecting a percussive envelope to a VCF control voltage input, and turning up the resonance. Just below self-oscillation, the filter can be made to oscillate for a short time by using the envelope to trigger the oscillation (Figure 2.2.39).

This 'ringing oscillator' is the basis of the designs for many drum machine sounds in the 1970s. See Section 2.4.5 and Figure 2.4.7.

Figure 2.2.40 Beats can be demonstrated by mixing together the outputs of two VCOs which have slightly different frequencies. The two waveforms will cyclically add together or subtract, and so produce an output that varies in level. The audible effect is an interesting 'chorus' type of sound for frequency differences of less than 2 Hz, and vibrato for 2–20 Hz.

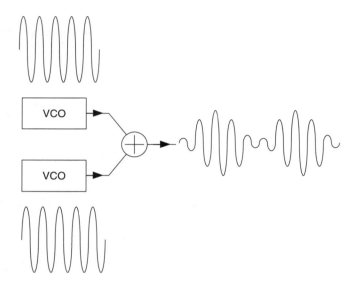

White noise filtered by a resonant low-pass filter changes from a hiss to a rumble as the cut-off frequency is reduced, because the filter is acting as a narrow bandwidth band-pass filter. With very narrow bandwidths, then the noise begins to produce a sense of pitch, and by connecting the keyboard voltage to the VCF so that it tracks the keyboard, then these 'pitched noise' sounds can be played with the keyboard. Keen experimenters might like to compare this with an alternative approach with audibly similar results: modulating the frequency of a VCO with noise.

Beats

Beats occur when two VCOs or audio signals are detuned relative to each other. The interference between the two signals produces a cyclic variation in the overall level as they combine or cancel each other out repeatedly (Figure 2.2.40). The time between the cancellations is related to the difference in frequency between the two audio signals or VCOs. Using two VCOs with a beat frequency of 1 Hz or less produces a 'lively', 'rich' and interesting sound.

PWM, uses an LFO to cyclically change the width of a pulse waveform from a single VCO. The result has many of the audible characteristics of two VCOs beating together.

Vibrato versus tremolo

- **Vibrato is FM:** The frequency of the audio signal is changed. Using an LFO to modulate the frequency of a VCO produces vibrato.
- **Tremolo is AM:** The level of the audio signal is changed. Using an LFO to modulate the level of an audio signal using a VCA produces tremolo.

Modulation summary and the cyclic variations of vibrato and tremolo are shown in Table 2.2.2 and Figure 2.2.41, respectively.

Table 2.2.2 Modulation summary

	Constant	Cyclic change
FM	Amplitude, pulse width	Frequency
AM	Frequency, pulse width	Amplitude
PWM	Amplitude, frequency	Pulse width

Figure 2.2.41 Vibrato is a cyclic variation in the frequency of a sound, whilst tremolo is a cyclic variation in the level of a sound.

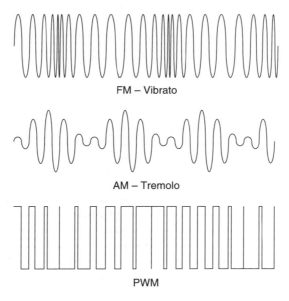

FM – Vibrato

AM – Tremolo

PWM

2.3 Additive synthesis

Whereas subtractive synthesis starts out with a harmonically rich sound, and 'subtracts' some of the harmonics, additive synthesis does almost the exact opposite. It adds together sine waves of different frequencies to produce the final sound. Because large numbers of parameters need to be controlled simultaneously, the user interface is usually much more complex than that of a subtractive synthesizer.

2.3.1 Theory: additive synthesis

Additive synthesis is based on the work produced by Fourier, a French mathematician from the nineteenth century. In 1807, Fourier showed that the shape of any repetitive waveform could be reproduced by adding together simpler waveforms, or alternatively, that any periodic waveform could be described by specifying the frequency and amplitude of a series of sine waves. The restriction that the wave-shape must repeat is imposed to keep the mathematics manageable. Without the restriction it is still possible to convert any waveform into a series of sine waves, but since the waveform is not constant, then the sine waves which make it up are not constant either.

One useful analogy is to think of trying to describe writing to someone, who has never seen it, over the telephone. You might start out by

Table 2.3.1 Harmonics, frequencies and overtones

Frequency	Harmonic	Overtone
f	fundamental	fundamental
$2f$	2	1
$3f$	3	2
$4f$	4	3
$5f$	5	4
$6f$	6	5
$7f$	7	6
$8f$	8	7
$9f$	9	8
$10f$	10	9

describing how the words are broken up into letters, and these letters are made up out of lines, dots and curves. This works perfectly well as long as the words you might try to describe stay fixed, but if they change, then you would have to keep updating your description. You could still convey the information about the shape of the letters which make up the words, but you would have to provide lots more detailed description as the letters change. The simplest example of synthesising a waveform using Fourier synthesis is a sine wave. A sine wave is made up of just one sine wave, at the same frequency! In terms of harmonics, a sine wave contains just one frequency component, at the repetition rate of the sine wave: its frequency.

More complicated waveshapes can be made by adding additional sine waves. The simplest method involves using simple integer multiples of the fundamental frequency. So, if the fundamental is denoted by f, then the additional frequencies will be $2f$, $3f$, $4f$, etc. These are the frequencies which occur in some of the basic waveshapes: sawtooth, square, etc., and are known as harmonics. Because the numbering of the harmonics is based around their position above the fundamental or first harmonic, with a frequency of f, then the second harmonic has a frequency of $2f$. The second harmonic is also sometimes called the first overtone (Table 2.3.1).

2.3.2 Harmonic synthesis

So far additive synthesis seems to be based around producing a specific waveform from a series of sine waves. In practice, the 'shape' of a waveform is not a good guide to its harmonic content, since minor changes to the shape can produce large changes in the harmonic content. Conversely, simple changes of phase for the harmonics can produce major changes in the shape of the waveform. In fact, although the human ear is mainly concerned with the harmonic content, the relative phase of the harmonics can be very important at low frequencies. For frequencies above 440 Hz, you can change the phase of a harmonic and thus alter the resulting shape of the waveform, but the basic timbre will sound the same. Control over phase is thus useful under some circumstances, and is found in some additive synthesizers.

Figure 2.3.1 (i) A triangle waveform constructed from six sine wave harmonics is very different from a sine wave, even though the fundamental is by far the strongest component. (ii) A combination of equal amounts of the first 12 harmonics produces a waveform which looks (and sounds) like a type of pulse waveshape.

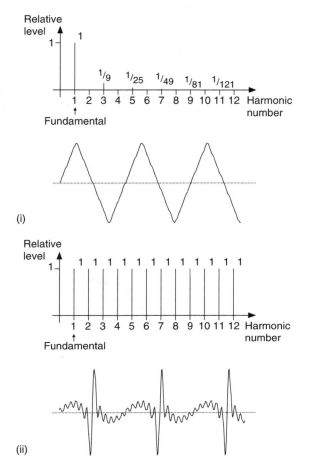

The harmonic content of waveshapes is a useful starting point for examining this relationship between shape and perception.

The simplest waveshape is the sine wave. Sine waves sound clean and pure, and perhaps even a little bit boring. Adding in small amounts of odd-numbered harmonics produces a triangular waveshape, which has enough harmonic content to stop it sounding quite as pure as the sine wave (Figure 2.3.1).

A square wave contains only odd harmonics. It has a characteristic 'hollow' sound, and the absence of the second harmonic is particularly noticeable if a square wave is compared with a sawtooth wave. A square wave which has been produced with a phase change in the second harmonic no longer looks like a 'square' wave, and yet the harmonic content is the same (Figure 2.3.2).

A sawtooth wave contains both odd and even harmonics. It sounds bright, although many pulse and 'super-sawtooth' waveshapes can contain greater levels of harmonics. Again, a sawtooth wave with a phase change in the second harmonic does not look like a sawtooth, although it still sounds like one to the ear (Figure 2.3.3).

Figure 2.3.2 (i) A square waveform constructed from six sine wave harmonics has a close approximation to the ideal waveshape. (ii) Changing the phase of the third harmonic radically alters the shape of the waveform.

Pulse waves contain more and more harmonics as the pulse width narrows (or widens). A 10% pulse has the same spectrum as a 90% pulse, and it also sounds the same to the ear. One special case is the square wave, where the even harmonics are missing completely. Pulse widths of anything other than 50% include the second harmonic, and this can usually be clearly heard as the pulse width is varied away from the 50% value.

Finally, there is the 'even harmonic' wave. If a sawtooth contains both odd and even harmonics and a square wave contains just the odd harmonics, then what does a wave containing just the even harmonics look like? Actually, it is just another square wave, but one octave higher in pitch, and with a fundamental frequency of $2f$!

In practice, adding together sine waves produces waveforms which have some of the characteristics of the mathematically perfect ideal waveforms, but not all. Producing square edges on a square wave would require large numbers of harmonics – an infinite number for a 'perfect' square wave. Using just a few harmonics can produce

Figure 2.3.3 (i) A sawtooth waveform constructed from 12 sine wave harmonics has a close approximation to the ideal waveshape. (ii) Changing the phase of the second harmonic radically alters the shape of the waveform.

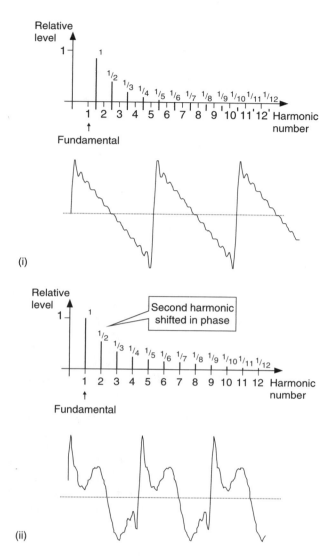

waveforms which have enough of the harmonic content to produce the correct type of timbre, even though the shape of the waveform may not be exactly as expected.

2.3.3 Harmonic analysis

In order to produce useful timbres, an additive synthesizer user really needs to know about the harmonic content of real instruments, rather than mathematically derived waveforms. The main method of determining this information is Fourier analysis, which reverses the concept of making any waveform out of sine waves and uses the idea that any waveform can be split into a series of sine waves.

The basic concept behind Fourier analysis is quite simple, although the practical implementation is usually very complicated. If an audio

Figure 2.3.4 Sweeping the centre frequency of a narrow band-pass filter can convert an audio signal into a spectrum: from the time domain to the frequency domain.

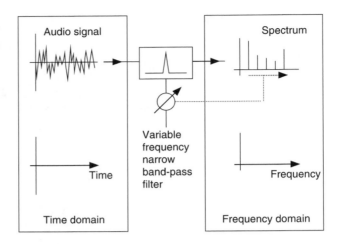

signal is passed through a very narrow band-pass filter that sweeps through the audio range, then the output of the filter will indicate the level of each band of frequencies which are present in the signal (Figure 2.3.4). The width of this band-pass filter determines how accurate the analysis of the frequency content will be: if it is 100 Hz wide, then the output can only be used to a resolution of 100 Hz, whereas if the band-pass filter has a 1-Hz bandwidth, then it will be able to indicate individual frequencies to a resolution of 1 Hz.

For simple musical sounds which contain mostly harmonics of the fundamental frequency, the resolution required for Fourier analysis is not very high. The more complex the sound, the higher the required resolution. For sounds which have a simple structure consisting of a fundamental and harmonics, then a rough 'rule of thumb' is to make the bandwidth of the filter less than the fundamental frequency, since the harmonics will be spaced at frequency intervals of the fundamental frequency. Having 1-Hz resolution in order to discover that there are five harmonics spaced at 1-kHz intervals is extravagant. Smaller bandwidths require more complicated filters, and this can increase the cost, size and processing time, depending on how the filters are implemented. Fourier analysis can be achieved using analogue filters, but it is frequently carried out by using digital technology (see Section 5.8 on Analysis–synthesis).

Numbers of harmonics

How many separate sine waves are needed in an additive synthesizer? Supposing that the lowest fundamental frequency which will be required to be produced is a low A at 55 Hz, then the harmonics will be at 110, 165, 220, 275, 330, 385, 440 Hz … The 32nd harmonic will be at 1760 Hz, and the 64th harmonic is at 3520 Hz.

An A at 440 Hz has a 45th harmonic of 19 800 Hz. Most additive synthesizers seem to use between 32 and 64 harmonics (Table 2.3.2).

Table 2.3.2 Additive frequencies and harmonics

Frequency (Hz)				Harmonic
55	110	220	440	fundamental
110	220	440	880	2
165	330	660	1320	3
220	440	880	1760	4
275	550	1100	2200	5
330	660	1320	2640	6
385	770	1540	3080	7
440	880	1760	3520	8
495	990	1980	3960	9
550	1100	2200	4400	10
605	1210	2420	4840	11
660	1320	2640	5280	12
715	1430	2860	5720	13
770	1540	3080	6160	14
825	1650	3300	6600	15
880	1760	3520	7040	16
935	1870	3740	7480	17
990	1980	3960	7920	18
1045	2090	4180	8360	19
1100	2200	4400	8800	20
1155	2310	4620	9240	21
1210	2420	4840	9680	22
1265	2530	5060	10120	23
1320	2640	5280	10560	24
1375	2750	5500	11000	25
1430	2860	5720	11440	26
1485	2970	5940	11880	27
1540	3080	6160	12320	28
1595	3190	6380	12760	29
1650	3300	6600	13200	30
1705	3410	6820	13640	31
1760	3520	7040	14080	32
1815	3630	7260	14520	33
1870	3740	7480	14960	34
1925	3850	7700	15400	35
1980	3960	7920	15840	36
2035	4070	8140	16280	37
2090	4180	8360	16720	38
2145	4290	8580	17160	39
2200	4400	8800	17600	40
2255	4510	9020	18040	41
2310	4620	9240	18480	42
2365	4730	9460	18920	43
2420	4840	9680	19360	44
2475	4950	9900	19800	45
2530	5060	10120	20240	46
2585	5170	10340	20680	47
2640	5280	10560	21120	48
2695	5390	10780	21560	49
2750	5500	11000	22000	50
2805	5610	11220	22440	51
2860	5720	11440	22880	52
2915	5830	11660	23320	53
2970	5940	11880	23760	54
3025	6050	12100	24200	55
3080	6160	12320	24640	56
3135	6270	12540	25080	57
3190	6380	12760	25520	58
3245	6490	12980	25960	59
3300	6600	13200	26400	60
3355	6710	13420	26840	61
3410	6820	13640	27280	62
3465	6930	13860	27720	63
3520	7040	14080	28160	64

Harmonic and inharmonic content

Real-world sounds are not usually deterministic: they do not contain just simple harmonics of the fundamental frequency. Instead, they also have additional frequencies which are not simple integer multiples of the fundamental frequency. There are several types of these unpredictable 'inharmonic' frequencies:

- noise
- beat frequencies
- sidebands
- inharmonics.

Noise has, by definition, no harmonic structure, although it may be present only in specific parts of the spectrum: coloured noise. So any noise which is present in a sound will appear as random additional frequencies within those bands, and whose level and phase are also random.

Beat frequencies arise when the harmonics in a sound are not perfectly in tune with each other. 'Perfect' waveshapes are always assumed to have harmonics at exact multiples of the fundamental, whereas this is not always the case in real-world sounds. If a harmonic is slightly detuned from its mathematically 'correct' position, then additional harmonics may be produced at the beat frequency, so if a harmonic is 1 Hz too high in pitch relative to the fundamental, then a frequency of 1 Hz will be present in the spectrum.

Sidebands occur when the frequency stability of a harmonic is imperfect, or when the sound itself is frequency modulated. Both cases result in pairs of frequencies which mirror around the 'ideal' frequency. So a 1-kHz sine wave which is frequency modulated with a few hertz will have a spectrum which contains frequencies on either side of 1 kHz, and the exact content will depend on the depth of modulation and its frequency.

Inharmonics are additional frequencies which are structured in some way, and so are not noise, but which do not have the simple integer multiple relationship with the fundamental frequency. Timbres which contain inharmonics typically sound like a 'bell' or 'gong'.

Many additive synthesizers only attempt to produce the harmonic frequencies, with perhaps a simple noise generator as well. This deterministic approach limits the range of sounds which are possible, since it ignores the many stochastic, probabilistic or random elements which make up real-world sounds.

2.3.4 Envelopes

The control of the level of each harmonic over time uses EGs and VCAs. Ideally, one EG and one VCA should be provided for each harmonic. This would mean that the overall envelope of the final sound was the result of adding together the individual envelopes for each

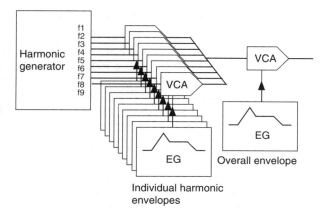

Figure 2.3.5 Individual envelopes are used to control the harmonics, but an overall envelope allows easy control over the whole sound which is produced.

of the harmonics, and so there would be no overall control over the envelope of the complete sound. Adding an overall EG and VCA to the sum of the individual harmonics allows quick modifications to be made to the final output (Figure 2.3.5).

In order to minimise the number of controls and the complexity, the EGs need to be as simple as possible without compromising the flexibility. Delayed ADR (DADR) envelopes are amongst the easiest of EGs to implement in discrete analogue circuitry, since the gate signal can be used to control a simple capacitor charge and discharge circuit to produce the ADR envelope voltage. DADR envelopes also require only four controls (delay time, attack time, decay time and release time), whereas a DADSR would require five controls and more complex circuitry. If integrated circuit EGs are used, then the ADSR envelope would probably be used, since most custom synthesizer chips provide ADSR functionality.

Control grouping and ganging

With large numbers of harmonics, having separate envelopes for each harmonic can become very unwieldy and awkward to control. The ability to assign a smaller number of envelopes to harmonics can reduce the complexity of an additive synthesizer considerably. This is only effective if the envelopes of groups of harmonics are similar enough to allow a 'common' envelope to be determined. Similarly, ganging together controls for the level of groups of harmonics can make it easy to make rapid changes to timbres – altering individual harmonics can be very time-consuming. Simple groupings like 'all of the odd', or 'all of the even' harmonics, can be useful starting points for this technique.

A more advanced use for grouping involves using keyboard voltages to give pitch dependent envelope controls. This can be used to create the effect of fixed resonances or 'formants' at specific frequencies.

Filter simulation/emulation

Filters modify the harmonic content of a sound. In the case of an additive synthesizer, there are two ways that this can be carried out: with a filter or with a filter emulation. As with the overall envelope control mentioned above, there are advantages to having a single control for the combined harmonics, and a VCF could be added just before the VCA. Such a filter would only provide crude filtering of the sound, in exactly the same way as in subtractive synthesis.

Filter emulation uses the individual EGs for the harmonics to 'synthesise' a filter by altering the envelopes. For example, if the envelopes of higher harmonics are set to have progressively shorter decay times, then when a note is played, the high harmonics will decay the first (Figure 2.3.6). This has an audible effect which is very similar to a low-pass filter being controlled by a decaying envelope. The difference is that the 'filter' is the result of the action of all the envelopes, rather than one envelope. Consequently, individual envelopes can be changed, which then allow control over harmonics that would not be possible using a single VCF.

As with the envelope control ganging and grouping, similar facilities can be used to make filter emulation easier to use, although the implementation of this is much easier in a fully digital additive instrument.

2.3.5 Practical problems

Analogue additive synthesis suffers from a number of design difficulties. Generating a large number of stable, high-purity sine waves

Figure 2.3.6 By using different envelopes for each harmonic, a filter can be 'synthesised'. This example shows the equivalent of a low-pass filter being produced by a number of different decaying envelopes.

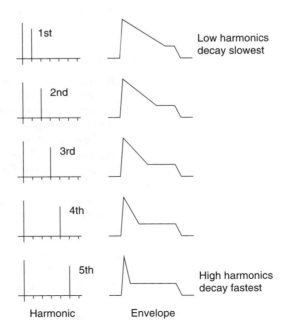

1st — Low harmonics decay slowest

2nd

3rd

4th

5th — High harmonics decay fastest

Harmonic   Envelope

simultaneously can be very complex, especially if they are not harmonically related. Providing sufficient controls for the large number of available parameters is also a problem. Depending on the complexity of the design, an additive synthesizer might have the following parameters repeated for each harmonic:

- frequency (fixed harmonic or variable inharmonic)
- phase
- level
- envelope (DADR, DADSR or multi-segment: four or more controls).

For a 32-harmonic additive synthesizer, these eight parameters give a total of just over 250 separate controls, ignoring any additional controls for ganging and filter emulation. Although it is possible to assemble an additive synthesizer using analogue design techniques, practical realisations of additive synthesizers have tended to be digital in nature, where the generation and control problems are much more easily solved.

Spectrum plots

The subtractive and additive sections in this chapter have both shown plots of the harmonic content of waveforms, showing a

Figure 2.3.7 A spectrum is a plot of frequency against level. It thus shows the harmonic content of an audio signal. In most of the examples in this book, the horizontal axis is normally shown with harmonic numbers instead of frequencies – the 55-Hz sine wave spectrum shows the correspondence with frequency. When a spectrum changes with time, then a 'mountain' graph may be used to show the changes in the shape.

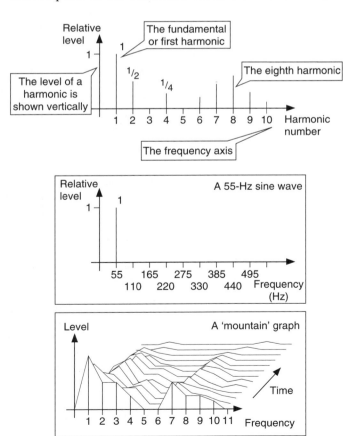

frequency axis plotted against level. This 'harmonic content' graph is called a spectrum, and it shows the relative levels of the frequencies in an audio signal. Whereas a waveform is a way of showing the shape of a waveform as its value changes with time, a spectrum is a way of showing the harmonic content of a sound. The shape of a waveform is not a very good indication of the harmonic content of a sound, whereas a spectrum is – by definition.

Spectra (the plural of the Latin-derived word 'spectrum') are not very good at showing any changes in the harmonic content of a sound – in much the same way that a single cycle of a PWM waveform does not convey the way that the width of the pulse is changing over time. To show changes in spectra, a 'waterfall' or 'mountain' graph is used, which effectively 'stacks' several spectra together. The resulting 3D-like representation can be used to show how the frequency content changes with time (Figure 2.3.7).

2.4 Other methods of analogue synthesis

2.4.1 *Amplitude modulation*

AM is a variation on one method used to transmit radio broadcasts. AM radio works by using a high frequency signal as the 'carrier' of the audio signal as a radio wave. The carrier signal on its own conveys no information – it is the modulation of the carrier by the audio signal that provides the information by changing the level of the carrier. In the simplest case, a sine wave audio signal is used to change (or modulate) the level of the carrier signal. The resulting output signal contains the original carrier frequency, but also the sum and the difference of the carrier and audio frequencies; these are called sidebands, because they are on either side of the carrier.

For audio AM, the two frequencies are both in the audio range, but the same principles apply – the output consists of the carrier frequency and the sum and difference frequencies (Figure 2.4.1). So with a carrier of 1000 Hz, and a modulator of 750 Hz, the output sideband frequencies will be 1000, 1750 and 250 Hz. Notice that the modulating frequency is not present in the output. For 100% modulation, the sidebands have half the amplitude of the carrier.

For AM with waveforms other than sine waves, each component frequency is treated separately. So for a sine carrier and a non-sinusoidal wave modulator, there are actually the equivalent of several modulator frequencies: one for each harmonic in the modulator. For a sawtooth modulator wave this means that there will be integer multiples of the modulator frequency, at decreasing levels. Each of these harmonics will produce sidebands around the carrier. The carrier frequency of 1000 Hz will also be present in the output. Again, with 100% modulation, the sidebands will have half the amplitude of the carrier.

With a non-sinusoidal carrier of 1000 Hz, and a sine wave modulator of 750 Hz, it is the equivalent of several carrier frequencies, and each

Figure 2.4.1 AM with two sine waves produces outputs at the sum and difference of the two input frequencies.

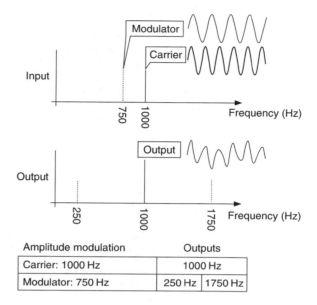

Amplitude modulation	Outputs	
Carrier: 1000 Hz	1000 Hz	
Modulator: 750 Hz	250 Hz	1750 Hz

carrier produces its own set of sidebands from the modulation frequency. For a sawtooth carrier, this means that there will be the equivalent of a carrier at each integer multiple of the carrier frequency, and each will produce sidebands from the modulator frequency. With 100% modulation the sidebands will have half the amplitude of the carrier. All of the harmonics in the carrier wave will also be present in the output (Figure 2.4.2).

For the case of two non-sinusoidal waves, AM produces a set of sidebands for each carrier harmonic, using each modulator harmonic. AM is thus a simple way of producing complex sounds with a number of harmonics which are not related to the fundamental (inharmonics) (Figure 2.4.3).

In an analogue synthesizer, AM is produced by connecting a VCO to the modulation control input of a VCA which is processing the output of another VCO. If the modulating frequency is lower than about 25 Hz, then AM is called tremolo, and it is perceived as a rapid cyclic change in the amplitude.

2.4.2 Frequency modulation

FM also employs another method which is normally used for the transmission of radio broadcasts. FM radio again uses a high frequency signal as the 'carrier' of the audio signal. The modulation of the carrier signal by the audio signal 'carries' the information by changing the frequency of the carrier. The simplest case is where a sine wave audio signal is used to change (or modulate) the frequency of the carrier signal. The amount of frequency change is called the deviation, Δf_c, and instead of producing just one pair of sideband

Figure 2.4.2 If the modulator is a non-sinusoidal waveform, then each of the harmonics of the modulator produces a pair of sum and difference frequencies in the output.

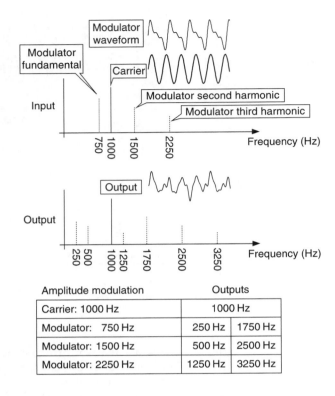

Amplitude modulation	Outputs	
Carrier: 1000 Hz	1000 Hz	
Modulator: 750 Hz	250 Hz	1750 Hz
Modulator: 1500 Hz	500 Hz	2500 Hz
Modulator: 2250 Hz	1250 Hz	3250 Hz

Figure 2.4.3 If the carrier is a non-sinusoidal waveform, then each carrier harmonic appears in the output, and also produces a pair of sum and difference frequencies.

Amplitude modulation	Outputs	
Carrier: 1000 Hz	1000 Hz	
Modulator: 750 Hz	250 Hz	1750 Hz
Modulator: 1500 Hz	1250 Hz	2750 Hz
Modulator: 2250 Hz	2250 Hz	3750 Hz

frequencies, FM can produce many sidebands, where the extra sidebands are similar to the harmonics in the sawtooth AM case described in Section 2.4.1, and this is just for sine wave carrier and modulator frequencies.

The number of sidebands which are produced can be determined by using the modulation index, which is a measure of the amount of modulation which is being applied to the carrier. The modulation index is given by dividing the deviation by the modulator frequency, f_m:

$$\text{Modulation Index} = \frac{\Delta f_c}{f_m}$$

Notice that the modulation index is dependent not only on how much the carrier frequency is changed, but also on the modulator frequency. The resulting output signal contains the original carrier frequency, but also the sum and difference sidebands for each of the multiples of the modulator frequency.

For audio FM with two sine waves, the output consists of the carrier frequency and sidebands made up from the sum and difference frequencies of the carrier and multiples of the modulator frequency. The number of sidebands depends on the modulation index (Figure 2.4.4), and a rough approximation is that there are two more than the modulation index. The modulating frequency is not present in

Figure 2.4.4 FM depends on the depth of modulation as well as the input frequencies. The number of sidebands that are produced depends on the modulation index.

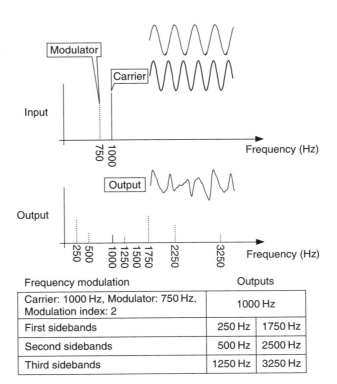

Frequency modulation	Outputs	
Carrier: 1000 Hz, Modulator: 750 Hz, Modulation index: 2	1000 Hz	
First sidebands	250 Hz	1750 Hz
Second sidebands	500 Hz	2500 Hz
Third sidebands	1250 Hz	3250 Hz

the output. The amplitudes of the sideband frequencies are determined by a set of curves called Bessel functions (Chowning and Bristow, 1986).

For FM with waveforms other than sine waves, each component frequency is treated separately. So for a sawtooth carrier and a sine wave modulator the output is similar to the sawtooth AM case, but there are many more sidebands produced. FM is thus a very powerful technique for producing complex spectra, but in an analogue synthesizer it suffers from problems related to the frequency stability of the carrier and modulator VCOs, and the response of the carrier VCO to FM at audio frequencies.

In an analogue synthesizer, FM is produced by connecting one VCO to the frequency control input of another VCO. If the modulating frequency is lower than about 25 Hz, then FM is known as vibrato, and it is perceived as a cyclic change in pitch. FM is described in more detail in Section 5.1.

2.4.3 Ring modulation

Ring modulation takes two audio signals and combines them together in a way that produces additional harmonics. It uses a circuit known as a 'balanced modulator' to produce a single output from two inputs: the output consists of the sum of the two input frequencies and the difference between the two input frequencies. The original inputs are not present in the output signal (Figure 2.4.5).

This is similar to AM, except that it is only the additional frequencies that are generated which are present at the output: only the sidebands are heard, not the carrier or the modulator. This means that ring

Figure 2.4.5 Ring modulation produces only the sum and difference frequencies – neither the carrier nor the modulator frequencies are present at the output.

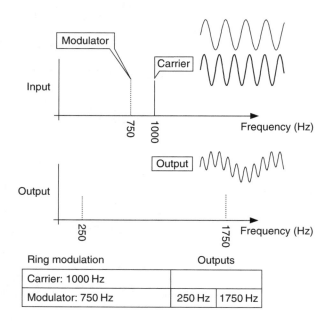

Ring modulation	Outputs	
Carrier: 1000 Hz		
Modulator: 750 Hz	250 Hz	1750 Hz

modulation can be useful where the original pitch information needs to be lost, which makes it useful for pitch transposition, especially where one of the sets of extra frequencies can be filtered out.

In an analogue synthesizer, ring modulation is produced by a special modifier circuit.

Modulation summary

Modulation summary is given in Table 2.4.1.

Table 2.4.1 Modulation summary

AM	Carrier in output	Simple sidebands for sine waves
FM	Carrier in output	Multiple sidebands for sine waves
RM	No carrier in output	Simple sidebands for sine waves

2.4.4 Formant synthesis

Formant synthesis is intended to emulate the strong resonant structure of many real instruments, where the spectrum of the output sound is dominated by one or more formants. Some analogue synthesizers have a simple high-pass filter after the low-pass filter to give some additional control over the bandwidth of sounds, and thus a simple type of formant.

In a formant synthesizer, this extra filtering is extended further: a graphic equaliser or complex filter is used to provide control over the bandwidth of the sound in addition to a VCF and VCA. Several parallel sections may be used to enable more detailed control over the individual formant areas of the sound (Figure 2.4.6).

2.4.5 Damped oscillators and ringing filters (drum sounds)

Circuits which have a strong resonance at a specific frequency can be made to oscillate if a sudden input causes them to self-oscillate. This 'ringing' is usually a sine wave and it dies away at a rate

Figure 2.4.6 A formant synth is intended to emulate the resonance found in real instruments. This can be achieved by using formant filters in addition to VCFs and VCAs.

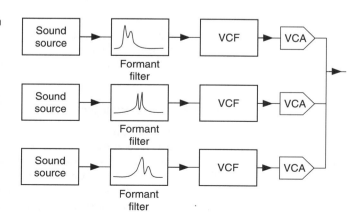

Figure 2.4.7 (i) A resonant circuit can produce some ringing when a trigger pulse is applied.
(ii) When a resonant circuit is placed in the feedback loop of an amplifier with a gain of less than one, then the ringing of the resonant circuit is enhanced.
(iii) If the gain of the amplifier is greater than one, then the circuit will oscillate at the frequency of the least attenuation in the resonant circuit.

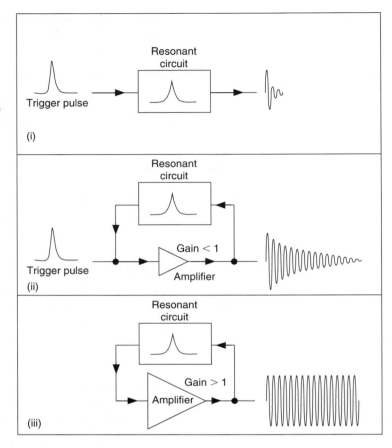

which is dependent on how close to self-oscillation the circuit is. The nearer it is to oscillating, the longer the ringing will last. Some VCFs can be made to self-oscillate if their Q or resonance is high enough, and at Q values just below this, they will ring. Conversely, an oscillator can be 'damped' so that it does not self-oscillate, but it will then ring. Filters and oscillators are just different applications of resonant circuits.

Decaying sine waves are very useful for producing percussive sounds, and many of the drum sounds produced by rhythm machines in the 1970s and early 1980s were produced by using ringing circuits (Figure 2.4.7).

2.4.6 Organ technologies

Most traditional organs are based around additive synthesis techniques, where a large number of sine waves are produced from a master oscillator, and then individual notes select mixes of sine waves via drawbar or other controls for the harmonic content (Figure 2.4.8). Unlike additive synthesizers, until the middle of the 1980s, organs tended not to have envelope control over the individual harmonics

Figure 2.4.8 Organs typically produce sounds by the addition of sine waves. The methods of producing the sine waves can be mechanical, electro-mechanical and electronic.

Figure 2.4.9 Simple 'piano'- and 'string'-type sounds can be produced by gating and filtering pulse waveforms which are derived from a master oscillator.

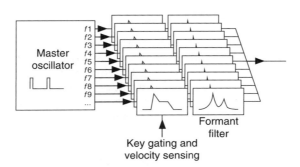

which make up the sounds. The advent of digital technology and sampling has made organs much more closely related to sample and synthesis synthesizers. Chapter 3 gives further details of digital master oscillators, whilst Chapter 4 describes sample and synthesis in more detail.

2.4.7 Piano technologies

Before digital sampling technology, piano-type sounds were produced by taking square or rectangular waveforms, often derived from a master oscillator by a divider technique, and then applying a percussive envelope and filtering. This produces a completely polyphonic instrument, although the sound suffers from the same lack of dynamic individual harmonic control as organs of the same time period. By using narrow pulse waveforms and different envelopes, the same techniques can be used to produce string-like sounds, and this was used in many 1970s 'string machines'. Section 3.3.2 describes 'beehive noise', a side effect of this sound generation technique.

By the mid-1980s, separate 'stand-alone' dedicated string machines had been replaced by polyphonic synthesizers, with the typical electronic piano becoming a specialised sample replay device by the end of the 1980s (Figure 2.4.9).

2.4.8 Combinations

Some analogue synthesizers use a combination of synthesis techniques, for example where several oscillators are used (additive-style) to provide the sound source, although this is then followed by a conventional subtractive synthesis modifier section. Ring modulation is another method which sometimes appears in otherwise straightforward implementations of subtractive synthesizers – perhaps because it is relatively simple to implement, and yet allows a large range of bell-like timbres which contrast well with the often more melodic subtractive synthesis timbres. Some 'string machines' in the 1970s added a VCF and ADSR EG section to provide 'synth brass' capabilities. Such combinations can provide additional control and creative potential, although their additions rarely become adopted generally.

2.4.9   Tape techniques

Perhaps the most straightforward method of analogue synthesis is the use of the tape recorder. By recording sounds onto magnetic tape, they can be stored permanently for later modification and manipulation. The raw sounds used can be either natural or synthetic. Chapter 4 details the use of tape as a recording medium, whilst Chapter 1 outlines some of the creative possibilities of using tape as a synthesis tool.

2.4.10   Optical techniques

Whilst tape offers a large number of possibilities for manipulating sound once it has been recorded on the tape, it does not allow the user to generate or control a sound directly. The audio signals are recorded onto the tape as changes to the magnetic fields stored on the iron oxide coating of the plastic tape, and so cannot be seen or changed, other than by recording a new sound over the top.

In contrast, by using the optical soundtracks which are often used in film projectors, it is possible to directly input the raw sound itself. Modern film projectors can use magnetic or digital techniques as well, but the basic method uses a light source and an optical sensor on either side of the film. When the soundtrack is clear, all the light passes through the film to the sensor, and conversely, when the film is dark, then no light passes to the sensor. By varying the amount of light that can pass through the film to the light sensor, the output of the sensor can be controlled. If the film soundtrack varies at a fast enough rate, then audio signals can be produced at the output of the sensor. Film soundtracks usually control the amount of light by altering the width of the clear part of the film – the wider the gap, the more light passes through to the sensor. The part of the film used to record this 'sound' track is by the side of the picture, and looks much like an oscilloscopic view of an audio signal, except that it is mirrored around the long axis (Figure 2.4.10).

By taking film which has no sound recorded onto it, and then drawing onto the film soundtrack with an opaque ink, it is possible to create sounds which will only be heard when the film is played. Sounds can thus be drawn or painted directly onto film. Although this sounds like an effective marriage of art and science, it turns out that the process of

Figure 2.4.10 Film sound-track uses the amount of light passing through the film to represent the audio waveform.

Audio waveform

Film soundtrack (optical)

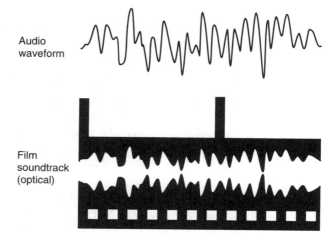

drawing sounds by hand is a slow and tedious one, and the precision required to obtain consistent timbres is very high. The rough '30-dB rule of thumb' that says that a drawn audio waveform represents only the most significant 30 dB of the harmonics is very relevant here.

Combining the drawing skills of optical sound creation with the tape manipulation processes of music concrete can offer a much more versatile technique. In this case, only short segments of film soundtrack need to be drawn, since the resulting short sounds can be recorded onto tape, copied many times to provide longer sounds, and then manipulated using tape techniques.

2.4.11 Sound effects

Perhaps the ultimate 'analogue' method of synthesising sounds is the work of the 'sound effects' team in a film or television studio. Using a floor covered with squares containing various surfaces, and a large selection of props, 'Foley' artistes produce many of the everyday sounds that accompany film and television programmes. For more unusual 'spot' effects, specialised props or pre-recorded sound effects may be used. Choreographing the sound effects for a detailed scene can be a very complex and time-consuming task, very similar to controlling an orchestra!

2.4.12 Disk techniques

Using a turntable, slip mat and a robust cartridge can also be a flexible and versatile analogue method of sound generation. Since the 1980s, the use of the vinyl disc as a source of complex sound effects, rhythms and musical phrases has become increasingly significant, and this has happened alongside the use of samplers (see also Chapter 8).

2.5 Topology

How do the component parts of a synthesizer fit together? This section starts by looking at typical arrangements of VCOs, VCFs, VCAs and EGs. It then looks at categorising types of synthesizers: the main divisions in type are between monophonic and polyphonic synthesizers, performance and modular synthesizers, and alternative controllers. This section looks at the topology of the modules that make up a typical synthesizer – how they are arranged and ordered. Although this information is fundamental to the actual construction of analogue synthesizers, the theory behind it is also relevant to some digital instruments, even though digital synthesizers often have no physical realisation of the separate modules at all.

2.5.1 Typical arrangements

The most common arrangement of analogue synthesizer modules is based on the 'source and modifier' or 'excitation and filter' model. This uses one or more VCOs plus a noise generator as the sources of the raw timbre. It then uses a VCF and VCA controlled by one or more EGs to shape and refine the final timbre. An LFO is used to provide cyclic modulation: usually of the VCO pitch (Figure 2.5.1).

This basic arrangement of modules is used so often by manufacturers that it has become permanently hard-wired into many designs, even some modular systems! The use of 'normalised' jack sockets allows for this type of preset wiring where the insertion of a plug into the socket opens the switch and removes the hard-wiring and thus allowing it to be over-ridden and replaced. 'Hard-wiring' is also used in many digital designs where there is no need for a rigid arrangement of modules because they are implemented in software.

One alternative method of subtractive synthesis replaces the single VCF with several. This enables more specific control of portions of the sound spectrum, and is often associated with the use of band-pass rather than low-pass filters. Because having separate filters for the oscillators enables them to be used as components of the final sound, rather than as a single source processed through a single modifier, this paralleling of facilities can be much more flexible in its creative possibilities. It is often used in formant synthesis, where the aim is to emulate the peaks in frequency response which characterise many

Figure 2.5.1 The basic synthesizer patch uses one or more VCOs and a noise generator as the sound source, with an LFO to provide vibrato modulation. The modifier section comprises a VCF and VCA, both controlled by one or more EGs.

real-world instruments, and particularly the human voice. Additive synthesis is an extension of this formant synthesis technique, where additional VCOs, VCFs and VCAs are added as required. Ganging of EGs by using voltage control of the EG parameters can make the control easier.

By using one VCO to modulate another, FM synthesis can be used, although the limitations of the VCO tuning stability and scaling accuracy limit its use. By using VCFs to process the outputs of each VCO, then the FM can be dynamically changed from using sine waves to using more complex waveshapes by increasing the cut-off frequency of the VCF on the output of the modulation VCO. This is something which most commercial digital FM synthesizers cannot do!

The basic synthesizer patch varies between monophonic and polyphonic synthesizers. It is often simplified for use in polyphonic synthesizers: only one VCO and VCA, and often less controllable parameters. Custom 'synth-on-a-chip' integrated circuits are often used to implement polyphonic synthesizer designs, and these chips are based on a minimalistic approach to the provision of modules and parameters.

2.5.2 *Monophonic synthesizers*

Monophonic synthesizers tend to be performance-oriented instruments designed for playing melodies, solos or lead lines. Despite the name, many monophonic analogue instruments can actually play more than one note at once: many have a duophonic note memory which allows two different note pitches to be assigned to two VCOs. With only one or two notes capable of being played simultaneously, an assignment strategy is required so that any additional notes played can be dealt with in a predictable way. Two common schemes are last-note and low-note priorities. Last-note priority is a time-based scheme: it always assigns the most recently played note to the synthesizer's voice circuitry, whilst low-note priority is a pitch-based scheme which always assigns the lowest pitched note to the voice circuitry. Low-note priority can be a powerful performance feature; for example, the performer can play legato 'drone' notes with the thumb of their right hand and use the rest of the fingers to play runs on top, with staccato playing dropping back to the 'drone' note. This technique is most effective with envelopes which are not retriggerable; that is, they do not restart the attack segment each time a new key is pressed on the keyboard. See Chapter 7 for more details on keyboard design and note assignment.

Portamento is a gliding effect which happens between notes. On a monophonic synthesizer it is normally used as a performance effect to give a contrast between the sudden pitch transition between notes, and the slow change of a portamento. The portamento circuits in analogue synthesizers work by restricting the rate at which a control voltage can change. Normally, the pitch control voltage from a keyboard will change rapidly when a new note is selected. A portamento circuit changes the slope of the transition between the two voltages. It thus takes time for the note to move from the existing pitch to the new pitch (Figure 2.5.2).

131

Figure 2.5.2 Portamento provides a smooth transition between successive pitches from the VCOs. The time taken for the keyboard control voltage to change from the previous value to the new value is called the portamento time.

Keyboard control voltage

Output of portamento circuit

Keyboard note on triggers

Portamento time

Glissando is a rapid movement from one note to another where the pitch changes chromatically through all the notes in between. At fast speeds, glissandos sound similar to portamento.

Monophonic synthesizers normally arrange the front panel controls so that they form a logical arrangement, often mimicking the topology of the modules inside. The front panel is normally arranged so that sources and controllers are on the left, with modifiers and the final output on the right. Early analogue monophonic synthesizers, and most modular systems, do not have any form of memory for the positions and settings of the front panel controls, and so a clear and functional arrangement of controls can aid the user in remembering settings. The process of using such a synthesizer requires a lot of practice to become thoroughly familiar with the workings of the instrument. Recalling a sound is often achieved iteratively, with adjustments of the controls gradually homing in on the required sound. Individuals who have mastered a synthesizer in this way have many similarities to a classically trained instrumentalist, where the way to produce a sound from the instrument requires dexterity, skill and a degree of coaxing.

Oberheim's OB1 monophonic synthesizer from 1978 had memories, but it is more famous for allegedly inspiring a character name from the first 'Star Wars' movie.

By the end of the 1970s, memory stores for the rapid recall of front panel settings had begun to appear, and by the end of the 1980s almost all monophonic synthesizers were equipped with memories. Front panels began to reflect this change by concentrating more on simplifying both the recall of memories, and of making simple minor edits to them. Many synthesizers became simply replay machines for preset sounds, and for many users, their programming changed from being part of the performance art to being an unwanted chore. By the late 1990s, live editing of sounds had become fashionable again, and synthesizer design reflected this with an increasing number of designs that included more controllers. In the first years of the 21st century, a number of manufacturers released synthesizers which were modern recreations of their own instruments of 20 or so years before, but with memories and additional performance controls.

The performance controls on monophonic analogue synthesizers are mono-oriented: *pitch-bend* (often set to an interval of a fifth or an octave); *octave switch* (up or down one or two octaves: often to compensate for a small keyboard span); *modulation* (normally vibrato) and, occasionally, *after-touch* (normally controlling vibrato). For those instruments which do have front panel controls, then they can be used as an additional method of control: real-time changes to sounds can be

Figure 2.5.3 A summary of the main features of a typical analogue monophonic synthesizer of the 1970s.

made 'live'. This usage of front panel and performance controls arises from the monophonic nature of the keyboard. For a right-handed player, the right hand is used to play the keyboard, whilst the left hand is used to provide additional expression by manipulating the pitch-bend and modulation wheels. 'Classical' two-handed static position techniques for playing monophonic melodies are rarely seen, and instead, a flowing right hand movement with lots of crossovers is used, thus freeing the left hand for the performance controls. Left-handed versions of monophonic synthesizers are very rare indeed: the placement of the performance controls is invariably on the left side of the keyboard (Figure 2.5.3).

2.5.3 Polyphonic synthesizers

Polyphonic analogue synthesizers are often implemented as several monophonic synthesis 'engines' or 'voices' connected to a common polyphonic keyboard. Each of these 'voices' receives monophonic note-pitch voltage, gate and trigger information, and performance controller information. It is usual for each voice to produce the same sound or timbre: multi-timbrality is normally commonly found in digital instruments only. The assignment of the voices to the keys which are played on the keyboard is carried out by key assignment circuitry or software in the polyphonic keyboard. This deals with the reassignment of notes which are playing (note stealing) and the method of assigning notes to the voices (last-note priority, etc.). (More details of keyboards can be found in Chapter 7.)

Controlling portamento on a polyphonic synthesizer is much more complex than on a monophonic synthesizer. The transitions between several notes can be made using several portamento algorithms. These are often named according to the effective polyphony which they produce, although in practice, only short portamento times are used to give a slight movement of pitch at the beginning of notes: this is frequently used for vocal, brass and string sounds.

Longer portamento times do not suit polyphonic keyboard technique, except for special effect usage with block chords, and often a glissando is more musically useful – where all the notes in between the last-note played and the next are played in sequence.

The Roland Juno-6 and Juno-60 memory version illustrate this well, since the follow-up model, the Juno 106, was only available with memories.

Memory stores seem to be more widespread in early polyphonic analogue synthesizers than in their monophonic equivalents. Initially, some manufacturers produced low-cost polyphonic synthesizers without memories, but these were not very popular in comparison to their more expensive memory-equipped versions. The designers of polyphonic synthesizers seem to have placed more emphasis on the accessibility of the memory recall controls than on the front panel controls for programming the synthesizer voices. The Yamaha CS-80 demonstrates this principle in its design: the programmable memories are hidden underneath a flap, and have tiny controls, whilst the large, colourful memory recall buttons are handily placed right at the front of the control panel.

The performance controls on polyphonic analogue synthesizer tend to be optimised for polyphonic playing techniques. Pitch bend is normally only a semitone, and can often be applied to only the top note, or the last note which has been played on the keyboard. The modulation wheel is often replaced or parallelled by a foot pedal, and often controls timbre via the VCF cut-off rather than vibrato. After-touch is almost invariably used to control vibrato or tremolo, and some instruments provide polyphonic after-touch pressure sensing instead of the easier-to-implement global version. Some instruments have an LFO which is common to all the voices, and so vibrato or tremolo modulation is applied at exactly the same frequency and phase to all the voices. In contrast, instruments which use separate LFOs for each voice circuit will have slightly different frequencies and phases, and this can greatly improve string and vocal sounds.

Real-time changes to the timbres are normally made using additional controllers: foot pedals, foot switches and breath controllers.

Figure 2.5.4 A summary of the main features of a typical analogue polyphonic synthesizer of the 1980s.

Manipulating front panel controls whilst playing with both hands on a polyphonic keyboard seems to be unpopular, and if front panel controls are used, then the playing technique used often reverts to monophonic usage, as described above – although polyphonic keyboards almost always have retriggered envelopes, which restricts some performance techniques. The performance controls are placed on the left hand side of the keyboard, just as with monophonic synthesizers (Figure 2.5.4).

2.5.4 *Performance versus modular synthesizers*

Performance synthesizers (monophonic or polyphonic) need a simplified and ordered control panel in order to make them usable in live performance. For this reason they usually have a fixed topology of modules: VCO, VCF and VCA.

Analogue modular synthesizers are not arranged in a logical order because there is no way to anticipate what they will be used for, except for the simplest cases. The most usual arrangement has the oscillators and other sound sources grouped together, usually on the top or on the left, with the modifiers (filters and amplifiers) in the centre or middle, and the EGs on the right or bottom.

Performance instruments have memories which can be used to store and recall sounds or timbres quickly. They are often used as replay machines for a series of presets.

Modular synthesizers normally have no memory facilities, or very simple generic ones which do not have the immediacy of those found in polyphonic instruments (Figure 2.5.5).

It has been said that modular synthesizers are the ultimate synthesizers, and that it is only time that limits people's use of them. Actually, modular synthesizers are severely limited, by a combination of the design and the user. The design is limited by the problems of trying to cope with patch-leads and lots of controls underneath. The user is fully occupied trying to hold everything about what is happening in their head: a simple VCO–VCF–VCA setup with a couple of EGs can be spread over more than a dozen modules and twenty or more patch-leads. The limitations are all too evident: no programmability, a confusing and obscure user interface and lots of scribbled sheets noting down settings and patches. They are also write-only devices – once the user has produced a patch, coming back 3 months later and trying to figure out what is happening is almost impossible. It is often much faster to start all over again.

Modular synthesizers are very good for appearance. Large panels covered with knobs, switches and patch-cords can look very impressive on stage. In reality, modular synthesizers are very good at producing lots of variations on a very specific set of sounds, and not very much outside of that set. Some FM-type sounds can be produced, but not very usefully, since the VCO modulation at audio frequencies is often less than ideal. Filter sweeps have a nasty habit of getting boring too,

Figure 2.5.5 (i) A performance-oriented synthesizer is designed to rapidly recall stored sounds and allow detailed performance effects to be applied with a range of specialised controllers. (ii) A modular synthesizer provides a wide range of modules which provide great flexibility, but at the expense of complexity and ease-of-use.

and it is very, very easy to fall into the 'lots of synth brass sounds' cliché. And do not forget that beyond about 20 patch-cords, most people lose track of what is connected to what! Modular synthesizers can be considered as almost 'write-only' devices, where trying to work out what a patch does can be very difficult, especially if someone else did the patching. It is also often forgotten that despite the large number of modules which are available in many modular synthesizers, their polyphony is very limited: two or three notes, and frequently only one note! Modular synthesizers are really not designed for polyphonic use, and trying to keep several separate sets of modules with anything like the same parameter settings is almost impossible. Although sampling the sounds that are produced using a modular synthesizer can be one way of producing a polyphonic sound and having programmability, given the synthesis power of many hardware and software samplers, the modular synthesizer is almost redundant even for this application.

Perhaps the most persuasive argument for the limited timbre palette of modular synthesizers is stored forever in recordings of the early 1970s. The problems of keeping track of patch-cords, the very limited polyphony, trying to avoid sweeping filter clichés, attempting to stay in control of the sound, the complete lack of any memory facilities and other limitations all conspire to make modular synthesizers an expensive chore.

Of course, from a very different viewpoint, modular synthesizers are very collectable, and may well be very sought after in the future as 'technological antiques'.

2.5.5 Keyboards versus other controllers

Most synthesizers come with a keyboard. Most expander modules are equipped with musical instrument digital interface (MIDI) input, which is a strongly keyboard-oriented interface. Many of the controls on a typical synthesizer are monophonic keyboard-oriented: pitch-bend, modulation, keyboard-tracking, after-touch, key-scaling …

How to do individual vibrato on specific notes? Take two synthesizers and set them to the same sound. Now play the non-vibrato notes on one keyboard, and the notes requiring vibrato on the other, using after-touch to bring in the vibrato (or just set the modulation wheel with a preset amount of vibrato). Simple but effective, and a challenging test of two handed playing technique. The inventive reader is encouraged to find other 'two-keyboard' solutions to 'And you can't do *that* on a synthesizer!' challenges.

Alternative controllers often have different parameters available which are not keyboard related. Stringed instruments like violin and cello have control over the pressure of the bow on the string in a way which is analogous to velocity and after-touch combined. Guitars enable the performer to use vibrato on specific notes: something which is very difficult on most keyboard-based synthesizers. Woodwind instruments have a number of performance techniques which do not have a keyboard equivalent – like pitch bending, changing the timbre or producing harmonics, all by using extra breath pressure and lip techniques. (Additional information on controllers can be found in Chapter 7.)

2.6 Early versus modern implementations

Electronics is always changing. Components, circuits, design techniques, standards and production processes may become obsolete over time. This means that the design and construction of electronic equipment will continuously change as these new criteria are met. The continuing trend seems to be for smaller packaging, lower power, higher performance and lower cost but at the price of increasing complexity, embedded software, difficulty of repair and rapid obsolescence. Over the last 25 years, the basic technology has changed from valves and transistors towards microprocessors and custom integrated circuits.

2.6.1 Tuning and stability

The analogue synthesizers of the late 1960s and early 1970s are infamous for their tuning problems. But then so are many acoustic instruments!

In fact, it was only the very earliest synthesizers which had major tuning problems. The first Moog VCOs were relatively simple circuits built at the limits of the available knowledge and technology – no one had ever built analogue synthesizers before. The designs were thus refined prototypes which had not been subjected to the rigorous trials of extended serious musical use. It is worth noting that the process of converting laboratory prototypes into rugged, 'road-worthy' equipment is still very difficult, and at the time, valve amplifiers and electromechanical devices like tape echo machines were very much the technology. Modular synthesizers were the first 'all-electronic' devices to become musical instruments that actually left the laboratory.

The oscillators in early synthesizers were affected by temperature changes because they used diodes or transistors to generate the required exponential control law and these change their characteristics with temperature (you can use diodes or transistors as temperature sensors!). Once the problem was identified it was quickly realised that there was a need for temperature compensation. A special temperature compensation resistor called a 'Q81' was frequently used – they have a negative temperature coefficient which exactly matches the positive temperature coefficient of the transistor.

Eventually circuit designers devised methods of providing temperature compensation which did not require esoteric resistors, usually based around differential pairs of matched transistors. Developments of these principles into custom synthesizer chips have effectively removed the need for additional temperature compensation. Unfortunately, the tuning problems had created a characteristic sound, which is one reason why the 'beating oscillator' sounds heard on vintage analogue synthesizers are emulated in fully digital instruments that have excellent temperature stability.

Tuning problems fall into four categories:

- overall tuning
- scaling
- high frequency tracking
- controllers.

Because of the differences in the response of components to temperature, the tuning of an analogue synthesizer can change as it warms up to operating temperature. This can be compensated for manually by adjusting the frequency control voltage, or automatically using an 'auto-tune' circuit (see below). Some synthesizers used temperature controlled chips to try and provide elevated but constant temperature conditions for the most critical components: usually the transistors or diodes in the exponential converter circuits. These 'ovens' have been largely replaced in modern designs by careful compensation for temperature changes.

Tuning polyphonic synthesizers requires patience and an understanding of the way that key assignment works (see Section 6.5.3). The tuner needs to know which VCO is making the sound (sometimes indicated by a light emiting diode (LED), or by a custom circuit add-on), as well as how to cycle through the remaining VCOs – often by holding one note down with a weight or a little wedge and then pressing and holding additional notes.

Temperature drift of the octave interval is the problem that most people mean when they say that analogue synthesizers go out of tune. Trying to match two exponential curves means that two interdependent parameters need to be changed: the offset and the scaling. The offset sets the lowest frequency that the VCO will produce, whilst the scaling sets the octave interval get the doubling of frequency for each successive octave. On a monophonic instrument this is not so hard, and any slight errors only help to make it sound lively and interesting. For polyphonic analogue synthesizers this process can be very time-consuming, and very tedious. With lots of VCOs to try and adjust, the problem can begin to approach piano-tuning in its complexity.

One method used to provide an 'automatic' tuning facility for polyphonic analogue synthesizers was introduced in the late 1970s. A microprocessor was used to measure the frequencies generated by each VCO at several points in its range, and then work out the offset and scaling correction control voltages. Because of the complexity of this type of tuning correction, and its dependence on a closed system, it has never been successfully applied to a modular synthesizer. (Auto-tuning is covered in more detail in Section 3.3 on DCOs.)

High frequency tracking is the tendency of analogue VCOs to go 'flat' in pitch at the upper end of their range. This is normally most noticeable when two or more VCOs are tuned several octaves apart, and although often present in single VCO synthesizers, is only apparent when they are used in conjunction with other instruments. Most VCOs use a constant current source and an integrator circuit to generate a rising voltage, and resetting the integrator when the output reaches a given voltage. This produces a 'sawtooth' waveform. The higher the current, the faster the voltage rises, and the sooner it will be reset: which produces a higher frequency sawtooth waveform. At low frequencies, the time it takes to reset the integrator is not significant in comparison with the time for the voltage to rise. But at high frequencies the reset time becomes more significant until eventually the waveform can become triangular in shape, which means that only one part of the waveform is actually controlled by the current source, and so the oscillator is not producing a high enough frequency (Figure 2.6.1). Some VCO designs generate a triangular waveform as the basic waveform and so do not suffer from this problem.

Controllers are another source of tuning instability. The stability of the pitch produced by a VCO is dependent on the control voltages that it receives. So anything mechanical that produces a control voltage can be a source of problem. Slider controls are one example of a mechanical control which can be prone to movement with vibration, whilst pitch-bend devices with poor detents can cause similar 'mechanical' tuning problems. The detent mechanism varies. One popular method involves using the pitch-bend wheel itself – it has two of the finger notches opposite to each other. One is used to help the user's fingers grip the wheel, whilst the other is used to provide the detent – a spring steel cam follower clicks into place when it is in the detent, and pops out again when the wheel is moved. This can wear, and produce wheels which do not click into position very reliably, which can mean that the whole instrument is then put out of tune.

Figure 2.6.1 At low frequencies, the rising part of the sawtooth waveform is much longer than the fixed reset time. But at higher frequencies, the reset time becomes a significant proportion of the cycle time of the waveform and so the frequency is lower than it should be. This high frequency tracking problem needs to be compensated for in the control voltage circuitry of the VCO.

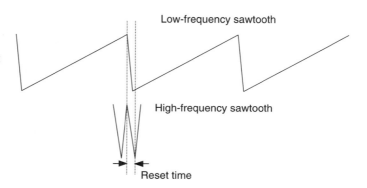

Low-frequency sawtooth

High-frequency sawtooth

Reset time

2.6.2   Voltage control

As has already been mentioned several times, despite the name, most of the electronic circuitry used in synthesizers is actually controlled by currents, not voltage! The voltages which are visible in the patch cords in the outside world are converted into currents inside the synthesizers, and the control is achieved using these currents.

Two 'standards' are in common usage:

- 1 volt/octave.
- exponential.

1 volt/octave

The 1-volt/octave system uses a linear relationship between the control voltage and pitch, which in practice means that there is a logarithmic relationship between voltage and frequency. This means that small changes in voltage become more significant at higher frequencies – just where small changes in pitch might become significant and audible as tuning problems. A 0–15-volt control signal can be used to control a pitch change of 15 octaves.

Exponential

The exponential system uses a linear relationship between the control voltage and the frequency. Because this method provides more resolution at high frequencies, it can be argued that it is a superior method to the 1-volt/octave system, since minor tuning errors at low frequencies are less objectionable. If the highest control voltage is 15 volts, then one octave down is 7.5, then 3.25, 1.875, 0.9375, 0.46875, 0.234375, 0.1171875, then 0.05859375 and so on, halving each time. Notice that just eight octaves down a voltage change of 58 millivolts is equivalent to an octave of pitch change.

Despite the apparent advantage of the exponential system, the most popular method was the 1-volt/octave system. Conversion boxes which enabled interworking between these two systems were available in the 1970s and 1980s, but they are very rare now.

2.6.3   Circuits

VCO

The basic oscillator circuit for a VCO uses a current to charge a capacitor. When the voltage across the capacitor reaches a preset limit, then it is discharged, and the charging process can start again. This 'relaxation' oscillator produces a crude sawtooth output, which can then be shaped to produce other waveforms (Figure 2.6.2). By varying the current which is used to charge the capacitor, then the time it takes to reach the limit changes, and so the frequency of the oscillator changes. By using a voltage to control the current, perhaps with a transistor, then the oscillator becomes voltage controlled. This type of circuit forms the basis of many VCOs.

Figure 2.6.2 A relaxation oscillator circuit consists of a capacitor which is charged by a current, *i*, and discharged by a switch when the voltage across the capacitor reaches the point at which the comparator triggers. Two output waveforms are available: the sawtooth voltage from the capacitor and the reset pulses from the comparator.

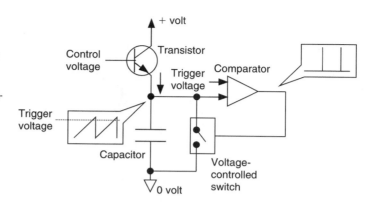

VCF

Simple low-pass filters use RC networks to attenuate high frequencies. By making the resistor variable, it is possible to alter the cut-off frequency. This RC network forms a single-pole filter, which has poor performance in terms of cut-off slope. Two- or four-pole filters improve the performance, but require more resistors and capacitors. This requires separate buffer stages, and multiple variable resistors. One way to produce several variable resistors uses the variation in impedance of a transistor or diode as the current through it is varied. By arranging a cascade of RC networks, where the transistors or diodes have the 'voltage controlled' current flowing through them, it is possible to make a low-pass filter whose cut-off frequency is controlled by the current which flows through the chain of transistors. This is the principle behind the 'ladder' filters used in Moog synthesizers.

The basic Moog-type filter uses two sets of transistors or diodes in a 'ladder' arrangement (Figure 2.6.3). The important parts of the filter are the base-emitter junction resistance, and the capacitors which connect the two sides of the ladder. Current flows down the ladder, and the input signal is injected into one side of the ladder. Since the resistance of the junctions is determined by the current which is flowing, the RC network thus formed changes its cut-off frequency as the current changes. This gives the voltage (actually current) control over the filter.

Another type of filter which is found in analogue synthesizers is the 'state variable' filter. This configuration had been used in analogue computers since valve days to solve differential equations. Once op-amps were developed, making a state variable filter was considerably easier, and by using field effect transistors (FETs) or transconductance amplifiers the cut-off frequency of the filter could easily be changed by a control voltage.

A typical state variable filter is made in the form of a loop of three op-amps (Figure 2.6.4). It is a constant-Q filter. Three outputs are available: low-pass, high-pass and band-pass (a band-reject can be produced by adding a fourth op-amp. Other types of multiple op-amp filters can be made: the bi-quad is one example whose circuit looks

141

Figure 2.6.3 A typical 'ladder' filter. The current flows through the control voltage transistor, TR1, and then through the two chains of diodes. The diodes and the connecting capacitors form RC networks which produce the filtering effect, with the diodes acting as variable resistors. The op-amp amplifies the difference between the two chains of diodes and feeds back this signal, thus producing a resonance or 'Q' control.

Figure 2.6.4 A typical 'two-pole' state variable filter. This produces three simultaneous outputs: high-pass, band-pass and low-pass.

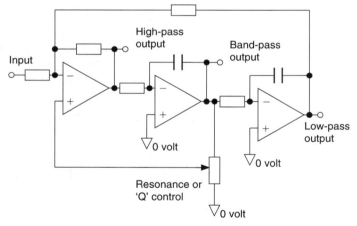

similar to a state variable, but the minor changes make it a constant-bandwidth filter and it only has low-pass and band-pass outputs.

2.6.4 Envelopes

It has been said that the more complex the envelope, the better the creative possibilities. The history of 'the envelope' is one of continuous evolution. The beginning lie with organ technology, where RC networks were used to try and damp out the clicks caused by keying sine waves on and off, and then the clicks ended up being generated deliberately so that they could be added back in as 'key click'.

Figure 2.6.5 Envelope scaling using voltage control.
(i) An ADSR envelope.
(ii) The same envelope with the attack, decay and release times reduced proportionally.
(iii) The same envelope with just the attack time reduced.
(iv) The same envelope with just the decay time reduced.
In order to produce each of these envelopes, a voltage controlled EG would need both ganged (all time altered equally) and individual controls.

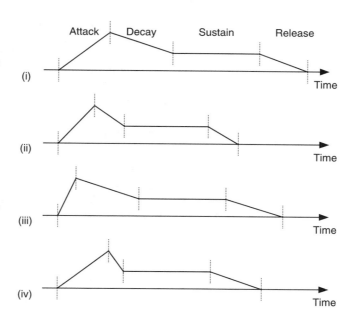

Trapezoidal waveform generators followed, which provided control over the start and finish of the envelope. ADSR-type envelopes, and their many variants, were used for the majority of the analogue synthesizers of the 1970s and 1980s. The advent of digital synthesizers with complex multi-segment EGs has made the ADSR appear unsophisticated, and analogue synthesizers designed in the 1990s have tended to emulate the multi-segment envelopes by adding additional break-points to ADSR envelopes.

The suitability of an envelope has very little to do with the number of segments, rates, times or levels. Instead, it is connected with the way that things happen in the real world. There are several things to consider:

- Many instruments have envelopes with exponential attacks rather than the much easier to produce linear slopes which many analogue synthesizers use. One solution to this is to add in two or more attack segments and so produce a rough approximation of an exponential envelope. This is much easier to achieve in a digital instrument than in analogue circuitry.
- Envelopes often change their shape and their timing in ways that are related to the note's pitch and the velocity with which it was played. Most modular and monophonic analogue synthesizers are not velocity sensitive, and so instruments which depend on this sort of performance technique tend to suffer (pianos, for example). Changing the attack times with pitch can be quite complex in an analogue synthesizer – you need an envelope generator with voltage controlled time parameters, and this can require a large number of additional patch-cords and control knobs (Figure 2.6.5).

Sophisticated multi-segment envelopes suffer from being harder for the user to visualise the shape of the envelope being produced. Probably the best compromise is an ADSR with a couple of attack, decay and release segments, and control over the slopes: 'function' generators meeting this sort of design criteria are beginning to appear. EG design research is still ongoing.

2.6.5 Discrete versus integration

Early analogue synthesizers used individual transistors to build up their circuits. This 'discrete' method of construction was gradually replaced by integrated circuits, usually op-amps for the majority of the analogue processing. Custom chips began to integrate large blocks of circuitry into single chips: a VCO or VCF, for example. Finally, by the mid-1980s, complete VCO, VCF, VCA, LFO and EG circuits could be placed on a single 'voice' chip intended for use in polyphonic analogue synthesizers. In the 1990s, the VCO would probably be replaced by digital generation techniques, with analogue filtering and enveloping from VCF and VCA chips. The specialist chips which are used can become collectors' items, particularly some of the older and rarer designs.

2.6.6 Pre- and post-MIDI

The development of MIDI signalled a major change in synthesizer technology (Rumsey, 1994). At a stroke, many of the incompatibility problems of analogue synthesizers were solved. Control voltages; gates and trigger pulses were replaced by digital data. The note-control and control parameters, sound data; pitch-bend and modulation controls were now standardised, and instruments could be easily interconnected.

Before MIDI, manufacturers were relatively free to use any method to provide interconnections between the instruments they produced, if at all. Commercial interests dictated that if you used a different control voltage, gate and trigger pulse system, then purchasers would only be able to easily interconnect to other products in your range. As a result, with a few exceptions, any interfacing between synthesizers from different manufacturers would require the conversion of voltages or currents. In addition, the performance controls were not fixed. Some manufacturers provided pitch-bend controls and multiple modulation controls, whilst others only had switched modulation: on or off. If an instrument was programmable, then the sound data was normally stored on data cassettes – again in proprietary formats.

MIDI was intended to enable the interchange and control of musical events with and by electronic musical instruments. It replaced the analogue voltages, currents and pulses with digital numbers, and so provided a simple way to assemble simple instruments into a larger unit. The layering of one sound with another changed from requiring two tracks on a multi-track tape recorder, to being a simple case of connecting two instruments together with a MIDI cable.

The introduction of MIDI had a profound and lasting effect on synthesizer design. Because the MIDI specification included a standard set of performance controllers, it effectively froze the pitch-bend and modulation wheel permanently into the specification of a synthesizer. MIDI is also biased towards a keyboard-oriented way of providing control: monophonic pressure is one example of this. MIDI also provided a standardised way of saving sound data by using system exclusive messages, and the possibility of editing front panel controls remotely.

The uniformity of many aspects of synthesizer design post-MIDI has meant that the emphasis has been placed onto the method of sound generation, rather than the functional design of the instrument. Although this has provided a wide variety of sounds, it has also meant that alternative controllers for synthesizers have tended to be largely ignored: the guitar synthesizer being one example.

2.6.7 Before and after microprocessors

The adoption of MIDI was also accompanied by a consolidation in the use of microprocessors. Microprocessors had begun to be used in polyphonic synthesizers to provide memory functions for storing sounds, but MIDI made the use of a microprocessor almost obligatory.

Before microprocessors, analogue synthesizers did not typically have auto-tune facilities, or memories for sounds. Interfacing was via analogue voltages and the complexity meant that only two or three instruments would be connected together. Front panel controls actually produced the control voltages that controlled the synthesizer sound circuitry.

Once microprocessors were incorporated in synthesizer designs, then auto-tuning was introduced for polyphonic synthesizers. Memories for sounds, and storage on floppy disk, data cassette or via MIDI system exclusive messages were possible. MIDI cables could be used to connect many instruments together. Front panel controls were scanned by the microprocessor to determine their position and thus produce a control voltage, or the front panel controls were replaced by a parameter system using buttons and a single control to select a parameter and edit it.

The changes in synthesizer design post-MIDI and post-microprocessors are most evident in rackmounting synthesizer modules, which have very little in common with the exterior appearance of analogue synthesizers of the late 1970s: no keyboard, few or no control knobs, no data cassette, no control voltage sockets and no performance controls – MIDI is totally essential to their production and control of sounds.

2.7 Example instruments

Moog modular (1965)

The Moog modular synthesizers comprise a number of modules which are placed in a frame which provides their power. Connections between modules are made using quarter inch front panel jack connectors. Models were available where the number and choice of

Figure 2.7.1 Moog modular.

modules was pre-determined, or the user could make their own selection. The system shown here (Figure 2.7.1) provides enough facilities for a powerful monophonic instrument, although producing polyphonic sounds does require a large number of modules, and can be very awkward to control.

Note the logical arrangement of the panels: the controls are at the top, the sockets at the bottom. Although with two rows of modules, the patch cords do tend to obscure the lower set of mostly VCO controls.

Mini Moog (1969)

The Mini Moog was intended to provide a portable monophonic performance instrument (the Sonic Six repackaged similar electronics in a different case for educational purposes). It provides a hard wired arrangement of synthesizer modules: VCOs, VCF, VCA, with two ADS EGS. This topology has since become the *de facto* 'basic' synthesizer 'voice' circuit, and can be found in many monophonic and polyphonic synthesizers, as well as custom 'synth-on-a-chip' integrated circuits (Figure 2.7.2).

Yamaha CS-80 (1978)

The Yamaha CS-80 was an early polyphonic synthesizer made up from eight sets of cards, each comprising a dual VCO/VCF/VCA/ADSR type of synthesizer 'voice' circuit. Comprehensive performance

Figure 2.7.2 Mini Moog.

Figure 2.7.3 CS-80.

controls made this a versatile and expressive instrument, if a little bulky and heavy. Preset sounds were provided, and these could be layered in pairs. Four user memories were provided, these used miniature sliders and switches which echoed the arrangement of the front panel controls, which provided another two user memories. The presets could be altered only by changing the resistor values on a circuit board inside the instrument (Figure 2.7.3).

Sequential Prophet 5 (1979)

The Prophet 5 was essentially five 'Mini Moog'-like synthesizer voice cards connected to a polyphonic keyboard controller. The major

Figure 2.7.4 Prophet 5.

Figure 2.7.5 SH-101.

innovation was the provision of digital storage for sounds, although the ability to use one VCO to modulate the other, called 'poly-mod' by sequential, allowed the production of unusual FM sounds (Figure 2.7.4).

Roland SH-101(1982)

The SH-101 was intended for live performance, and contained a simplified basic synthesizer 'voice' circuit. The instrument casing was designed so that it could be adapted for on-stage use by slinging it over the shoulder of the performer – a special hand grip add-on provided pitch-bend and modulation controls (Figure 2.7.5).

Oberheim Matrix-12 (1985)

The Matrix-12 (and the smaller Matrix-6) was a modular synthesizer in a case which was more typical of a performance synthesizer. The front panel extends the use of displays which was pioneered in earlier OB-X models – this time using green cold-cathode displays to provide reassignable front panel controls. The wide range of

Figure 2.7.6 Matrix-12.

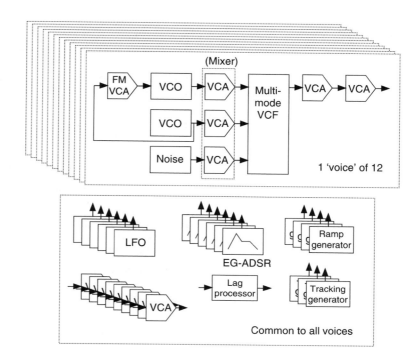

processing modules made this a versatile and powerful instrument. Only one voice from the 12 available is shown in Figure 2.7.6.

2.8 Questions

1. Name three ways of producing sound electronically using analogue synthesis, and briefly outline how they work.
2. Describe the 'source and modifier' model for sound synthesis.
3. What are the basic analogue synthesizer source waveforms and their harmonic contents?
4. What are the four major types of filter response curve? What effect do they have on an audio signal?
5. What are the main parts of an envelope? Include examples of the envelopes of real instruments.
6. How do vibrato and tremolo differ?
7. Why is it difficult to construct an analogue additive synthesizer?
8. What are the differences between AM, FM and ring modulation? Draw a spectrum for a 1-kHz carrier and 100-Hz modulator for each type of modulation.
9. Compare and contrast monophonic and polyphonic synthesizers.
10. Outline the effect of MIDI on synthesizer design.

Time line

Date	Name	Event	Notes
1500s	Barrel Organ	The barrel organ. Pipe organ driven by barrel covered with metal spikes	The forerunner of the synthesizer, sequencer and expander module!
1700s	Orchestrion	Orchestrions made in Germany. Complex combinations of barrel organs, reeds and percussion devices. Used for imitating orchestras	
1700	J.C. Denner	Invented the Clarinet	Single-reed woodwind instrument
1804	Leonard Maelzel	Leonard Maelzel invented the Panharmonicon, another mechanical orchestral imitator	
1807	Jean Baptiste Joseph Fourier	Fourier published details of his theorem, which describes how any periodic waveform can be produced by using a series of sine waves	The basis of additive synthesis
1870s	Gavioli	Fairground organs from Gavioli begin to use real instruments to provide percussion sounds	
1876	Alexander Graham Bell	Invented the telephone	Start of the marriage between electronics and audio
1877	Thomas Edison	Thomas Alva Edison invented the cylinder audio recorder, the 'phonograph'. Playing time was a couple of minutes!	Cylinder was brass with a tin foil surface, replaced with metal cylinder coated with wax for commercial release
1878	David Hughes	Moving coil microphone invented	
1887	Torakusu Yamaha	Torakusu Yamaha built his first organ	
1888	Emile Berliner	First demonstration of a disk-based recording system, the 'gramophone'	Disk was made of zinc, and the groove was recorded by removing fat from the surface, and then acid etching the zinc
1896–1906	Thaddeus Cahill	Invented the telharmonium, which used electromagnetic principles to create tones	Telephony
1896	Thomas Edison	Motion picture invented	
1903	Double-sided record	The Odeon label release the first double-sided record	Two single-sided records stuck together?
1904–1915	Valve	Development of the valve	The first amplifying device – the beginnings of electronics
1915	Lee de Forest	The first valve-based oscillator	
1920s	Cinema organs	Cinema organs, using electrical connection between the console keyboard and the sound generation	Also start to use real percussion and more: car horns, etc. Mainly to provide effects for silent movie accompaniment
1920s	Microphone recordings	First major electrical recordings made using microphones	Previously, many recordings were 'acoustic', using large horns to capture the sound of the performers
1920	Lev Theremin	The Theremin, patented in 1928 in the US. Originally called the 'Aetherophone'	Based on interfering radio waves

Date	Name	Event	Notes
1928	Maurice Martinot & Ondes	Invented the Ondes Martenot, an early synthesizer	Controlled by a ring on a wire, finger operated.
1930s	Baldwin, Welte, Kimball & others	Opto-electric organ tone generators	
1930s	Bell Telephone Labs	Invented the Vocoder – a device for splitting sound into frequency bands for processing	More musical uses than telephone uses!
1930s	Record groove direction	Some dictation machines record from the centre out instead of edge in.	This pre-empted the CD 'centre out' philosophy
1930s	Ondes	Ondioline, an early synthesizer	A relaxation oscillator was used as a sound source
1930s	Run-in Grooves	Run-in grooves on records invented	Previously, you put the needle into the 'silence' at the beginning of the track
1930	Friedrich Trautwein	Invented the Trautonium, an early electronic instrument	Wire pressed onto metal rail. Original was monophonic. Later duophonic
1933	Stelzhammer	Electrical instrument using electromagnets to produce a variety of timbres	
1934	John Compton	UK patent for rotating loudspeaker	
1934	Laurens Hammond	Hammond 'Tone Wheel' organ uses rotating iron gears and electromagentic pickups	Additive sine waves
1935	AEG, Berlin	AEG in Germany used iron oxide backed plastic tapes produced by BASF to record and replay audio	Previously, wire recorders had used wire instead of tape
1937	Tape recorder	Magnetophon magnetic tape recorder developed in Germany	The first true tape recorder
1939	Hammond	Hammond Novachord, first fully electronic organ	Uses 'master oscillator plus divider' technology to produce notes
1939	Hammond	Hammond Solovox, monophonic 'synthesizer'	British Patent 541911, US Patent 209920
1945	Metronome	First pocket metronome produced in Switzerland	
1945	Ronald Leslie	Patents rotating speaker system	
1947	Conn	Independent electromechanical generators used in organ	
1948	Baldwin	Blocking divider system used in organ	
1949	Allen	Organs using independent oscillators	
1950	John Leslie	Reintroduction of Leslie speakers	This time they are a success
1951	Hammond	Melochord	
1954	Milton Babbitt, H.F. Olsen & H. Belar	RCA Music Synthesizer mark I	Only monophonic
1957	RCA	RCA Music Synthesizer mark II	Used punched paper tape to provide automation
1958	Charlie Watkins	Charlie Watkins produced the CopyCat tape echo device	

Date	Name	Event	Notes
1958	RCA	RCA announced the first 'cassette' tape, a reel of tape in an enclosure	Not a success
1959	Yamaha	First 'Electone' organ	
1960s	Clavioline	Clavioline	British Patent 653340 and 643846
1960s	Mellotron	The Mellotron, which used tape to reproduce real sounds	Tape-based sample playback machine
1960s	Wurlitzer, Korg	Mechanical rhythm units built into home organs by Wurlitzer and Korg	
1963	Don Buchla	Simple VCO, VCF and VCA-based modular synthesizer – 100 Series: 'The Black Box'	Not well publicised
1963	Herb Deutsch	First meeting with Robert Moog. Initial discussions about voltage controlled synthesizers	
1963	Philips	Philips in Holland announced the 'Compact Cassette', two reels plus tape in a single case	A success well beyond the original expectations!
1965	Paul Ketoff	Built the 'Synket', a live performance analogue synthesizer for composer John Eaton	Commercial examples like the Mini Moog and Arp Odyssey, soon followed
1965	Robert Moog	First Moog synthesizer was hand-built	Only limited interest at first
1966	Don Buchla	Launched the Buchla Modular Electronic Music System, a solid-state, modular, analogue synthesizer	Result of collaboration with Morton Subotnick and Ramon Sender
1966	Rhythm machine	Rhythm machines appeared on electronic organs	Non-programmable and very simple rhythms
1968	Ikutaro Kakehashi	First stand-alone drum machine, the 'Rhythm Ace FR-1'	Designed by the future boss of Roland
1968	Walter Carlos	'Switched On Bach', an album of 'electronic realisations' of classical music, became the best seller	Moog synthesizers suddenly change from obscurity to stardom
1969	Peter Zinovief	(Electronic music studios) EMS produced the VCS3, the UK's first affordable synthesizer	The unmodified VCS3 is notable for its tuning instability
1969	Robert Moog	Mini Moog was launched. Simple, compact monophonic synthesizer intended for live performance use	Hugely successful, although the learning curve is very steep for many musicians
1970	ARP Instruments	ARP 2600, 'Blue Meanie' modular-in-a-box released	
1970	ARP Instruments, Alan Richard Pearlman	ARP 2500, very large modular studio synthesizer	Uses slider switches – a good idea, but suffer from cross-talk problems
1971	ARP Instruments	The 2600, a performance-oriented modular mono-synth in a distinctive wedge-shaped box	
1972	Roland	Ikutaro Kakehashi founded Roland in Japan, designed for R&D into electronic musical instruments	First products are drum machines
1972	Roland	TR33, 55 and 77 preset drum machines launched	
1978	Electronic Dream Plant	Wasp Synthesizer launched. Monophonic, all-plastic casing, very low-cost, touch keyboard, but it sounded much more expensive	Designed by Chris Huggett and Adrian Wagner

Date	Name	Event	Notes
1978	Roland	Roland launched the CR-78, the world's first programmable rhythm machine	
1978	Yamaha	Yamaha CS series of synthesizers (50, 60 and 80), the first mass-produced successful polyphonic synthesizers	Korg, Oberheim and others also produced polyphonic synthesizers at about the same time
1978	Korg	PS-3200	Programmable polyphonic synthesizer in modular-style casing
1978	Oberheim	OB-1	Programmable mono-synth. Claimed to be the inspiration for a character name in 'Star Wars'
1979	Sequential Circuits	Sequential Circuits prophet 5 synthesizer, essentially five Mini Moog-type synthesizers in a box	A runaway best seller
1979	Roland	Boss 'Dr. Rhythm' programmable drum machine	
1979	Roland	Roland Space-Echo launched. Used long tape loop and has built in spring line reverb and chorus	A classic device, used as the basis of several specialist guitar performance techniques (for example, Robert Fripp)
1980	Roland	Jupiter-8 polyphonic synthesizer	Eight note polyphonic, programmable poly-synth
1980	Roland	Roland TR-808 launched. Classic analogue drum machine	
1981	Moog	Robert Moog was presented with the last Mini Moog at NAMM in Chicago	The end of an era
1981	Roland	Roland Jupiter-8. Analogue eight-note polyphonic synthesizer	
1982	Moog	Memory Moog, six-note polyphonic synthesizer with 100 user memories	Cassette storage! Six Mini Moogs in a box!
1982	Roland	Jupiter 6 launched, first Japanese MIDI synthesizer	Very limited MIDI specification. Six-note polyphonic analogue synth
1982	Sequential	Prophet 600 launched, first US MIDI synthesizer	Six-note polyphonic analogue synth, marred by a membrane numeric keypad
1983	Oxford Synthesizer Company	Chris Huggett launches the Oscar, a sophisticated programmable analogue monophonic synthesizer	One of the few mono-synths to have MIDI as standard
1984	Sequential	Sequential launch the Max, an early attempt at mixing home computers and synthesizers	A complete failure, too early for the market
1984	Sequential Circuits	SixTrak. A multi-timbral analogue synthesizer with a simple sequencer	The first 'workstation'?
1985	Oberheim	OB-8	Classic analogue poly-synth keyboard
1986	Roland	JX-10 polyphonic synthesizer	Descended from the JX-3 via the JX-8P

Date	Name	Event	Notes
1988	Oberheim	Matrix-1000	Rackmount analogue poly-synth. Some critics have said that it has 1000 memories, but not 1000 different sounds
1990s	Europe	Many rackmount analogue synthesizers	From TB-303-influenced monophonics to generic polyphonics
1994	Oberheim	OB-MX	Multi-timbral programmable analogue poly-synth expander module
1995	Novation	Bass Station	Programmable analogue mono-synth with a TB-303 emulation mode
2000	Jomox	Sunsyn	Multi-timbral analogue poly-synth expander module
2001	Alesis	Andromeda analogue poly-synth	Polyphonic true-analogue keyboard with distinctive front panel graphics
2002	Robert Moog	Mini Moog Voyager released	Robert Moog's 21st century rework of the model D Mini Moog
2003	Dave Smith	Evolver, tabletop analogue poly-synth expander module, audio filter box and 16-step sequencer	Radical minimalistic matrix-based interface
2003	Cwejman	Cwejman S1	Analogue rackmount modular mono-synth with mini-jacks and patch-cords
2003	Phobos	Red Square	Jack-patchable analogue mono-synth and audio filter box

3   Hybrid synthesis

Hybrid synthesis is the name usually associated with methods of synthesis which are not completely analogue or digital. These borderline methods were most important during the changeover from analogue to digital sound generation in the early 1980s, but the underlying techniques have also become part of the all-digital synthesis methods. With the continuing increase of interest in 'analogue' synthesis that began in the 1990s, it is intriguing to note that very few of the instruments which are now being designed are truly analogue; in many ways they are actually either hybrids or are completely digital!

Synthesis methods which combine more than one techniques or methods of synthesis to produce a composite sound are described as 'layered' or 'stacked', and are covered in Chapter 6. Although these methods are sometimes called 'hybrid' methods, I prefer the word 'composite' synthesis.

It is possible to divide hybrid synthesizers into different classes. One possible division is based on the roles of the digital and analogue parts:

- **Digital control of the parameters of analogue synthesis**, as used in many programmable analogue monophonic synthesizers.
- **Digital control of the oscillator** (in other words, digitally controlled oscillator, DCO) with the remainder of the instrument analogue, perhaps with digital control of the parameters.
- **Digital oscillator with analogue modifiers**, and with digital control of the analogue parameters. These are the forms which many of the mid-1990s 'retro' analogue synthesizers used.

Another classification might be made on the method used to produce the sound. This section divides hybrid synthesizers using this method. Wavecycle, wavetable and DCO technologies are all discussed.

The predominant method of hybrid synthesis uses digital sound generation and control of parameters with analogue filtering and enveloping. This uses the technology which is the most appropriate for the task. Hybrid synthesizers are often characterised by a more sophisticated raw sound as the output from the 'source' part, which is due to the use of digital technology, especially in wavetable-based synthesizers. In fact, this availability of additional waveshapes, beyond the 'traditional' analogue set of sine, triangle, sawtooth and rectangular waveforms, could be seen as the major differentiator between analogue and hybrid instruments.

Although the idea of mixing analogue with digital was pioneered in the late 1970s, most notably with the Wasp synthesizer, it still forms the basis of many of the most successful hybrid (and hybrid masquerading as analogue) instruments. In fact, the recent trend of using digital circuitry to replace traditionally analogue functions like filters or oscillators follows on from hybrid synthesis. Also, the hybrid design philosophy of using a complicated oscillator and conventional modifiers also forms the basis of all sample and synthesis (S&S) instruments.

3.1 Wavecycle

A wavecycle is another word for waveform, although it emphasises the word cycle, which is very significant in this context. It is used here to emphasise the difference between the 'static, sample-based' replay-oriented wavecycle oscillators, and the 'dynamic, loop-based' wavetable oscillators.

Analogue synthesizers incorporate voltage controlled oscillators (VCOs) or oscillators which can typically produce a small number of different waveforms with fixed waveshapes, where each cycle is identical to those before and after it. The one exception to this is a pulse width modulation (PWM) waveform, where the shape of the pulse can be changed using a control voltage, often from a low frequency oscillator (LFO) for cyclic changes of timbre.

Hybrid synthesizers which use wavecycle-based sound generation can use this single-cycle mode, but they can also produce additional waveshapes, and use more complex schemes where more than one cycle of the waveform is used before the shape is repeated. The logical conclusion to this is for there to be a large number of cycles, each different, and with no repetition at all; the result is then called a sample. The only really important differences between these examples are the length of the audio sample and the amount of repetition. Each method has its own strengths and weaknesses.

3.1.1 Single-cycle

Single-cycle oscillators produce fixed waveforms, somewhat like an analogue synthesizer, although the selection of waveshapes is often much larger. The method of producing the waveform is often a mixture of analogue and digital circuitry, although as digital technology has fallen in price, so the use of predominantly digital circuit design has increased.

Possibly the simplest method of controlling a waveshape is the pulse width control, which is sometimes found on analogue synthesizers. With a single control, a variety of timbres can be produced: from the 'hollow'-sounding square wave with the missing second harmonic, through to narrow pulses with a rich harmonic content and a thin, 'reedy' sound.

Pulse waveforms usually only have two levels, although there are variants which have three, where the pulses are positive and negative

Figure 3.1.1 This eight slider scanner circuit runs a counter at eight times the required frequency. The counter causes a '1 of 8' multiplexer to sequentially activate each of the eight slider controls, which produce a voltage dependent on their position. The slider outputs are summed together to produce the output waveform.

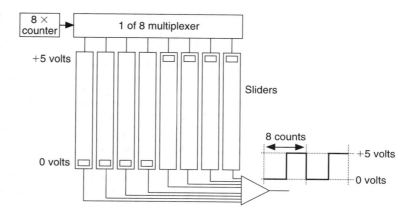

with respect to a central zero value. Multiple levels were used in one of the first 'user-programmable' waveforms, called 'slider scanning'. In this method, the oscillator runs at several times the required frequency, and is used to drive a counter circuit, which then controls an electronic switch called a *multiplexer*. The multiplexer 'scans' across several slider controls, and thus creates a single waveform cycle where the voltage output for each or the stages is equivalent to the positions of the relevant slider. By setting half of the sliders to the maximum voltage position, and the remainder to the minimum, a square waveform is produced, but a large number of other waveforms are also possible (Figure 3.1.1).

Slider scanning oscillators are limited by the number of sliders which they provide. Eight or sixteen sliders are often used, and this means that the oscillator is running at 8 or 16 times the frequency of a VCO producing the same note conventionally. The counter is normally arranged so that it switches in each slider in turn, and when all of the sliders have been scanned, it returns to the first slider again. The sliders thus represent one cycle of the waveform. This type of counter is called a *Johnson counter*; although it is possible to use counters which scan back and forth along the sliders, in which case the relationship between the sliders and the cycle of the waveform is less obvious; although it appears to be only half of a cycle, the reversal of the scan direction merely adds in a time-reversed version of the sliders, and this sounds like a second cycle of a two-cycle waveform. The pitch is thus unaltered. Slider scanners thus provide single-cycle (or two-cycle) waveforms where the shape of the waveform is static (unless the slider positions are changed) and repeated continuously.

For detailed control over the waveform, the obvious solution is to add more sliders. Providing many more than 16 separate sliders quickly becomes very cumbersome, and it makes rapid selection of different waveforms almost impossible. One alternative is to provide pre-stored values for the slider positions in a memory chip (read-only memory (ROM) or random-access memory (RAM)), and to use these values to produce the output waveform. In this way many more 'sliders' can be used, and the waveform can be formed from a large number of separate values, instead of just 8 or 16. The difficulty lies in producing the values to put in the memory – preset values can be provided – but sliders or other means of user control of the values are preferable. A minimalistic approach might be to provide two displays: one for the 'slider' number and another for the value at that slider position.

The Fairlight Computer Musical Instrument (CMI) allowed users to draw waveforms on a screen using a light-pen. Unfortunately, the waveform is not a good guide to the timbre that will be produced: see Figure 2.3.2.

Drawing the values on a computer display screen has been used as one method of providing a more sophisticated user interface to large numbers of sliders, but this can be very tedious to use and difficult to achieve the desired results. Trying to set the positions of several hundred sliders on a screen to produce a particular timbre is also hampered by the relationship between the shape drawn and the timbre produced, which is not intuitive for most users, and requires considerable practice and experience before a specific timbre can be quickly set up. The simpler oscillators which scan through values in a memory chip are very economical in their usage of memory, and a large number of waveforms can be made available in this way. As with analogue synthesizers, selecting a waveform is easier if they are arranged in some sort of order: either a gradually increasing harmonic content, or else in groups with variations of specific timbres: pulses, multiple sine waves added together, etc.

Scanning across a series of voltages, which are set by slider positions, using a multiplexer is straightforward. Replacing the slider positions with numbers then requires some way of converting from a number to a voltage. This is achieved using a circuit called a 'digital-to-analogue converter', normally abbreviated to DAC, which converts a digital number into a voltage. By sequentially presenting a series of numbers at the input to the DAC, a corresponding set of output voltages will be produced (Figure 3.1.2).

Figure 3.1.2 When a wavecycle is stored in memory, then a counter can be used to successively read each of the values and output them to the DAC, and so produce the desired waveform. This repeats for each cycle of the waveform. The location in the memory which is selected by the 'cycle select' logic determines the cycle shape. The 8-bit values are shown here only for brevity: 16-bit representations became widely adopted in the late 1980s.

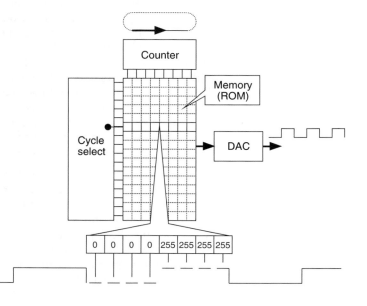

By storing the numbers in ROM, a large number of preset waveforms can be provided, merely by sending different sets of numbers from the ROM to the DAC. This is easily accomplished by having additional control signals which set the area of the memory which is being used. One simple way to achieve this is to use the low order bits from the counter to cycle through the memory, whilst the high order bits can be used to access different cycles, and thus, the different stored waveforms. For user-definable waveforms, RAM memory chips are used, where values can also be stored in the chip instead of merely recalled. Often a mixture of ROM and RAM is used to provide fixed and user-programmable waveforms, but there are alternatives to using memory chips. By using the output of a counter as the input to the DAC, a number of different waveforms can be produced: it depends on the way that the counter operates. Assuming that the DAC converts a simple binary number representation, then a simple binary counter would produce a sawtooth-like staircase waveform. There are a large number of types of counter that could be used for this purpose, although dynamically changing the type of counter is not straightforward. Some other types which might be used include the up–down and Johnson counters mentioned earlier, and Gray-code counters.

Whilst ingenious synthesis techniques sometimes find their way into commercial instruments, the long-term trend has always been for straightforward metaphors and user interfaces, especially when referenced to 'real-world' circuits. Digitally modelled analogue synthesis is one example of how strong this bias is.

By using digital feedback between the stages of a counter, it is possible to produce a counter which does not just produce a short sequence of numbers in sequence, but a very much longer sequence of numbers in a fixed but relatively unpredictable order. These are called *pseudo-random sequence generators,* and they can be used to produce noise-like waveforms from a DAC. Actually this is a multi-cycle waveform (see below) rather than a single cycle. But by deliberately using the wrong feedback paths (or by resetting the count, it is possible to shorten the length of the sequence so that it produces sounds with a definite pitch, where the length of the sequence is related to the basic pitch that is produced. In effect, the length of the sequence becomes the length of one cycle of the waveform. Slight changes in the feedback paths or the initial conditions of the counter can produce a wide variety of waveforms from a relatively small amount of circuitry with simple (if unintuitive) controls, and no memory is required (unless the paths and initial conditions need to be stored, of course!).

Since digital circuitry is concerned with only two values: on and off or one and zero, the square or pulse waveform is a basic digital waveform, in much the same way as sine waves are the underlying basis of analogue. By taking several square or pulse waveforms at different rates, and adding them together, they can be used to produce waveshapes in much the same way as adding several sine waves together (see Chapter 4). These are called Walsh functions, and although conceptually very different from the systems which read out values from a memory chip, the simplest method of producing them is to calculate the values and store them in memory. Providing a user interface to a Walsh-function-driven waveform generator would require comprehensive control over many sine waves, and as with drawing a waveform on a screen, it suffers from the same problems of complexity and

Figure 3.1.3 Walsh functions combine square or pulse waveforms to produce more complex waveforms. In this example, four square waves are added together to produce a crude 'triangle' type of waveform. Each position in the output is produced by adding together the level of each of the component waveforms.

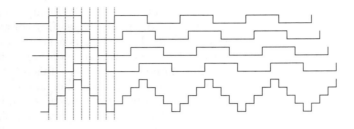

detail, without any intuitive method of determining the settings (Figure 3.1.3).

Filtering of outputs

All of the methods of producing arbitrary waveshapes using digital circuitry described above produce outputs which tend to have flat segments connected by sudden transitions. Since these rapid changes produce additional (often unwanted) harmonics, the outputs need to be filtered so that the final output is 'smooth'; usually with a low-pass filter whose cut-off frequency is set to be at the highest required frequency in the output. Because the frequency of the output waveform from an oscillator can change, then the filter needs to track the changes in frequency, which means that the VCO needs to be coupled to a voltage controlled filter (VCF), and set up so that the cut-off frequency follows the oscillator frequency, usually by connecting the same control voltage to the VCO and VCF. In some circumstances, the additional frequencies are deliberately allowed to pass through the filter. Because these frequencies are linked to the oscillator frequency, they are actually harmonics of it: albeit high harmonics. Removing the filter means that a waveform, which might appear to be a sine wave from the slider positions, is actually a sine wave with extra harmonics. The ability to switch the filter in and out, together with knowledge of how the waveforms are being produced, is very useful if the most is to be made of the potential of single-cycle oscillators (Figure 3.1.4).

Waveshapes

In summary, single-cycle oscillators normally have a selection of waveshapes based on the following types:

- **Mathematical shapes**: sine, triangle, square, sawtooth.
- **Additions of sine wave harmonics** (organ 'drawbar' emulations).
- **Additions of square or pulse waves** (Walsh functions).
- **Random**: single-cycle 'noise' waveforms from pseudo-random sequence generators tend to be very non-white in character, and often have large amounts of high harmonics.

Single-cycle oscillators have a characteristic fixed timbre – each cycle is the same as the previous one and the next one. This means that subsequent processing through modifiers is often used to make the sound more interesting to the ear.

Figure 3.1.4 Filtering the output of a generated wavecycle waveform can smooth out the abrupt transitions and produce the required shape.

Figure 3.1.5 By addressing two (or more) different parts of the waveform ROM memory, the output waveform can be changed on a 'per cycle' basis. Here, 8-bit representations of a square and sawtooth waveforms are present in the ROM memory, and are sequenced cyclically to form a composite output waveform.

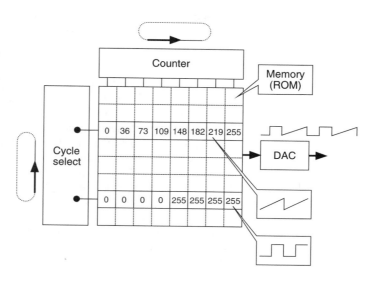

3.1.2 Multi-cycle

The important thing about multi-cycle oscillators is that although they can have many cycles of waveforms which they output in sequence, the same set of cycles repeats continuously. This is different to a sample, where the sample may only be played through in its entirety once.

The technology for producing multi-cycle waveforms is very similar to single-cycle oscillators, although the user interface is often restricted to merely choosing the specific waveforms for each cycle, rather than providing large numbers of slider or other controls. The basic method uses a memory chip and DAC, just as with single-cycle oscillators. The difference is that the area of memory which is being cycled can also be controlled dynamically. A simple example might have two areas of ROM memory: one containing a square waveform and the other a sawtooth waveform (Figure 3.1.5). By setting the ROM to output first the square, then the sawtooth, and then repeating the process, the output will be a series of interspersed square and sawtooth cycles. This two-cycle waveshape has a harmonic structure which incorporates some elements from each of the two types of waveform, but also has additional lower frequency harmonics which are related to half of the basic cycle frequency. This is because the complete cycle repeats at half the fundamental cycle rate.

By concatenating more waveforms together, more complex sequences of waveforms can be produced. As the length of the sequence increases, then the extra low frequency harmonics also drop in frequency.

To take an extreme example, imagine one cycle of a square waveform followed by three cycles of a silence waveform. The equivalent is a pulse waveform with a frequency of a quarter of the square wave cycle: two octaves below the pitch which was intended. With eight cycles in the sequence, then the lowest harmonic component will be three octaves down, and with 16 cycles, the frequency will be four octaves down. If the length of the sequence is not a square of two, then the frequency that is produced may not be related to the cycle frequency with intervals of octaves.

For example, if a square wave cycle is followed by two cycles of silence, then the effective frequency of the pulse waveform is a third of the basic cycle frequency, which means that the lowest frequency will be an octave and a fifth down, and the ear will interpret this as the fundamental frequency, and so the oscillator has apparently been pitch shifted by an octave and a fifth. With longer sequences of single-cycle waveforms, this pitch change can be harmonically unrelated to the basic pitch of the oscillator. With more than one cycle of non-silence, the resulting harmonic structures can be very complicated. This can produce sounds which have complex and often unrelated sets of harmonics, which gives a bell-like or clangorous timbre. The pseudo-random sequence generators mentioned earlier are one alternative method of producing sequences of single-cycle waveforms, and exactly the same length-related pitch-shifting effect happens. Chapter 4 looks at these effects in more detail.

For very long sequences of cycles, the pitch-shift can become so large that the frequency becomes too low to be heard, and it is then only the individual cycles which are heard. This means that by concatenating a series of pulse waveforms which gradually change their pulse width, it is possible to produce a repeated multi-cycle waveform which sounds like a single-cycle PWM waveform. (By altering the number of repeated cycles of each different pulse width waveform, it is possible to change the effective speed of PWM. In fact, this is exactly how wavetable oscillators work; see Section 3.2.)

There are two methods for reading out the values in a wavecycle memory. When the values are accessed by a rising number, then the shape is merely repeated, whilst by accessing the same values using a counter which counts up and down, then each alternate cycle is reversed in time. This can be a powerful technique for producing additional multi-cycle waveforms from a small wavecycle memory. If repeats of the cycles can be inverted as well, then even more possibilities are available. All of these variations on a single cycle can produce changes in the spectrum of the sound: from minor detail through to major additional harmonics where the transition between the repeats is not smooth (see Figure 3.1.6 for more details).

Fixed sequences of single-cycles can be thought of as short samples, the PWM waveform is one example where a complete 'cycle' of PWM is repeated to give the same audible effect as a pulse waveform which is being modulated. Many other dynamically changing multi-cycle waveforms can be produced.

Figure 3.1.6 Pairs of wavecycles can be arranged in different ways by exploiting the symmetry (or lack of) the waveshapes. These four examples show a wavecycle followed by the four possibilities of reversing or inversion. The transitions between the wavecycles can be smooth or abrupt depending on the shape of the wavecycle.

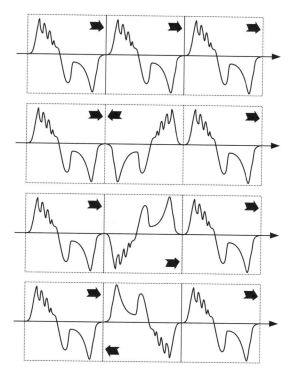

Multi-cycle oscillators can also be likened to granular synthesis, since they both concatenate cycles of waveforms, although granular synthesis normally works on groups of cycles rather than individual cycles (see Chapter 5). Roland's 'RS-PCM', and many other soundcards and computer sound generators all often use loops of multi-cycle 'samples' to provide the sustain and release portions of an enveloped sound; and sometimes even the attack portion. This technique is equivalent to changing the waveform of a multi-cycle oscillator dynamically, which is an advanced form of wavetable synthesis (see the next section).

Multi-cycle oscillators normally have a selection of single-cycle waveshapes plus the following additional types:

- concatenations of mathematical shapes meaning sine, triangle, square and sawtooth cycles in sequences
- symmetry variations of mathematical and other shapes
- PWM waveshapes which change their harmonic content with time
- waveshapes which change their harmonic content with time, but not in a regular sequence (i.e. not progressively as in PWM)
- shapes with additional non-harmonic frequencies (clangs, chimes and vocal sounds)
- noise in other words, more cycles mean that the noise produced can be more 'white' in character than from single-cycle oscillators.

Interpolation is a method of producing gradual changes from one wavecycle shape to another, rather than the abrupt changes which occur when wavecycles are concatenated. Section 3.2 deals with this in more detail.

Sample

Memory size is closely related to date. In the 1980s, a megabyte was large: the first external small computer system interface (SCSI) hard drive for the 128-Kbyte RAM Macintosh Plus computer had a capacity of 20 Mbytes.

For very long sequences of single cycles, the complete sequence may not repeat whilst a note is being played, and it then becomes a sample rather than a multi-cycle waveform. Samples are usually held in either ROM or RAM memory, and because of the length, the amount of memory used can be quite large. For example, for 16-bit values, where there are 44 100 values output per second (the same as a single channel of a compact disk (CD) player) then 705 600 bits (just over 86 Kbytes) are required to store just 1 second of monophonic audio. This means that it requires a megabyte of memory to store just below 12 seconds of monophonic audio sample. Obviously long samples are going to require large quantities of memory, and stereo samples will double the memory requirements. Because of this, most hybrid synthesizers of the 1970s and 1980s used very short samples, and it was only with the availability of low-cost memory in the 1990s that sampling techniques became more widespread, and this was in an all-digital form.

Trying to reduce the amount of memory which is required to store cycles affects the quality of the audio. Hybrid wavecycle synthesizers suffer from the resolution limitations of their storage. At low frequencies there are not enough sample points to adequately define the waveshape, whilst at high frequencies the circuitry may not run fast enough. For example, suppose that a single cycle of a waveform is represented by 1024 values. At 100 Hz, this means that the VCO needs to run at 1024 times 100 Hz, which is 102.4 kHz. But at 1000 Hz, the VCO needs to run at 1.204 MHz and at 10 kHz, the VCO is oscillating at 10.24 MHz. Accurate VCOs with wide ranges, good temperature stability and excellent linearity at these frequencies are more normally found in very high quality radio receivers. More importantly, affordable late 1970s memory technology began to run out of speed at a few MHz. Reducing the number of bits which are used to represent the waveform cycles also reduces the quality by introducing noise and distortion. The whole of the sampling process is covered in more detail in Chapters 1 and 4.

It is a common fallacy that the ability to produce complex waveshapes is all that is required to recreate any sound. In practice, most methods of producing a waveshape do not have the required resolution to provide enough control over the sound over time. Harmonics are often a more reliable guide to the sound, and it is normally the change of harmonics over time that provides the interest in most sounds. The effect of harmonics on waveshape is covered in more detail in Section 2.3 on Additive synthesis.

Modifiers

When the oscillator has produced the raw source waveform, then most hybrid synthesizers pass it through a modifier section which is typical of those found on analogue synthesizers. This is usually a VCF and voltage controlled amplifier (VCA), with associated envelope generator (EG) control. Curiously, whereas most hybrid synthesizers attempt to improve on the selection of available waveforms for the oscillator, the VCF is often still just a simple resonant low-pass filter. This puts a great deal of emphasis on the oscillator as the prime source of the timbre, and means that the possibilities for the changing of the timbre by the modifiers are just the same as an analogue synthesizer. This means that the resonant filter sweep sound remains an audio cliché for both analogue and hybrid synthesizers. Only on some all-digital S&S synthesizers are the filtering capabilities enhanced significantly.

In a historical perspective, hybrid synthesizers reached a peak of popularity in the early 1980s, after polyphonic analogue synthesizers and just before the all-digital synthesizers. The many 'analogue' synthesizers of the mid-1990s' 'retro' revival of analogue technology are often not truly analogue, but are actually more modern hybrids, where the 'VCOs' are actually sophisticated wholly digital DCOs which use the methods described above to produce their waveforms, but coupled with a conventional analogue modifier section in a standard hybrid synthesis way. These updated hybrids have, in turn, being incorporated into all-digital instruments which use a mixture of synthesis methods. By the start of the 21st century, stand-alone hybrid synthesis had been almost entirely replaced by digital synthesis, often using modelling techniques, although there was also a new 'retro' revival for 'pure' analogue.

3.2　Wavetable

Initially, wavetable synthesis might appear to be very similar to multi-cycle wavecycle synthesis. Both methods use sequences of cycles to produce complex waveshapes. The major difference lies in the way the cycles are controlled. In multi-cycle wavecycle synthesis, the chosen sequence of cycles is repeated continuously, whereas in wavetable synthesis, the actual waveform which will be used can be chosen on a cycle-by-cycle basis.

3.2.1　Memory

Wavetable synthesis is based around memory even more strongly than wavecycle synthesis where there are a few methods which do not use large quantities of memory, for example pseudo-random sequence-based waveform generators. But wavetable synthesis uses the memory as an integral part of the synthesis process, since the cycle being used is dynamically selected by controlling the memory.

Just as with single-cycle wavecycle synthesis, a cycle of a waveform is stored in a memory chip, and successive values are retrieved from the memory and sent to a DAC where they produce the output waveform (Figure 3.2.1). The values are retrieved in order by using a counter which steps through the memory in an ascending sequence of memory locations. Determining which values the counter steps

Figure 3.2.1 A wavetable synthesizer uses several wavecycle locations in the memory: accessing each in turn. In this example, the cycle select logic sequentially selects wavecycles 1, 2 and 3, and then repeats this continuously. The output thus consists of three concatenated wavecycles. For simplicity the values shown are just 0s and 1s, but they could be 8-, 12- or 16-bit values, depending on the required precision.

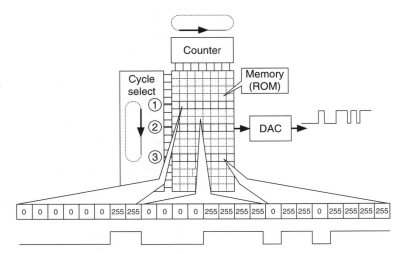

through is set by controlling which part of the memory is used. In a single-cycle wavecycle oscillator, the control signals are set to point to the specific single-cycle waveform, and the oscillator then outputs that waveshape continuously.

In a wavetable oscillator, the control signals which determine where the waveform information is stored can be changed dynamically as the oscillator is outputting the waveshape. Normally the changes are made as one cycle ends and another begins, so that the waveshape does not change mid-way through a cycle. The control signals which set where the cycle is retrieved from can be thought of as modulating the shape of the output waveform, although they are really just pointing to different parts of the memory. The name 'wavetable' comes from the way that the memory can be thought of as being a table of values, and so the control signals just point to cycles within that wavetable. There are two basic ways that the cycle being used in the table can be changed: *swept* and *random-access*.

Swept

One common usage for swept wavetables is to emulate a waveform that has been passed through an enveloped resonant filter.

By incrementing the pointer so that it points to successive cycles in the wavetable, the control signals effectively 'sweep' the resulting waveshape through a series of waveforms. The fastest rate at which this can happen is when only one cycle of each waveform is used before moving to the next waveform, although by omitting waveforms the sweep speed can be increased. Wavetables which are intended to be swept in this fashion are normally arranged so that the waveforms are stored in an order where similar sounding waveforms are close together. This produces a 'smooth' sounding change of waveshape and harmonics as the table is swept. Large changes of harmonics or sudden changes of waveshape, can produce rich sets of harmonics, and this is catered for by allowing sweeps to occur over the boundaries between these groups of similar timbres. For example, a series of added sine waves might be followed by a group of pulse

Figure 3.2.2 (i) A swept wavetable outputs each of the cycles between the start and finish points in the memory. (ii) A random-access wavetable only outputs the specific cycles which have been chosen.

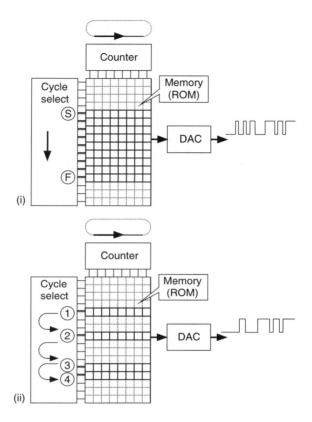

waves in the table, and a sweep which crossed over between the two groups would have large changes between the two sections.

Random-access

By allowing the pointer to be set to point to anywhere in the table for each successive cycle, any cycle can be followed by any other cycle from the wavetable. This is called *random-access*, since any random cycle in the table can be accessed. By supplying a series of pointers, the waveform can be swept, so a sweep is in fact a special case of random-access. More normally, a series of values are used to make the pointer access a sequence of waveform cycles. This can be a fixed sequence, in which case the wavetable behaves as a multi-cycle wavecycle oscillator, or a dynamically changing sequence of pointer locations, in which case the modulation of the waveform is characteristically that of a wavetable (Figure 3.2.2).

3.2.2 Table storage

The actual storage of the waveforms inside the wavetable can be of several forms. Some hybrid oscillators only have one of these types. Others provide two types or all three. The naming conventions differ with each manufacturer: wavesamples and hyperwaves are just two examples of names used for samples and multi-cycle waves, respectively. Some types are found in analogue/digital hybrids as well as digital instruments which emulate analogue hybrids, whilst the more complex types are only found in digital instruments. The major types of table storage are:

- Single-cycle wavetable oscillators provide large numbers of single-cycle waveforms, and can be implemented in hybrid or digital technologies. This method can be used to provide results which are similar to crude granular synthesis.
- Multi-cycle wavetable oscillators contain waveforms with more than one cycle, and can be implemented in hybrid or digital technologies, but are found mostly in digital instruments.
- Samples are just longer multi-cycles, although the implication is that the sample plays through once or only partially, whilst a multi-cycle waveform is usually short enough to be repeated several times in the course of a note being played. Some samples in wavetables are provided with multiple start points, which means that the sample can be played in its entirety, or that it can be started mid-way through. This can be used to provide a single sample which can be used as an attack transient sound with a sustain section following, or as just a sustained sample by playing the sample from the start of the sustain portion. The section on using S&S in this chapter contains more detail on these techniques.

The Korg Wavestation is an example of a synthesizer which allows samples to be sequenced.

- Sequence lists or wave sequences are the names usually given to the sequential set of pointers to cycles or samples in the wavetable. This list determines the order in which the cycles or samples will be replayed by the oscillator. Lists can automatically repeat when they reach the end, reverse the order when they reach the end, or merely loop the last cycle or sample. Some sequences allow looping from the end of the list to an arbitrary point inside the sequence list, which allows a set of cycles or samples to be used for the attack portion of the sound, whilst a second set of cycles or samples is used for the sustain and release portions of the sound. This interdependence of the oscillator and envelope is common in sample-based instruments, whereas in analogue synthesis the VCO and EG are normally independent.
- Mixed modes. Some hybrid wavetable oscillators allow mixtures of single-cycle, multi-cycle and sample waveforms to be used in the same sequence list. Additional controls like repetitions of single- or multi-cycle waveforms, or even the length of time that a sample plays, may also be provided.

Multi-samples

The term 'multi-samples' can be applied to the result of a sequence list which causes samples to be played back in a different order to that in which they were recorded. These samples may be looped, in which case some interaction with an EG is usually used to control the transitions between the individual looped samples. Roland's 'RS-PCM' and many other soundcards and computer sound generators all often use loops of multi-cycle 'samples' to provide the sustain and release portions of an enveloped sound, and sometimes even the attack portion. This technique is equivalent to changing the waveform of a multi-cycle oscillator dynamically, which is an advanced form of wavetable synthesis.

'Multi-samples' are also used to mean the use of several different samples of the same sound, but taken at different pitches.

Loop or wave sequences

A loop or wave sequence is the name for the sequence of samples which are used in a multi-sample. It provides the mapping between the envelope segments and the samples which are looped in that segment (Figure 3.2.3). Loop sequences are sometimes part of a complete definition which includes the multi-sample and loop-sequence information: for example, the musical instrument definitions in Apple's QuickTime and Roland's 'RS-PCM' are stored in this way.

When samples are looped as part of a sequence then the playback time will normally be set by the sequence timing, and can be made uniform across the keyboard span. So in a string sound consisting of a bowing attack sound, and a sustain vibrato sound, the bow scrape attack might last the same time regardless of the key played on the keyboard, even though the pitch will change. This is the exact opposite to how most samplers work: in a sampler, the pitch and length of

Figure 3.2.3 Loop sequences control the order in which looped wavecycles are replayed. In this example, the loop sequence is controlled by the envelope. Wavecycle 1 loops during the attack part of the envelope, followed by wavecycle 2 during the decay segment. The sustain segment is produced by looping wavecycle 3, and wavecycle 4 loops during the release part of the envelope.

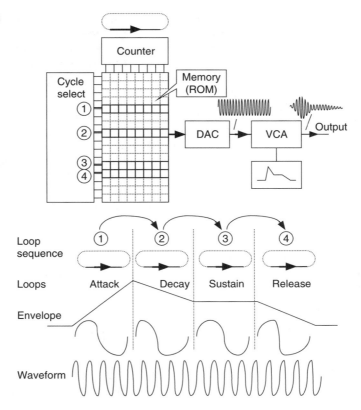

time which the sample pays for are normally directly related. So higher pitched sounds tend to have shorter attacks and decays. The solution in a sampler is to provide several different samples taken at different pitches. These are also called 'multi-samples'. At the end of the 20th century, sample-replay techniques which combined these two contrasting approaches were developed, and the link between pitch and time was largely removed.

Interpolation tables

Sequences of samples do not need to abruptly change from one sample to the next. By taking two differently shaped wavecycles or samples from a wavetable and gradually changing from one shape to another, the harmonic content can be dynamically changed, and the audible effect is like cross-fading from one sound to the other. The initial cycle will contain all of the values from one of the two cycles or samples, and the final cycle will contain only the values from the other cycle or sample (Figure 3.2.4). The process of changing from one set of values to another is called *interpolation*. Interpolation is mathematically intensive, but requires only small amounts of memory to produce complex changing timbres.

Interpolating between two waveshapes does not always produce a musically useful transition because the changes in harmonic content may not be too great, and the result does not sound smooth and predictable. The relationship between the shape of a waveform and its harmonic content is not a simple one, and minor changes in the shape of a waveform can produce large changes in the harmonic content. Interpolation can emphasise this effect by producing timbres which change from one sound to another, but which pass through many other timbres in the process, rather than the smooth 'evolution' which might be expected.

Rather than using interpolation as a mathematical transformation of information about a waveform, a much more satisfactory method is to use interpolation to produce the changes between one spectrum and another. This method does produce smooth changes of timbre (or very unsmooth changes, depending on the wishes of the user!). This sort of spectral transformation is described in the section on Analysis–synthesis techniques in Chapter 5.

Figure 3.2.4 Interpolation allows two waveforms to be defined as start and finish points, and the 'in-between' wavecycles are then calculated (or interpolated) to produce a smooth transition between the two waveforms.

Start waveform   Finish waveform

Interpolated waveform changes from the start to the finish shape

3.2.3 Additional notes

- Wavetable synthesis is a term used to cover a wide range of techniques, and as a result, there are as many definitions of wavetable synthesis as there are techniques.
- The differences between single-cycle wavetable synthesis and sampling are actually greater than the differences between multi-cycle wavecycle and sampling. Very few samplers have the facility to alter the order in which the parts of a sample are played back!
- Soundcard manufacturers tend to describe almost any hybrid technique as 'wavetable' synthesis.
- By loading a wavetable oscillator with a set of multi-cycle waveforms which have been generated from the addition of sine waves, an additive synthesis engine can be produced using hybrid digital and analogue techniques.

3.2.4 Sample sets

The samples which are provided in wavetable and wavesample-based hybrid synthesizers can have a great effect on the sound set which is possible to create. A determined programmer will see the exploitation of even the sparsest of sample sets as a challenge. So one of the first tasks for a synthesist, who wishes to explore the sound-making possibilities of a hybrid instrument, is to become familiar with the supplied sound resources.

Typical wave and sample sets come in very standardised forms, mainly due to the twin influences of general musical instrument digital interface (General MIDI, abbreviated as GM) and history. GM has meant that most instruments need to contain sufficient samples to produce the 128 GM sounds. This means that otherwise serious professional instruments may well have samples of birds, a telephone ringing, a helicopter and applause in their sample set. History dictates the inclusion of instruments like key-click organs, harpsichords, accordions and some now-unusual percussion sounds: all of which have become somewhat clichéd. Some manufacturers have used synthesis to create these sounds, but the results are often very different to the GM standard samples.

The provision of a good piano sound is also frequently obligatory, especially in a keyboard that is expected to have broad appeal, like a workstation. Only in a few 'pure' synthesizers has the piano sound ever been omitted. Unfortunately, a good piano sound requires large amounts of memory, often a significant proportion of the total sample set. It can be very difficult to utilise piano sounds as the raw material for anything other than piano sounds.

There are three main types of samples that are found in sample sets:

- Pitched samples are sounds which have a specific frequency component (the note that you would whistle).
- Residues, which are the non-pitched parts of a sound: the hammer thud, fret-buzz, string-scrape, etc. Often produced by processing the original sound to remove the pitched parts.
- Inharmonics: unpitched, noisy, buzzy or clangorous sounds.

Piano sounds will often start with several pitched multi-samples at different pitches. Piano residues will be one or more hammer thuds and harp buzz sounds. These can be useful for adding to other instrument sounds or for special effects when pitch shifted. Piano and electric piano residues can be good for sound effects; they sound like metal tapping and clunks.

'Classic' or 'analogue' samples will be the basic square sawtooth and pulse waveforms, possibly a sine, and sometimes some residues. These are intended to be used in emulations of analogue synthesizers.

Strings and vocal sounds will be looped sustained sounds, but some may have looping artefacts and cyclic variations in timbre: audition all of the samples and listen closely to the sustained sound as it loops. If a loop does exhibit a strong artefact, then consider using that as part of a rhythmic accompaniment.

Woodwind sounds can be used as additional waveforms for analogue emulations, as well as thickening string sounds.

Plucked and bass sounds can be pitch shifted up to provide percussive attacks, or shifted down for special effects.

Percussive sounds can be used as attack segments, or assembled into wave sequences to provide rhythmic accompaniment.

Digital waveforms come in either PWM wave sequences, or sequences with varying harmonic content. These can be used in cross-faded wave sequences to provide movement in the sound. Samples that are large enough to contain complete cycles of PWM beating are rare, whereas single cycles of waveforms are small.

Gunshots, sci-fi sirens, laughter and raindrops can be pitch shifted downwards to provide atmospheric backdrops.

3.3 DCOs (Digital controlled oscillators)

DCOs are the digital equivalent of the analogue VCO. DCOs have much in common with wavecycle synthesis. In fact, they can be considered to be a special case of the most basic wavecycle oscillator: one which only produces the 'classic' synthesizer waveforms (sine, square, pulse and sawtooth).

DCOs are combinations of analogue and digital circuitry and design philosophies; they are literally hybrids of the two technologies. They were originally developed in order to replace the VCOs used in analogue synthesizers with something which had better pitch stability. The simple exponential generator circuitry used in many early analogue VCOs was not compensated for changes in temperature, and so the VCOs were not very stable and would go out of tune. Replacing the VCOs with digitally controlled versions solved the tuning stability problems, but changed some of the characteristics of the oscillators because of the new technology. The DCO is also notable because it marked the final entry of the era of three character acronyms: VCO, VCF, LFO and VCA. Subsequent digital synthesizers moved away from acronyms towards more accessible terms like oscillator, filter, function generator and amplifier.

It is worth noting that although early VCO designs did suffer from poor pitch stability, the designs of the late 1970s gradually improved the temperature compensation and the custom chips developed in the early 1980s had excellent stability. But the damage to the reputation of analogue VCOs had been done, and DCOs replaced the VCO permanently for all but the most purist of analogue users. DCO-based synthesizers were also used by frequency modulation (FM) in the mid-1980s, and the DCO-based sample player is the basis of all S&S synthesizers. In a curious looping of technology, the modelled analogue synthesizers of the 2000s use DCO-based sample-replay to replay samples which are based on mathematical models of analogue VCOs of the 1970s.

3.3.1 Digitally tuned VCOs

The simplest DCOs are merely digitally tuned VCOs. A microprocessor is used to monitor the tuning of the VCOs and retune it when necessary. This process was usually carried out when the instrument was initially powered up, although it could be manually started from the front panel. The technique isolates the VCO from the keyboard circuitry (see Chapter 7), and then uses a timer to measure the frequency by counting the number of pulses in a given time period. This measurement process is carried out near the upper and lower frequency limits of the VCO by switching in reference voltages, with additional measurement points as required to check the tracking of the VCOs (see Chapter 2). From this information, the two main adjustments can be calculated:

- an 'offset' voltage can be generated to bring the VCO back into 'tune'
- a 'tracking' voltage can be used to set the tracking.

The offset and tracking voltages for all the VCOs in a synthesizer would be stored in battery-backed memory. This technique is often known as 'auto-tune' (Figure 3.3.1).

Some analogue poly-synths need to be left alone during auto-tuning. Moving the pitch-bend wheel on some 1970s' examples could put the entire instrument out of tune.

A variation of this technique is currently used in some analogue-to-digital converter (ADC) chips, where the circuit monitors its own performance and recalibrates itself continuously for each sample. In early auto-tune synthesizers, the time required to measure the frequencies at the various points meant that it was not possible to continuously tune the VCOs – it could take several minutes to retune all of the VCOs in a polyphonic synthesizer. It would be possible to arrange the voice allocation scheme so that a VCO that was out of tune could be removed from the 'pool' of available voices, but this would effectively reduce the polyphony by one. Some synthesizers allowed voices to be disabled in exactly this way if they could not be tuned correctly by the auto-tune circuitry, which explains the poor reliability reputation of the VCOs in some early polyphonic synthesizers.

Figure 3.3.1 In an 'auto-tune' system, a microprocessor sends a series of control voltages to the VCO, and compares the output frequencies with the ideal values. These numbers are then used to provide offset and tracking adjustments to the VCO so that its response matches the ideal curve.

Another combination of digital microprocessor technology with analogue VCOs occurs in synthesizers where the keyboard is scanned using a microprocessor and the resulting key codes are turned into analogue voltages using a DAC and then connected to the VCOs. Although suited to polyphonic instruments, this technique has been used in some monophonic synthesizers, particularly where simple sequencer functions are also provided by the microprocessor. The Sequential Pro-One is one example of a monophonic instrument which uses 'digital' storage of note voltages, and the use of a micro-controller chip allows it to also provide two short sequences with a total of up to 32 notes clocked by the LFO.

3.3.2 Master oscillator plus dividers

A nearer approach to an all-digital 'true' DCO uses ideas taken from master oscillator organ chips. A quartz crystal controlled master oscillator provides a high frequency clock, which is then divided down to provide lower rate clocks through a series of divider chips (Figure 3.3.2). By using a high rate of master clock and correspondingly large divider ratios, it is possible to generate all the frequencies which are required for all the notes for a synthesizer from just a few chips. These notes can then be gated from the keyboard circuitry, with the end result being a polyphonic oscillator which is derived from a stable crystal controlled master oscillator.

Rather than having separate stages of dividers for each note, an obvious design simplification is to have 12 dividers which produce the

Figure 3.3.2 A 'top octave' divider systems uses a high frequency master oscillator and dividers to provide all the required frequencies for all the notes on a keyboard. In this example, the master oscillator frequency is 500 kHz, and the division values required to produce the 12 top notes in an octave are shown. Each of these frequencies then needs to be further subdivided down to produce the lower octave notes.

highest required frequencies, and then divide each of these outputs successively by 2 to produce the lower octaves (this presupposes that the scale used will be fixed: usually equal temperament, and that each octave has identical ratios between the notes). This is called a 'top octave' method, and with the right division values and a high enough clock, it gives very good results. For example, with a master clock frequency of 500 kHz, the division required to produce a C#6 at 1108.73 Hz is 450.96, which is almost exactly 451, whilst for the next note, the D6 at 1174.66 Hz, the division is 425.65, and so an integer value of 426 will produce an output frequency which is slightly too low. In a real-world design, you might expect that the clock frequency and division values would be chosen to minimise the errors by setting real division values which are as near to integers as possible. In practice, a real-world custom top octave synthesizer chip, the General Instruments (GI) AY1-0212A used exactly these division values, as shown in Table 3.3.1.

As the table shows, using integer dividers only makes a slight difference in the output frequency. Using individual separate division stages only improves the accuracy slightly. Taking the values for the C1, C5 and C7 division values given above, if the 239 value for the C7 'top octave' division is then divided down by successive dividers, this is equivalent to doubling the effective division value: which would thus have the values of 956 for the C5 frequency and 15 296 for the C1 frequency. The 956 division value is identical to the one used, but the 15 296 is slightly too large, which means that output frequency will be too low – by about 0.0149 Hz or 0.00045 cents. The difference in division values thus produces only very slight differences in the output frequencies. For comparison purposes, the human ear can detect pitch changes of a minimum of about 5 cents, whilst the E-mu Morpheus has fine tuning steps of 1.5625 cents, the Yamaha

Table 3.3.1 DCO dividers

Note	Note frequency (Hz)	True divider	Integer divider (Hz)	Actual note frequency	Frequency error (%)	Frequency difference (Hz)
Clock 500 000 Hz						
C#0	17.3239	28861.84	28862	17.3238	−0.0006	9.8E−05
D0	18.354	27241.95	27242	18.354	−0.0002	3.6E−05
D#0	19.4454	25712.97	25713	19.4454	−0.0001	2E−05
E0	20.6017	24269.82	24270	20.6016	−0.0008	0.00016
F0	21.8268	22907.66	22908	21.8264	−0.0015	0.00033
F#0	23.1247	21621.95	21622	23.1246	−0.0002	5.6E−05
G0	24.4997	20408.4	20408	24.5002	0.00196	−0.0005
G#0	25.9565	19262.97	19263	25.9565	−0.0002	4.7E−05
A0	27.5	18181.82	18182	27.4997	−0.001	0.00027
A#0	29.1352	17161.35	17161	29.1358	0.00205	−0.0006
B0	30.8677	16198.16	16198	30.868	0.00098	−0.0003
C1	32.7032	15289.03	15289	32.7033	0.00017	−6E−05
Clock 500 000 Hz						
C#4	277.183	1803.865	1804	277.162	−0.0075	2.07686
D4	293.665	1702.622	1703	293.6	−0.0222	0.06524
D#4	311.127	1607.061	1607	311.139	0.00379	−0.0118
E4	329.628	1516.863	1517	329.598	−0.009	0.02967
F4	349.228	1431.728	1432	349.162	−0.019	0.06622
F#4	369.994	1351.372	1351	370.096	0.02751	−0.1018
G4	391.995	1275.525	1276	391.85	−0.0372	0.14591
G#4	415.305	1203.935	1204	415.282	−0.0054	0.02231
A4	440	1136.364	1136	440.141	0.032	−0.1408
A#4	466.16	1072.593	1073	465.983	−0.0379	0.17678
B4	493.88	1012.392	1012	494.071	0.03869	−0.1911
C5	523.25	955.5662	956	523.013	−0.0454	0.23745
Clock 500 000 Hz						
C#6	1108.73	450.9664	451	1108.65	−0.0074	0.08255
D6	1174.66	425.6551	426	1173.71	−0.081	0.95108
D#6	1244.51	401.7645	402	1243.78	−0.0586	0.72891
E6	1318.51	379.2159	379	1319.26	0.05694	−0.7512
F6	1396.91	357.9329	358	1396.65	−0.0188	0.26196
F#6	1479.98	337.8424	338	1479.29	−0.0466	0.69006
G6	1567.98	318.8816	319	1567.4	−0.0371	0.58188
G#6	1661.22	300.9836	301	1661.13	−0.0054	0.09043
A6	1760	284.0909	284	1760.56	0.032	−0.5634
A#6	1864.66	268.1454	268	1865.67	0.05422	−1.0116
B6	1975.53	253.0966	253	1976.28	0.03818	−0.7546
C7	2093	238.8915	239	2092.05	−0.0454	0.94979

Based on a GI AY1-0212A TOS chip.

SY99 has micro-tuning steps of 1.171875 cents, and the MIDI Tuning Standard has steps of 0.0061 cents.

Changes in the pitch of these types of 'master oscillator plus divider' DCO might be achieved by using a voltage controlled crystal oscillator to make minor changes in pitch for pitch-bend or vibrato effects, but it is difficult to change the frequency of a crystal oscillator enough for satisfactory pitch control. A more satisfactory method uses a rate

Figure 3.3.3 Dividers can be used as filters. In this example, a single pulse is missing from the 4.096-MHz clock. Subsequent divide-by-2 stages reduce the effect of the missing clock by 'averaging' out the frequency.

One clock pulse missing in a 4.096-MHz clock pulse stream

adapter, which is a counter-based circuit which removes just one clock occasionally from a continuous clock signal and so reduces the effective clock rate. This 'gapped' clock needs to be followed by the equivalent of a low-pass filter to remove the effects of the jitter in the clock pulses, but this type of DCO has just such filters in the form of divider circuits.

For a 4.096-MHz clock, removing just one clock pulse with a rate adapter can be thought of as changing the effective frequency to 2.048 MHz whilst the clock pulse is missing, but then the next clock pulse restores the frequency to 4.096 MHz again (Figure 3.3.3). The actual number of pulses per second (which is what frequency measures) varies depending on when and how the measurement is taken. If the frequency is measured by timing from the start of one clock pulse to the start of the next, then the frequencies of 4.096 and 2.048 MHz are correct. The brief change in frequency is large when measured in this way, but by measuring more than one clock pulse, the change in frequency reduces as the number of clock pulses used for the measurement increases. This process of averaging the frequency over several clock cycles is just what happens when a divider circuit is used to divide down the output of a rate adapter.

The missing clocks are an extreme form of the random variation in the time between successive clock edges in a digital system called *jitter*. Caused by unstable clock sources, or noise affecting the switching point of gates, jitter usually varies randomly around the true edge position: some clocks are slightly closer together, whilst others are slightly further apart. More complicated circuits like accumulator/divider circuits can provide very small frequency changes without the need for large numbers of divider stages to filter out the jitter.

The main indicator of the use of this type of 'master oscillator plus dividers' DCO is global pitch control, particularly for portamento.

Glissando can be produced by altering the way that the keyboard circuitry controls the gating of the note frequencies, and can be carried out on a 'per voice' basis; so the glissando can be polyphonic, with held notes being unaffected. If portamento is provided, then it can be implemented by producing a series of pitch bends from note to note, and this may well produce audible glitches in the transitions; but more importantly, because the pitch-bend rate adaption is done before the divider circuitry, it will affect all of the notes which are being played. This type of portamento is thus called 'monophonic' portamento. Because of this 'global' pitch change, synthesizers which have this type of DCO do not usually provide pitch envelopes which change the pitch of a note, since any new notes which are played would pitch bend any pre-existing held notes, which is not very useful musically. The pitch bend and vibrato normally affect all notes that are being played, and individual pitch control for pitch bend or vibrato on a 'per voice' basis is more unusual.

The upper frequency limit of many top octave synthesizer chips was limited; for a 500-kHz input, the chip mentioned above can only produce a C7 at 2093 Hz, which is an octave below the top note of an 88-note piano keyboard. Also, having lots of simultaneous frequencies produced by a large number of divider chips can induce a characteristic buzzing sound in the audio output if care is not taken with wiring layout and circuit board design – the commonly used onomatopoeic term for this problem is 'beehive' noise. 'Electronic pianos' and 'string machines' of the mid-1970s which used top octave synthesis were notorious for this extraneous noise.

Another useful hint that this type of DCO is being used is the lack of any 'detune' facility if there is more than one DCO provided in the voice. Since the only way to provide fine resolution pitch changes is by using the rate adapter (of which there is usually only one), which produces global pitch changes, then it is not possible to achieve the slight 'detuning' effects of two VCOs. By using two rate adapters and two sets of divider circuits, it is possible to produce detune, but this almost doubles the required circuitry. Many 'master oscillator plus divider' synthesizers provide 'sub-oscillators, which are merely the output of the gated notes divided by 2 or 4 to give extra outputs which are one or two octaves down in pitch from the main output. Chorus is often also provided to try and reproduce the effect of detuned VCOs (see Chapter 6).

3.3.3 Waveshaping

The basic output of most simple DCOs is a pulse or square wave at the required frequency. In order to emulate a conventional VCO, this needs to be converted into the 'classic' analogue waveforms: sine, square, sawtooth, triangle and pulse. This can be done using analogue electronics, but a much more flexible system can be achieved by using wavecycle/wavetable techniques. By setting the DCO to produce an output which is much higher in frequency, a look-up table can be used to store the values for each point in the waveform and a DAC can be

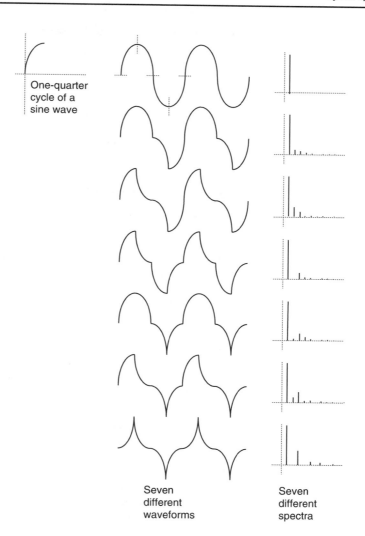

Figure 3.3.4 Symmetry can be used in a wavetable synthesizer to produce many waveforms from a small segment of a complete waveform. In this example, a quarter cycle of a sine wave is used to generate a sine waveform, plus six other waveshapes.

One-quarter cycle of a sine wave

Seven different waveforms

Seven different spectra

used to produce the required waveforms. The purity of the waveforms produced is limited only by the number of bits used to represent each point on the waveform, and the highest output frequency of the DCO, which sets the number of points that can be used (Figure 3.3.4).

Since several of the 'classic' analogue waveforms have lots of symmetry, the number of points which need to be stored in order to produce a single cycle can be minimised. For a square wave, it can be argued that only two values need to be stored, but this is inadequate for the remaining waveforms. Whilst the sine and triangle waveforms can be perfectly described with only a quarter of a cycle, the sawtooth and pulse waveforms require at least half a cycle. By using 256 points to define a half-cycle of the waveform, this is thus the equivalent of having 512 separate stored points which means that the DCO needs to run at 512 times the cycle frequency. By exploiting the symmetry or asymmetry of waveforms, a number of waveforms can be produced by using the same set of points.

179

Sometimes 8-bit values are used to store the waveform point values, but for only a doubling of the memory requirements, 16-bit values give a huge increase in the perceived quality (the doubling of the number of bits produces a disproportionately large increase in the audio quality, see Chapter 4). For high pitched notes, the whole of the wavetable needs not be used, since only one or two harmonics will be audible, and so less points are required in the table; this can be achieved by only using every other value, or perhaps even missing out three points, and only using every fourth value.

3.3.4 High resolution DCOs

By the 1990s, DCOs were using higher frequency oscillators and similar division techniques to those of the mid-1970s, but with much finer resolution: sufficient to provide frequency steps so small that they were almost inaudible. They also usually multiplexed the rate adapter and division circuitry so that each voice can have an effectively independent DCO. The multiplexing usually happened at a very high rate, often higher than the CD sample rate of 44.1 kHz: 48 or 62.5 kHz are frequently used for this 'sample rate' clock. These enhancements removed all the problems described above for the 'master oscillator plus divider' type of DCO, and gave a tone generation source which has almost ideal performance – limited only by the master clock rate and the precision of the dividers and rate adapters. Most of these improvements were due to the availability of faster chips rather than any major changes in design.

The frequency steps from realisable oscillators depend on the number of bits which are used to control the frequency changes. With 20 bits of divider resolution, it is possible to have frequency steps of 0.3% at 20 Hz, and 0.005% at 1 kHz, using a basic clock rate of 62.5 kHz. For comparison with the 500-kHz clocks of the 1970s, here are some mid-1990s' figures: the Roland D50 uses 32.768 MHz for its tone generator Application-Specific Integrated Circuits (ASICs), the Yamaha FB01 uses a 4-MHz clock and a 62.5-kHz sample rate, whilst the Yamaha SY99 uses 6.144-MHz clocks for its tone generator chips and a 48-kHz sample rate. In the 2000s, digital signal processing (DSP) chips and even general-purpose microprocessors were used as tone generators, with clock speeds of hundreds of megahertz. The sample rate has remained at 48 kHz, with some examples using 96 kHz.

3.3.5 Minimum frequency steps

The most important pointer to a good DCO design (apart from temperature stability) is the minimum step in frequency which can be made. This is most apparent when the pitch-bend control is used. Some DCOs have audible jumps or steps in pitch, which shows that insufficient frequency resolution is available. A more rigorous method of verifying the size of the frequency steps can be achieved by detuning two DCOs so that they beat together. The pitch differences required for slow beating are quite small. For example, if two frequencies are 1 Hz apart, then they will beat once every second. If they are 0.1 Hz apart, then the beat will cycle once in every 10 seconds. For a pitch difference of 0.01 Hz, the beat will take 100 seconds to complete one cycle. So, to measure the minimum frequency step, you leave one DCO unchanged, and apply the smallest pitch change that you can produce to the other; this is probably not going to be audible, but by listening to the beats you can hear when the two DCOs go from the same frequency (no beats) to slightly different frequency, when the beating will start. By timing the length of one cycle of the beat, you can work out the difference in frequency.

3.4   S&S (Sample and synthesis)

S&S is a generic term for the many methods of sound synthesis which use variations on a sample playback oscillator as the raw sound source for a VCF/VCA synthesis modifier section. The samples are normally stored in ROM memory using pulse code modulation, normally abbreviated to PCM. This is just a technical term for the conversion of analogue values to digital form by converting each sample into a number, but the acronym has become widely used in manufacturer's advertising literature. The source sample playback is much the same as for a DCO driving a large wavetable, whilst the modifier sections are usually based on the VCF/VCA structure of analogue synthesizers.

Although the use of the term 'S&S' has been introduced for instruments where the modifiers are digital emulations of the VCF and VCA section of an analogue synthesizer, S&S is not necessarily restricted to digital instruments. It can also be produced with analogue equipment, and in fact, instruments like the Mellotron, Chamberlin, and Birotron could be considered to be S&S synthesizers which use magnetic tape instead of solid-state memory. Many of the early wavetable instruments and samplers replayed digital samples and then processed them through analogue modifiers.

The availability of low-cost, high capacity ROM memory is one of the major factors in the change from simple wavecycle DCOs to sample-replay instruments with hundreds of sampled sounds. In the same way, advances in digital technology have allowed a gradual changeover from analogue modifiers to digital emulations. So S&S synthesizers start out as hybrid instruments with a DCO driving a sample replay, processed by analogue filters, but end up as completely digital instruments. A typical S&S synthesizer of the early 2000s mixes many of the features of a sampler with an emulated modifier section which has the processing capability of an analogue synthesizer of 20 years ago – complete with detailed emulations of resonant VCFs.

3.4.1   Samples

Unlike dynamic wavetable synthesizers, the samples that are provided with S&S instruments are normally replayed singly rather than being sequenced into an order. The only available source of the raw sound material for subsequent modification is thus a collection of preset sounds or timbres. S&S instruments then allow the processing of this raw sample 'source' of sound through one or more 'modifiers', and so allow different sounds to be synthesised. The modifiers are usually just some sort of filtering and enveloping control. The complexity of the processing varies a great deal – some have just low-pass filtering and simple envelopes, whilst others have complicated filtering that can be changed in real time, and loopable or programmable function generators instead of envelopes. In general, the most creative possibilities for making interesting sounds are provided by a combination of powerful sample replays options with elaborate processing functions.

Most S&S instruments have their sample sets held in ROM memory, which means that there is a fixed and limited set of available source sounds. Many GM and low-cost 'home' keyboards use S&S technology to produce their sounds – replaying the sounds is relatively straightforward and can provide high quality sounds. In a typical GM instrument, a large proportion of the memory is taken up with a multi-sampled piano sound, and the rest is almost entirely devoted to other orchestral or band instruments. These instrument samples are chosen because they have the correct characteristics for the instruments that they are intended to sound like; if they do not sound correct, then they fail to sound convincing. Unfortunately, because these audio fingerprints are so effective at identifying a sound as being of a particular type, it is not easy to make any meaningful modifications to the sample – a violin sample still tends to sound like a violin, regardless of most changes to the envelope and the filtering. The sample sets in most S&S instruments thus represent a pre-prepared set of clichés: all readily identifiable, and all very difficult to disguise. Rather like the audio equivalent of a 'fingerprint', in fact.

This fingerprint analogy can also be extended to the modifiers of the source sounds as well. If the only filtering available is a low-pass filter, then there will be a characteristic change of harmonics as the filter frequency is changed; and this can be just as distinctive as a specific sample. Filters with alternative 'shapes' like high-pass, notch, band-pass and comb filters can help to give extra creative opportunities for sound making. Again, the creative potential is reflected in the complexity of the available processing.

In order to avoid this fingerprinting problem, the programmer of an S&S synthesizer needs to have more control over the samples than just simple sample replay. Multi-sample wavetable hybrid instruments – like the Korg Wavestation – provide sample sequencing, cross-fading and wave-mixing facilities that enable samples to be manipulated in ways that can remove some of the more identifiable characteristics. Pitch shifting a violin residue and then using just part of it as the attack for another sound can produce some powerful synthesis capabilities in just the sound source.

There are many synthesizers and expander modules which use the S&S technique to produce sounds, and it has been very successful commercially for a number of reasons. It is comparatively easy to design an S&S instrument which incorporates sounds like the GM set, and it will have a broad range of applications, from professional through to home use. Because S&S instruments use pre-defined and fixed samples in ROM form, there is also considerable scope for selling add-ons like extra sample ROMs. Despite this, because many samplers also have the same sort of synthesizer processing and modification stages but their samples are held in RAM instead of ROM, the creative possibilities of a sampler are much wider!

3.4.2 Topology

Because S&S instruments have a 'pre-packaged' set of samples, they are sometimes described as merely sample-replay instruments, and not true synthesizers. For the case of a single sample being replayed by an S&S instrument, the only changes that can be made to the sample are restricted to the modifier section, which allows changes to the filtering and envelope of the sound. But almost all S&S instruments provide rather more than this 'basic' mode of replay: normally either two independent sets of 'sound source and modifier', or two separate sound sources processed by a single modifier. In addition, some instruments also allow more than one sound to be triggered from the same note event, and so several samples can be combined (Figure 3.4.1).

It should be noted that poyphony is traded against the complexity of the topology. Polyphony decreases as the number of sets of sound source and modifiers increases. For example, the polyphony would halve if the sound source and modifier resources required are doubled. Because of this, polyphony has tended to increase with time. A typical S&S synthesizer of the early 2000s may have 128-note polyphony or more, although the demands of typical sounds will reduce this to 32 or 16 notes.

The ability to trigger the playback of several different samples from one event opens up considerably more synthesis possibilities. Some early S&S instruments used an 'attack and sustain' model, where one sample was used to produce the attack portion of a sound, whilst a simplified 'subtractive synthesizer'-type section was used to produce

Figure 3.4.1 The basic S&S topology is a single sound source followed by one modifier section, as shown in (i). But most S&S synthesizers have either two sound sources which share a single modifier section (ii), or two separate sets of sound source and modifier, as in (iii).

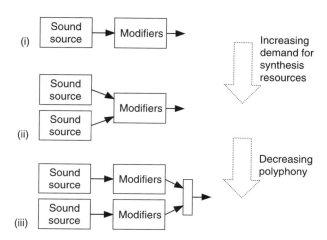

183

the sustained portion, with a cross-fade between the two portions of the sound. As the technology developed, two sample-replay sound sources could be used to produce the sound, and this allowed a more flexible division of their roles. Chapter 6 describes some of the ways in which two or more separate sound sources can be combined to produce composite sounds.

3.4.3   Counters and memory

The basic process for reproducing a sample from memory involves using a digital counter to sequentially access each sample value in the sample memory. The first sample is pointed to by the counter, which then increments to point to the next value, and this repeats until the entire sample has been read out. In practice, these retrieved values may be used as the input to an interpolation process, but the counter and memory structure remains the foundation of the replay technique. Samples are normally organised serially throughout the memory device – the end of a sample is followed by the start of the next sample. Some manufacturers deliberately order their samples so that successive samples are related in their harmonic content, which allows the sample memory to be used as a form of dynamic wavetable. But this is often complicated by the provision of multi-samples where the same instrument is represented by samples taken at different pitches (Figure 3.4.2).

In order to hold several different samples in one block of memory, pointers to the individual samples are required. There are many approaches to providing these pointers to the locations or addresses of the sample values. The simplest method specifies the start and stop addresses for each of the samples, where the start address can be used to pre-load the counter, and the stop address can be used to stop the counter when the end of the sample is reached. Alternatively, start and length parameters can be used when the counter merely adds an

Figure 3.4.2 Sample memory is often arranged as a single contiguous block of ROM or RAM (or a mixture of the two). Sample replay consists of setting pointers to the beginning and end of the required sample, and then loading a counter with the start location and incrementing a count until the finish location is reached. Often it is possible to set the start and finish pointers to encompass several samples. In the example shown only sample 15 will be replayed, but by moving the finish pointer then the multi-samples of sample 16 could also be included.

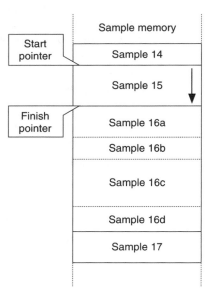

offset to the start address, since then the length parameter stops the counter when the count equals the length. By changing the length parameter, the playback time of the sample can be controlled.

If it is required to commence playback of the sample after the true start of the sample, then an offset parameter may be used to add an offset value to the start address which is loaded into the counter. Some instruments allow the length parameter to be set too longer than the sample, in which case the playback will continue into the following sample. Offsets can sometimes be used to provide similar control over the start address, and can cause the replay to commence from a different sample altogether. E-mu's Proteus series is one example of S&S instruments which implements start, offset and length parameters, as well as a partially ordered sample ROM to allow wavetable-like usage.

When offsets are applied to the start and end of sample replay, then the sample values at those points may be numerically large, which can produce clicks in the audio signal, especially if the sample is looped or concatenated with another sample. Some S&S instruments only allow start and stop addresses to be selected when the sample values at those addresses are close to zero; although this is useful to prevent clicks in the output, it can be a problem when trying to loop the sample. Most samples normally start and finish with values which are close to zero (Figure 3.4.3).

Looping samples can involve either the entire sample between the start and the stop addresses, or any portion in between. Some instruments allow loops to extend beyond a single sample – sometimes even through the whole of the sample memory. The obvious loop is to play the sample through to the end of the loop, then to return to the start of the loop, and then repeat the section between the start and end of the loop. Loops can be set to occur a number of times, or for a specified time period, or they may be controlled by the EG. Loops can be forwards, where the end of the loop is immediately followed by the start of the loop, or can alternately move forwards and backwards through the looped section of the sample. Alternate sample looping can help to prevent audible clicks when the sample values at the loop addresses are not at zero-crossing points. Another possibility is to invert the sample playback for each alternate repetition, again so that clicks are minimised.

Figure 3.4.3 Sample-replay parameters provide additional control over how the counter starts and loops whilst replaying a sample. An offset parameter allows the start of the sample to be later (or earlier) in the sample memory, whilst a length parameter allows the offset or start to be changed dynamically without altering the replay time of the sample.

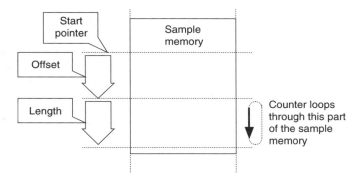

The predominant use of the loop is to provide a continuous sound when the EG is in the sustain portion of the envelope. These are called sustain loops. But it is also possible to have attack or release loops, where the start or end of notes can be extended without requiring long samples. Loops are a way of minimising the storage requirements for sounds which are required to have long attack, sustain or release envelopes. S&S techniques where each sample is closely connected to the EG, and so has separate attack, decay, sustain and release loops, are becoming rarer as the cost of memory reduces.

Storing the parameters required for playing back a sample requires two separate storage areas. A look-up table is required to map the samples to their addresses in the sample memory, and this can also contain details of the length of the sample, zero-crossing points for potential start and offset addresses, as well as default loop addresses. These default control parameters can often be replaced by values held as part of the complete definition of a sound.

3.4.4 Sample replay

Replaying a sample involves reading the individual sample values from a storage device, and then converting these numbers into an analogue signal. The conversion from digital to analogue is carried out by a DAC chip. There are two methods of replaying samples: variable frequency and fixed frequency.

Variable frequency playback

The easiest method of replaying a wavecycle or wavetable would be to output the sample values at a rate controlled by an oscillator or DCO. This is called *variable frequency playback*. The oscillator steps through the values which specify the waveform, and this is converted into an audio signal by a DAC. Although simple to understand conceptually, this technique has several major limitations:

- **The same number of sample values are replayed regardless of frequency**: Because the oscillator is merely stepping through a fixed series of values, the detail contained within the waveform is constant, but the same is not true for the spectrum: at low pitches all of the harmonics may be below the half-sampling frequency, whilst for high pitches then only one or two harmonics may be below the half-sampling frequency. The sample should thus ideally have more detail when it is used to produce low pitched sounds, because this is where the harmonic content is most important, whilst for higher pitched sounds less detail is required because less harmonics will be heard. One technique which can be used to provide the required detail in samples is to use different sample rates for different pitches (Figure 3.4.4).
- **The half-sampling frequency changes as the pitch changes**: Because using a DCO to control the replay rate means that the half-sampling rate tracks the playback pitch, then the reconstruction filter also needs to track the half-sampling rate. (Note that

Figure 3.4.4 Multi-sampling is often used to provide several different samples mapped onto a keyboard, but it can also be used to provide different degrees of detail in a given sample. This diagram compares two methods of shifting down by two octaves in pitch: using one sample and slowing down the replay rate provides the same number of sample points regardless of the output pitch, whilst the use of two multi-samples enables the same sample rate to be used for each sample, thus increasing the amount of detail which is available for the lower pitched sample compared to the single sample method.

Figure 3.4.5 Reconstruction filters are normally thought of as being used in the output stage of digital audio systems, but they can be required to process the output of a DCO if it uses the variable frequency method to provide different pitches. In this example, a digitally controlled filter (DCF) is used to track the DCO frequency so that any aliasing components are removed before any post-DCO modifiers process the audio. The DCF 'smooths' the DCO waveform so that no aliasing components are present.

early sample playback devices did not always do this. Instead they set the reconstruction filter so that it filtered correctly for the highest playback pitch, which meant that for lower pitches aliasing was present in the output signal.) Tracking means that a low-pass VCF is required to follow the changes in playback pitch, so that frequencies above the half-sampling rate are not heard in the output audio signal. Such VCFs have much more stringent design criteria than the VCFs found in analogue synthesizers: the 24 dB/octave roll-off slope of a typical analogue synthesizer VCF is not adequate for preventing aliasing, and slopes of 90 dB/octave or more are often required, with 90 dB or more of stopband attenuation (Figure 3.4.5).

Figure 3.4.6 (i) Variable frequency playback (4-note polyphonic) requires a separate DCO and counter to access the sample memory, followed by a DAC to convert the sample into an analogue signal for processing by the modifier section. (ii) Fixed frequency playback (4-note polyphonic) changes the pitch of the sample and allows the use of a digital modifier section, with a single DAC to convert the output to analogue.

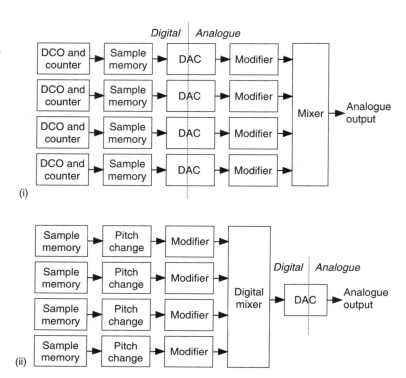

- **The playback is monophonic**: Because the sample replay rate is set by an oscillator or DCO, variable frequency playback requires a separate sample-replay circuit for each individual pitch that is playing. Each DCO in a polyphonic synthesizer can produce a differently pitched monophonic audio sample, and these analogue outputs are then processed by analogue filters in the modifier section.

Fixed frequency playback

Fixed frequency playback uses just one frequency for the sample rate, but changes the effective number of sample values which are used to represent the different pitches by calculating the missing values. It has the advantage that only one sample rate is used, and so a fixed frequency sample playback circuit can be polyphonic and can be connected to a digital modifier section, which allows a completely digital synthesizer to be produced where the digital-to-analogue conversion happens after all the digital processing. Fixed frequency playback is now almost exclusively used in digital synthesizers and samplers (Figure 3.4.6).

3.4.5 Interpolation and pitch shifting

Changing the playback sample rate is not the only way of changing the playback pitch of a sampled sound. Consider a sample of a sound: a single cycle of a given pitch will contain a number of sample points,

Figure 3.4.7 Interpolation is used to calculate missing or intermediate values in a sample. Linear interpolation draws straight lines between the sample points, whilst polynomial curve fitting attempts to match a curve to the sample points. In this example, a sample curve is shown, together with a linear interpolation based on five sample points, and a curve fitted interpolation. The linear interpolation misses some of the major features, whilst the curve fitting produces a much better fit.

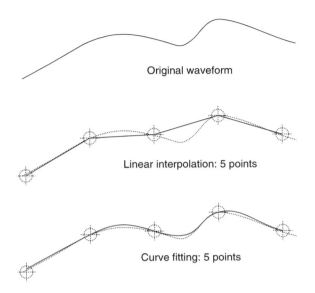

Original waveform

Linear interpolation: 5 points

Curve fitting: 5 points

where the number is related to the cycle time for the waveform, and the sample rate. So if 256 sample values represent a single cycle of a waveform at one pitch, then for a lower playback pitch more sample values would be required, whilst for a higher playback pitch less sample values are required. But it is possible to take the existing sample values and work out what the missing values are by a process called interpolation. This is used in fixed frequency sample playback.

Interpolation attempts to represent the waveform by a mathematical formula. If the sample values are thought of as points on a graph, then interpolation tries to join up those points. Once the points are joined up, then any sample values in between the available points can be calculated. The simplest method of interpolating merely joins the sample points with straight lines; this is strictly called linear interpolation, although it is often erroneously shortened to just interpolation. Although this is easy to do, real-world waveforms which consist of lots of straight lines joined together are rare! A better approach is to try and produce a curve which passes through the sample points (Figure 3.4.7).

One method which can achieve this uses polynomials: general-purpose algebraic equations which can be used to represent almost any curve shape. Polynomials are categorised by their degree, and in general, n points can be matched by an $(n-1)$th degree polynomial. So for two sample values, a first-degree polynomial is used which turns out to be the formula for a straight line. If more sample values are used to work out the shape of the curve, then higher order polynomials can be used: three points can be fitted by a quadratic (or second-degree) equation, whilst four points require a cubic equation. Manufacturers rarely reveal how they do their interpolation; in general, the lower cost the implementation, the lower the degree of polynomial which is used, and the poorer the resultant audio quality.

Interpolation using polynomials with degrees higher than 1 is sometimes called *differential interpolation* to distinguish it from linear *interpolation*. It is also possible to design filters which can interpolate, and these are used in many digital systems.

An alternative technique which can reduce the number of sample values at high frequencies is literally to remove samples, or conversely to add in extra sample values at low frequencies. The simplest way to do this for an octave shift up or down is to either miss every other sample, or to repeat each sample. This is called decimation, and it is crude but effective. Since there are no calculations involved, it is easier to implement than interpolation, but it can produce distortion in the output. Because of the relatively low cost of ROM, sampling at a higher rate than is required can be used. Known as 'over-sampling', the idea is to provide more points for the interpolation processing. Since the points are closer together, and there are more available for the calculations, the interpolation quality improves. Over-sampling can be at anything from twice the required rate to 64 times or more. The performance of the memory and the interpolation processing requirements limit the over-sampling rate.

3.4.6 Quality

Sample reproduction quality is determined by:

- sample rate in kilohertz (affects the bandwidth)
- sample size in bits (affects the signal-to-noise ratio, abbreviated as SNR)
- interpolation technique (linear or polynomial: affects the distortion when pitch transposing)
- the anti-aliasing and reconstruction filters (affect the distortion and SNR).

The CD sample rate of 44.1 kHz and the digital audio tape (DAT) sample rate of 48 kHz have become widely used in electronic musical instrument, with some instruments using even higher rates of 96 kHz or more. Samplers often have a range of available sampling frequencies, so that their memory usage can be maximised – sampling at 32 or 22.05 kHz can reduce the amount of storage which is required for sounds which have restricted bandwidths.

Sixteen-bit sample size has become the norm. Internal processing is often higher, but conversion chips designed for CD players (which are fundamentally based on 16-bit sample storage) are widely used in synthesizers and samplers. As higher resolution converters become available, they are likely to be incorporated, since the S/PDIF digital audio format has provision for 24-bit samples.

Interpolation techniques depend on the processing power which is available. Microprocessors and DSP chips continue to increase their performance, and so more sophisticated interpolation techniques will become possible, which should improve the quality of sample replay and transposition.

Analogue filter technology is almost at the theoretical limits, and so any improvements are likely to take place by adding in digital filtering. By increasing the sample rate inside the conversion chips, it is possible to augment the anti-aliasing and reconstruction filters with additional digital filtering using DSP chips. This allows enhanced performance, and yet outside the conversion chips, the samples can still be at sample rate of 44.1 or 48 kHz.

Synthesizers and samplers will continue to follow development in audio technology. Future developments are likely to include more digital processing, and less analogue electronics.

3.5   Early versus modern implementations

Section 3.3 discussed about the technology of DCOs, and mentioned the difference between the design of instruments in the mid-1970s, and those of the mid-1990s. This section summarises the differences between hybrid synthesizers since the 1970s.

The 1970s

In the 1970s, hybrid instruments were just developing. VCOs were gradually being enhanced by the addition of auto-tune and digital control features, as well as programmability of the complete synthesizer with the change in emphasis from 'live' user programming to instant access via large numbers of memories. There were two distinct types of keyboard synthesizers: versatile *monophonic* or *polyphonic synthesizers* with rather more limited functionality – all based on mixtures of analogue and digital circuitry, and fully polyphonic 'electronic pianos'/ 'string machines' and multi-instruments (string and brass) – all based on top octave chips plus dividers followed by simple filtering and enveloping circuits. The second category was already in decline: the polyphonic digital instruments of the 1980s would cause their complete disappearance.

The synthesizers had limited wavecycle waveforms, and if any controls were provided for the levels in the wavecycle, they would be on a 'one control per function' basis. The display would use light-emitting diodes (LEDs), or perhaps a discharge tube/fluorescent display. Waveform samples would be in 8 bits, and the sample rate would be between 20 and 30 kHz, giving an upper limit for frequency output of between 10 and 15 kHz. Control would be via 4- or 8-bit microcontrollers, adapted from chips intended for simple industrial control applications. The interfacing would be via analogue control voltages, gates and trigger pulses, or perhaps from a proprietary digital bus format.

The 1980s

The release of the Yamaha all-digital FM synthesizers in the early 1980s, saw all the other manufacturers trying to catch up and releasing hybrids whilst their development teams worked on the digital instruments which would begin to appear in the late 1980s.

These hybrids used digital enhancements to make the most of analogue oscillators, and eventually replaced the VCO completely with a digital equivalent. Portamento was the first casualty of this conversion, but by the end of the decade it had reappeared as the clock speed of chips made more sophisticated DCOs possible. Early designs used medium and large scale integrated circuits (ICs) containing tens, hundreds or thousands of digital gates. Wavecycle was joined by wavetable, usually with either 8- or 12-bit waveform samples.

The display gradually replaced the front panel knobs as the centre of attention during the programming process, although a 2 row by 16 character liquid crystal distal (LCD) display (which might be backlit) was not ideal. Individual controls were replaced by 'parameter access', where a single slider or knob was used to change the value of a parameter which was selected by individual buttons. The 8- and 16-bit microprocessors were used to control the increasingly complicated functionality, especially once MIDI became established. Interfacing polarised rapidly from proprietary interface busses to MIDI within a couple of years of the launch of MIDI in 1983.

The 1990s

The 1990s opened with a preponderance of all-digital instruments, and a consolidation of sampling. But this was quickly followed by a resurgence of interest in analogue technology, and some manufacturers began to rework older designs or even design completely new instrument from scratch. Although often labelled 'analogue', many of these instruments were actually hybrids; most often they use DCOs rather than VCOs. Even the 'pure' analogue instruments had considerable amounts of digital circuitry used for control and programming purposes.

DCOs used multiplexed circuitry to provide independent 'oscillators', and these used sophisticated accumulator/divider-type techniques to provide very fine resolution frequency control – typically on custom chips made for the individual manufacturer (ASICs). With plenty of processing power available, the wavecycle and wavetable generation techniques were joined by sampling, and wave sequencing, normally with 16-bit waveform samples and better than CD sampling rates (greater than 44.1 kHz sample rate). Displays increased in size, with 4 row by 40 character backlit LCDs (and larger) in common usage: some with dot-addressable graphics modes instead of just character-based displays. Allied to this was the increasing importance of a graphical user interface (GUI), sometimes with a mouse used as a pointing device, but almost certainly with softkeys or assignable buttons. Computer-based editing software helped to make the front panel display almost superfluous on some rackmounting instruments. Control functions were provided by 16- or 32-bit microprocessors, perhaps with a DSP for handling the more complex signal processing functions. Interfacing was via MIDI. The conversion from analogue to digital was almost complete – often only the VCFs and enveloping was analogue.

Figure 3.6.2 The Roland D50 mixes simple sample-replay technology with a basic DCO/DCF analogue synthesis emulation.

Figure 3.6.3 Waldorf MicroWave.

It also marks the end of the front panel as a guide to the operation of the synthesis method, and a change to mental models instead. The front panel clearly shows the influences at the time: diagrams influenced by FM synthesizers, joystick from vector synthesizers, and a large softkey-driven display to simplify the editing.

The D50 was the last of the hybrids, although very little analogue is present. Instead, it was designed to appeal as an alternative to the all-digital FM synthesizers, whilst appearing as analogue as possible. But the Korg M1 changed the rules and ended the hybrids forever.

Waldorf MicroWave (1989)

The Waldorf MicroWave is essentially a 'PPG Wave'-type of wavetable synthesizer, but redesigned to take advantage of the available electronics of the late 1980s. The minimalistic front panel design relied on a large data entry wheel and a few buttons (Figure 3.6.3).

195

3.7   Questions

1. What is a hybrid synthesis technique?
2. What are the differences between cycles, multi-cycles and samples?
3. Give four examples of single-cycle waveshapes.
4. What is the difference between multi-cycle wavecycle and wave-table synthesis?
5. How does a 'top octave' oscillator produce audio frequency outputs?
6. How does the frequency resolution of a DCO affect detuning and pitch bending?
7. Why is an early 1990s 'analogue' synthesizer really a hybrid?
8. How can dividers reduce the effect of jitter?
9. How does 'auto-tuning' work?
10. How are the contents of a wavetable converted into an audio signal?

Time line

Date	Name	Event	Notes
1960s	Clavioline	Clavioline	British Patent 653340 and 643846
1969	Philips	Digital master oscillator and divider system	
1970s	Ralph Deutsch	Digital generators followed by tone-forming circuits	The popularisation of the electronic organ and piano
1975	Moog	Poly Moog was released	More like a 'master oscillator and divider' organ with added monophonic synthesizer
1982	PPG	Wave 2.2, polyphonic hybrid synthesizer, was launched	German hybrid of digital wavetables with analogue filtering
1986	Sequential	Sequential launched the Prophet VS, a 'vector' synth which used a joystick to mix sounds in real time	One of the last sequential products before the demise of the company
1986	Ensoniq	ESQ-1	Digital sample-replay synth with analogue modifers (VCF and VCA)
1989	Waldorf	MicroWave, a digital/analogue hybrid based on wavetable synthesis	Effectively a PPG Wave 2.3 brought up to date
1996	Waldorf	Pulse	The Waldorf Pulse was a three VCO, VCF analogue mono-synth
1998	Waldorf	MicroWave XT	The MicroWave XT was an analogue wavetable and FM poly-synth with arpeggiator and FX
1998	Ensoniq	Fizmo	Advanced wavetable synth

4 Sampling

A sampler is the name given to a piece of electronic musical equipment which records a sound, stores it and then replays it on demand. There are thus three important functions:

- record the sound
- store the recording on some sort of storage medium
- replay the stored sound.

A sampler combines all of these functions into one unit, and this makes it very different from almost all of the other examples of synthesizers described in this book. Most synthesizers can fulfil the last two functions: store and replay, but the distinguishing feature of a sampler is its ability to record sounds.

This definition of a sampler in terms of its functionality is important because it enables a wide range of equipment to be classified as being samplers, whereas the commonly used term is often restricted to merely electronic music equipment which stores sounds in random-access memory (RAM). Using the functional description, the following can all be described as samplers:

- tape recorder
- cassette recorder
- video recorders
- digital audio tape (DAT) recorder (MiniDisc, etc.)
- echo effects unit
- computers with sound input and output facilities.

All of these 'samplers' represent ways to record, store and subsequently replay sounds. In some of these cases, the sounds will probably be naturally occurring sounds that can be recorded with a microphone, but this does not prevent the process of collecting the sounds, storing and manipulating them, and then replaying them from being called 'sound synthesis'.

Within this wider context, any of the techniques that have already been described can become part of a larger synthesis system by utilising sampling. The 'source and modifier' model can be used to describe the working of an analogue subtractive synthesizer, but it can also be used to describe the process of using a synthesizer merely as the source of sounds which are then recorded, stored, modified and finally replayed using a sampler which acts as the modifier of those sounds.

One popular usage of the word 'sampler' which is not covered by this type of definition is the recorded collections of material from more than one source, which are also called *samplers*.

Samplers thus form a bridge between the analogue and the digital synthesizer, since they span the two technologies with very similar instruments. Analogue sampling can be tape-based or chip-based, although analogue sound storage chips have been largely ignored since digital technology became available. Digital sampling has increasingly used the technology and approach of synthesis, and this has led to the convergence of sampling and synthesis.

4.1 Tape-based

Samplers based on tape recording techniques have a long history. The first 'tape' recorders did not use tape at all, but used wire instead. Plastic tape covered with a thin layer of iron oxide is much easier and safer to handle than reels of wire, and far easier to cut and splice!

4.1.1 Tape recording

The underlying idea behind how a tape recorder works is very simple. The sound signal is converted into an electrical signal in a microphone, and this signal is then amplified, converted into a changing magnetic field and stored onto tape. By passing this magnetised tape past a replay head, the changes in the magnetic field are picked up, amplified and converted back into sound again.

Magnetic tape is made up from two parts:

- a plastic material which is chosen for its strength, wear-resistance, and temperature stability
- magnetic coating which is chosen for its magnetic properties.

It is actually possible to record and replay sounds using a fine layer of iron rust placed onto the sticky side of adhesive tape, although this is not recommended as a practical demonstration. The commercial versions of recording tape are just more sophisticated versions of this 'rust on tape' idea.

A tape recorder is a mixture of mechanical and electronic engineering. The mechanical system has to handle long lengths of fragile tape, pulling it across the record and replay heads at a constant speed, and ensuring that the tape is then wound onto the spool neatly. This requires a complex mixture of motors, clutches and brakes to achieve. The pulling of the tape across the heads is achieved by pressing the tape against a small rotating rod called the *capstan*. The tape is held onto the capstan with a rubber wheel called the *pinch roller*. The spool which is supplying the tape is arranged so that it provides enough friction to provide sufficient tension in the tape to press it against the record and replay heads as it is pulled past. Once past the capstan and roller, the tape is then wound onto the other tape spool. When the tape is wound forwards or backwards, the pinch roller is moved away so that the tape no longer presses against the capstan or the heads, and the spools can then be moved at speed (Figure 4.1.1).

The electronic part of a tape recorder has two sections: record and replay. The record part amplifies the incoming audio signal, and then

Figure 4.1.1 A tape recorder/player pulls the tape past the record/replay head. The capstan revolves at a constant rate, and the tape is held against the capstan by the pinch wheel.

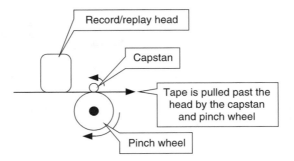

Record/replay head

Capstan

Tape is pulled past the head by the capstan and pinch wheel

Pinch wheel

drives the record head with the amplified signal plus a high frequency 'bias' signal. The combination of the two signals allows the response of the magnetic tape to be 'linearised'. Without the bias, the tape recorder would produce large amounts of distortion. The replay section merely amplifies the signal from the replay head (no bias is required for replay).

4.1.2 Mellotrons

The word 'Mellotron' is a trademarked name for one type of sample-replay musical instrument which uses short lengths of magnetic tape. The concept is simple, the practicalities are rather more involved. The basic idea is to have a tape replayer for each key on the keyboard. A long capstan stretches across the whole of the keyboard. Pressing a key pushes the tape down onto the capstan and pulls it across the replay head. The tape is held in a bin with a spring and pulley arrangement to pull it back when the key is released. The length of tape is thus fixed, and so the key can only be held down for a limited time. Loops of tape cannot be used because the start of the sound would not be synchronised with the pressing of the key; that is, by arranging for the tape to be pulled back into the bin each time the key is released, it automatically goes back to the start point of the sound, ready for the next press of the key. There have been several other variants on the same idea from other manufacturers, but the Mellotron is the best known (Figure 4.1.2).

Because the capstan is the same size for each key, the tape for each key needs to be recorded separately, with each tape producing just one note (although several tracks are available on each tape, with a different sound on each track). The tapes are thus multi-sampled at one-note intervals. Recording user samples for such a machine requires time, patience and attention to detail: the levels of the sounds must be consistent across all the tapes, for example. The 'frames' which contain complete key-sets of the tape bins can be changed, but this is not a quick operation. Because of the difficulty of recording your own sounds onto a tape, these tape samplers can almost be regarded as being sample-replay instruments rather than true samplers.

4.1.3 Tape loops

By looping a piece of tape around and joining the end to the beginning with splicing tape, it is possible to create a continuous loop of

Figure 4.1.2 A side view (i) and top view (ii) of a tape sample-replay instrument. The capstan spans the whole of the keyboard, and revolves continuously. When a key is pressed down, this presses the tape against the capstan, which pulls the tape across the replay head.

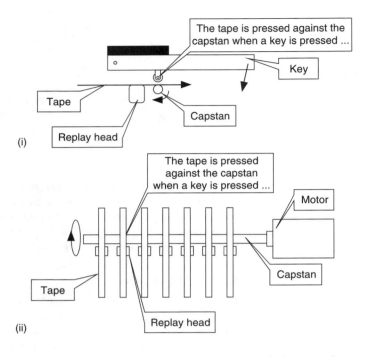

tape which will play the same piece or recorded material repeatedly. The only limitation on the size of the tape is physical: short loops may not fit around the tape recorder head and capstan, whilst long loops can be difficult to handle as they can easily become tangled.

The Watkins (WEM) CopyCat echo unit consists of a loop of tape and several replay heads, but the addition of the record head changes the function!

The repetition of a sequence of sounds produces a characteristic rhythmic sound, which can be used as the basis of a composition. As with the Mellotron tape player, synchronising the playback of the start of a loop is difficult, and synchronising two loops requires them to be exactly the same length. Tape loops are thus usually used for asynchronous sound generation purposes.

4.1.4 Pitch changes

Analogue tape recorders have one fundamental 'built-in' method of modifying the sound: speed control. Changing the speed at which the tape passes through the machine alters the pitch of the sound when it is played back. This can be either during the record or the replay process. For example, if a sound is recorded using 15 inches per second (ips), and then replayed at 7.5 ips, then it will be played back at half the speed, and thus will be shifted down in pitch by one octave. Conversely, sounds which are recorded at 7.5 ips and replayed at 15 ips will be played at twice normal speed, and will thus be shifted up in pitch by one octave. Note that the pitch and time are linked: as the pitch goes up, the time shortens, whilst lower pitch means longer time. The 'length' of a sound is exactly the length of the piece of tape on which it is recorded. If the tape is played back faster, then the tape passes over the replay head faster, and so the sound lasts for a shorter time. (The same is not necessarily true for digital samplers …)

This 'pitch and time doubling' was used to great effect by guitarist Les Paul in the 1950s. Using the technique of recording at a slow tape speed, and then replaying at a faster tape speed, he was able to achieve astonishingly fast and complex performances on guitar. The same technique is still a powerful way of changing the pitch of sounds, or for enabling virtuoso performances at slow tempos.

4.2 Analogue sampling

Analogue sampling covers any method which does not use tape or digital methods to store the audio signals.

The most common technology which meets these requirements is the 'bucket-brigade' delay line, or analogue delay line. This uses the charge on a series of capacitors to represent the audio signal, rather than the magnetic field used in tape systems or the numbers used in digital systems. The sampling process is merely the opening of an electronic switch to charge up the first capacitor in the delay line. The size of the voltage determines the amount of charge which is transferred to the capacitor: the higher the voltage which is being sampled, the more the charge which is stored in the capacitor. Effectively, the capacitor acts as a store for the voltage, since the presence of the charge in the capacitor is shown by the voltage across the capacitor. The switch then opens and the charge is held in the capacitor since there is no significant leakage path. Another switch is then used to transfer the charge to the next capacitor in the delay line, where it again produces a voltage. The original capacitor is then available to sample the next point on the incoming audio signal. This process continues, with the sample voltages moving along the delay line formed by the capacitors; hence the term 'bucket-brigade' delay lines (Figure 4.2.1).

Because each section of the delay line is just a capacitor and some electronic switches, it is easy to fabricate, and so several thousand can be placed on a single integrated circuit (IC) chip. The sampling and transfer of charges requires a high frequency clock signal, but the control circuitry is straightforward. This simplicity of control and application made analogue 'bucket-brigade' delay lines popular in the 1970s and early 1980s for producing echo, chorus and reverberation effects. At least one monophonic sampler was produced using analogue delay lines in the early 1980s, but it was rapidly superseded by digital versions.

The limitations of the analogue delay line technique are two-fold: firstly, the capacitors are not perfect, so some of the charge leaks away causing signal loss, distortion and noise; but more importantly, the high frequency clock signals tend to become superimposed on the output audio signals and this degrades the usable dynamic range of the delay line. Because of these problems, digital sampling technology has replaced analogue delay lines in the majority of cases; although recent advances in the technology mean that some analogue delay lines continue to be used in specialist applications like audio storage for telephone answering machines.

Figure 4.2.1 An analogue delay line moves charge along a series of capacitors connected by switches. The input voltage is stored on the first capacitor (i). The charge is then transferred to the next capacitor (ii). This repeats for the entire chain, and so the input voltages move along the capacitors.

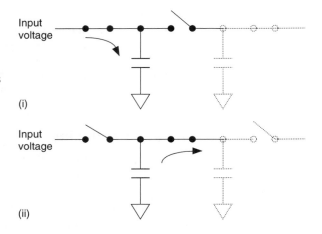

4.2.1 Delay lines

An alternative to bucket-brigade delay lines moving charge around is to use metal springs or metal plates to carry the sound signals acoustically. Sounds are transferred to the metal using modified loudspeaker drivers, and the delayed sound signals are recovered with contact microphones. The physical size of these acoustic delay lines is large, and the 'spring lines' and 'plate echoes' of the 1960s and 1970s have again been largely replaced by digital techniques. Acoustic delay lines have the advantage that they are not a sampling system, but are more suited to reverberation effects than pure sampling – they are not suited to storing a sound and subsequently replaying it.

4.2.2 Optical

One alternative sampling method uses a technique which is similar in principle to tape recording. Optical film soundtracks are a light-based variation of tape recording. Instead of storing the audio as a changing magnetic field, the film soundtrack uses the amount of light passing through the film to store the audio signal. This is normally achieved by arranging for large audio signal levels to allow a large amount of light to pass through the film, whilst small signals allow less light through. A photodetector and lamp are used to convert the transmission of light into an audio signal.

This modulation of light by an audio signal is normally achieved by using the audio waveform to control the width of a slot, and so the amount of light which passes through the film. Variable density (opacity) film can also be used, but this is rare for film use, although it has been used for experimental systems where film is used to produce sound by literally painting onto it to control the amount of light which passes through it at any given instant. By passing the resulting film through a lamp and photodetector, the optical version of the audio can be converted into sound. Although flexible, the complexity of producing the required degree of detail is enormous and very time-consuming. At least one manufacturer produced an optical sample-replay machine in the 1980s, but as with all analogue methods, this was not a success against the digital competitors.

4.3   Digital sampling

Digital sampling is based around three electronic devices:

- analogue-to-digital converters (ADCs)
- memory devices (RAM, flash electrically programmable read-only memory (EPROM), etc.)
- digital-to-analogue converters (DACs).

These three devices carry out the three major sampling functions:

- the ADC records the sound
- the memory devices store the recording
- the DAC replays the stored sound.

A sampler works in three modes: record, edit/store and replay. The record mode is used to convert signals from a continuous analogue form into a numeric digital representation. The digital data which represents the sound is then held in RAM memory inside the sampler, and this is edited in the second mode: edit/store. When an audio signal is recorded by a sampler, the start of recording is normally set to before the actual start of the sound, so that the initial attack part of the sound is captured. Once in RAM memory inside the sampler, the sample data needs to be edited so that the start of the sound is at the start of the sample data. This ensures that when the sample is replayed, it will start playing without any time delay. Once edited, the sample data is then stored in some sort of permanent storage like hard disk. This sample data can be reloaded into the sampler's RAM memory when it is required. The replay mode takes the sample data in the sampler memory and converts it back to an analogue audio signal.

The major division between a samples and synthesis (S&S) instrument and a sampler can be considered to be the type of memory: S&S instruments use fixed ROM memory and so the samples cannot be edited, whilst samplers use volatile RAM memory where the samples can be edited.

The actual process of sampling sounds is often forgotten because of the vast range of pre-recorded material which has become available. But at the heart of almost all samplers is the capability to record sounds. The sources of sounds are not as apparent as might be imagined.

Singing

Recording a vocal line into a sampler can be useful in several ways. If recorded early in the song creation process, then it can provide a guide vocal for building up the arrangement, or for working on vocal harmonies, or it can serve as a baseline recording for later improvement by the singer. As with any recording or composition process, the 'sleep-on-it' test can be harsh, but very useful: a sound, vocal performance or song that sounds perfect one day can seem rather less perfect the following day when it is auditioned again.

Real instruments

Sampling real instruments is a difficult and exacting task which requires skilled performance ability, determination, patience and time. Unless the instrument is not readily available as samples already, then this process is not recommended. Trying to record clean, correctly pitched samples of different notes with similar intonations at similar volumes, and with smooth transitions between multi-samples across the keyboard range, is much harder than it sounds, and looping those sounds so that they have inaudible looping artefacts can be very challenging.

Other electronic musical instruments

Sampling other electronic musical instruments is easier than real instruments. The pitching is more repeatable in most cases, the noise floor and maximum output level (MOL) set the limits to the dynamic range and by using a sequencer to send consistent musical instrument digital interface (MIDI) velocity values, then the intonation can be made consistent. Assembling large stacks of keyboards and lots of effects units can indeed produce very big and complex sounds, but these are not always useful in all musical contexts. Using factory presets sounds on some of the keyboards is not a good idea, which means that the sounds all need to be custom programmed in order to ensure that the samples are unique and that the factory presets are not recognisable. Whenever a factory programmer produces a complex and clever special effect sound from a new synthesizer it will automatically become well known and almost unusable in a real performance context. Producing sounds which are distinctive and useful is considerably more challenging.

Real-world

Real-world sounds are sometimes found in unexpected places. The author has an ancient oven where the grill door hinge requires lubrication, but has not been lubricated because the sound it makes when it is opened is very similar to the sound made by Klingon ships when they fire their main phase disruptor weapon.

The real-world is an astonishing source of sounds, but these present great challenges when they are going to be reproduced by a sampler. Wind noise, aircraft noise and general background noises can be either unwanted distractions, or the sounds being recorded, but usually the former. Sounds which are pitched are unlikely to be tuned to a note based on A-440, and changing the pitch of many real-world sounds can completely destroy their characteristic timbre. The triangle is one example: a pitch-shifted triangle sounds nothing like a triangle.

The usual technique when capturing real-world sounds is to use a portable DAT recorder and a detailed notebook. Making a safety backup copy of the source recording immediately before starting any sampling of the DAT tape is strongly recommended.

CD-ROMs

Pre-recorded samples on compact disk-ROMs (CD-ROMs) are a popular and potentially expensive source of sounds. Because the sounds have already been sampled, looped and assigned to notes on the keyboard with smooth transitions between the multi-samples, then these are an almost ideal source of sounds that other people have produced. If the required sounds are not available on CD-ROM, then they are less than ideal.

CDs

Many samplers include facilities which are designed to ease the recording of sounds from this type of media, although these facilities are normally intended for use with sample CDs, where the sounds are presented in sequence of pitch, intonation, etc. Sampling audio CDs other than specially licensed sample CDs will require permission from the copyright owner.

DVDs

Movies are noted for sound-bites: short pithy phrases or sentences that capture a mood or express an emotion. Sampling these from movie soundtracks also requires permission from the copyright holder, although there are a number of ways of producing close emulations of the originals, ranging from actors and actresses who specialise in sound-alikes, through to specialist sample CDs.

Once the sample has been recorded, then it will probably need to be edited …

4.3.1   Editing

The most important editing function which is required by a sampler is the trimming of the unwanted portions of the raw samples – 'before' and 'after' the wanted sample. This trimming or 'topping and tailing' process allows the sampler user to set the start and the end of the sound. This can be especially important if the raw sample is noisy, because the start of the sound may not be apparent, and a compromise may be to be made between finding the true start of the sample and hearing noise. Some samplers provide automatic functions which will trim a sample using criteria which can be adjusted to suit the user. Although this trimming function is of great importance to a user who produces their own samples, the majority of sampler users merely use the sampler to replay pre-prepared samples, and so the trimming function is not as important as might be supposed. But the ability to manipulate segments of audio is essential for the user who wishes to use a sampler as a synthesizer rather than merely a sample-replay device.

Most S&S synthesizers have only limited sample manipulation facilities and no sample editing facilities – the sample are in ROM memory, and so the only manipulations which are possible are changes in direction, start point and loop points. S&S instrument thus rely almost entirely on their synthesizer modifier section to make changes to the timbre of the samples. In contrast, samplers normally have a powerful sample manipulation and editing section as well as a synthesizer-type modifier section. The synthesis modifier facilities are thus less critical to the operation of the sampler, and in fact, many samples of filter sweeps and related modifier sounds are available so that the modifier section is less important. By having the sample editing facilities available, changes can be made to individual samples rather than the global filtering which is available in a modifier section.

Three of the most powerful of these sample editing techniques are looping, stretching and re-sampling.

Looping

Looping consists of implementing the equivalent of a loop of tape, but in the digital domain instead of with a physical loop of magnetic tape. In the simplest case, this is merely a repetition of the same portion of the sample, but it may also be controlled by an envelope generator (EG) so that the loop does not stay at a fixed volume. The transition from the end of the loop to the beginning of the loop is the equivalent of the splice in a physical tape loop, but the control of this transition can be much more sophisticated because the sample is stored in a digital form.

The basic method of joining the end of the loop to the beginning is to splice the two points together. If the end and start of the loop do not have the same level, then the resulting 'glitch' will be audible as a click in the loop. There are several approaches to avoid this problem. It is possible to arrange for the splice to be made only when the two levels meet one or more criteria:

- same level
- same slope
- same rate of change of slope.

The 'same level' criteria is often refined to when the audio waveform crosses the zero axis; this is called *zero crossing*. Since the zero axis is normally the effective 'silence' level of the sample, splicing at zero crossings can produce splices without clicks, although this is not guaranteed. Better techniques take into account the shape of the waveform at the transition. Matching the slopes of the two portions of the waveforms can reduce the level of any click, whilst matching the two waveforms so that the splice point occurs at similar points on both can minimise the click, although this restricts the available splicing points. By reversing the direction of playback at the splice point, some of the problems of matching the levels or slopes can be avoided, but this works only for short samples where the backwards and forwards playback of the loop will not be heard – long loops can sound very unusual if they are looped using this technique, although it is useful as a special effect. The length of the loop also affects the perceived pitch of the sound as it is replayed. In extreme cases, the looped section can shift its pitch markedly from the original pitch before looping. This is most obvious for short loops, especially where a single cycle of the waveform is being looped (Figure 4.3.1).

Even if clicks are not produced by the splice in a loop, the start and end of the loop may not have the same timbre. This produces a

Figure 4.3.1 Splicing loops. (i) Choosing the same level does not guarantee a good splice. (ii) Matching the level and the slope can give a good splice. (iii) Even a good splice can alter the frequency of the loop if the same cycle time (or an integer multiple of it) is not maintained.

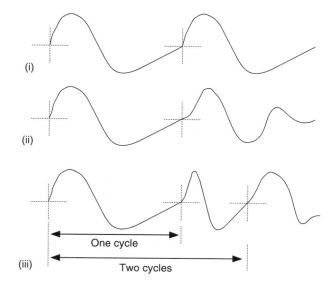

(i)

(ii)

One cycle

Two cycles

(iii)

sudden change in the timbre which can be almost as noticeable as a click. The abrupt change in spectrum of the sound is often interpreted by the listener as a click even though examination of the waveform at the spice point shows no obvious mismatch of level or slope. Cross-fading between the start and end of the loop can be used to reduce the effect of a sudden change into a smoother transition between the two contrasting timbres. Unfortunately, even a cross-faded loop can still produce a cyclic variation in timbre which is apparent when the loop repeats. The subject of producing usable looped samples of sounds is an entire subject, and is often covered in detail in the literature and support material produced by the manufacturers of samplers.

Once looped, the timbre of a sample is fixed. In real instruments, the timbre tends to change rapidly during the initial attack and decay portions of the envelope, and changes more slowly during the sustain and release portions. Looped samples are thus most frequently used for the sustain and release parts of sounds, with cross-fading between two samples or a modifier filter used to provide a changing timbre.

As with advanced S&S and wavetable synthesis techniques, it is possible for samples to have multiple loops: each with an envelope, pitch modulation and velocity switching facilities. The transitions between multi-samples can also be modified with velocity and note-position cross-fades, which can help to minimise the abrupt changes in timbre which can be present between multi-samples. The hardware and software of the sampler itself determines exactly which methods can be applied to the samples. The sound manipulation possibilities which are opened by these techniques should not be underestimated – most samplers provide powerful synthesis capability even without using the 'synthesis' modifier section.

Looping a sound can markedly reduce the amount of storage which is required. For a 10-second 'CD quality' stereo sound without any looping, then nearly 2 Mbytes of storage is required. If 7 seconds can be produced by looping part of the sample during the sustain segment, then this can be reduced to half an Mbyte. Reducing the storage requirements can reduce the amount of RAM memory required in the sampler which may reduce the manufacturing cost, or can allow more samples to be stored in the RAM memory.

Stretching

Stretching is the name given to the process of independently adjusting the timing or pitch characteristics of a sample. Transposition changes both the pitch and the time of a sample: if the sample is shifted up by an octave, then it plays back twice as fast and so a '1 second' sample lasts only half a second. In contrast, stretching aims to change the time without changing the pitch, or the pitch without altering the timing.

To change the timing, this involves analysing the existing sample and either removing or adding sections, depending on whether the sample is being lengthened or shortened. For pitch changes, the sample is either lengthened or shortened by the pitch change, and so sections again need to be added or removed. The length of the sections which are repeated or removed is normally quite short: at least one cycle, but short enough that the repeated sections are not heard as repeats. The transitions between the original and new sections suffer from the same problems as loop splice points, except that there are many more of them. This makes time and pitch stretching prone to audio quality problems – although this can be used as an effect on rapidly changing material to give a result which is similar to granular synthesis.

Re-sampling

Re-sampling is the name for using the sample record facility of a sampler to record the output of a sample replay. In a digital sampler it is the digital signals which are used, and so the loss in quality is not dependent on the analogue-to-digital or digital-to-analogue conversions. Re-sampling allows the sample rate of a sample to be changed, or for a low frequency oscillator (LFO) modulated sound to be stored as a sample, or for a filter sweep to be stored as part of a sample. It enables 'snapshots' to be made of the output of the sampler, and then the reuse of these sample snapshots as the raw material for further sounds by reprocessing the sample snapshots through the sample manipulation and modifier sections.

Because the sampling process does not capture all of a sound nor reproduce it perfectly, then re-sampling can reduce the audio quality. This can be especially noticeable if electrical hum or noise is introduced into the re-sampled version. Pitch-shifting also produces distortions, typically because of the limitations of the interpolation techniques used. Each time when re-sampling occurs, any degradation will accumulate.

4.3.2 Multi-sampling

Multi-sampling is normally used merely to provide changes in sound across the keyboard. It is typically used for instruments which have marked differences in their harmonic structure for high and low pitches, most notably the piano. Samples are taken of the source sound played at different pitches, normally all at the same sample rate. The limiting case of multi-sampling is that when each note on a music keyboard is sampled separately, then each note will be reproduced using a different set of sample values. This uses large amounts of memory, but provides the potential for the most accurate reproduction. Most multi-sampling is less extravagant than this, with samples being transposed to provide spans of an octave or perhaps a fifth rather than individual notes.

For multi-sampling where two or more samples are used across the whole keyboard range, the transition between samples can be important. As an example, consider two samples made of a piano: one from each extreme of the keyboard. The low pitched sound would be rich in harmonics, whilst the high pitched sound would be a sine wave plus a 'plink' transient hammer noise from the hitting of the string. The changeover from one sample to the other in the middle of the keyboard is likely to be very noticeable to a listener! Most multi-sampling does not use two samples taken from extreme ends of the range of an instrument. Instead, the aim is to provide enough samples to capture the characteristic sound of the instrument whilst minimising the unwanted effects of transposing samples. Extreme transpositions of samples produce an effect called 'munchkinisation', where the changes of pitch and timing emphasise the pitch change and give it a comic effect. This is particularly apparent on the spoken word or singing, although many instruments change their character noticeably when they are transposed by a large amount.

Since most playing of an instrument concentrates on the middle portion of the range of the instrument, most multi-sampling schemes involve having the most detail in this area. The transpose range of the multi-samples is thus small where the detail and transitions between multi-samples are important, but increases at the extremes of the range. This can be observed in many piano multi-sample sets. The bass notes often use a single transposed sample, whilst the high notes use a single 'plink' sample, with the smallest multi-sample ranges being present in the middle area of the keyboard. The transitions between these two extreme samples and those used in the central area are often the most striking, since this is where the largest compromise is made between choosing a suitable sample and ensuring a smooth transition between adjacent samples.

For instruments with large changes in timbre across their range, producing multi-sample sets can be complex and exacting work. Instruments which have a restricted range can also be a problem because most samplers will enable the playback of a sample over the complete range of the sampler, even if the source instrument cannot! This often means that a single violin multi-sample set is utilised as a violin, viola, cello and even double-bass, which although useful for providing synthetic textures which have some of the characteristics of real instruments, it is not suitable as a means of emulating real instruments. The most extreme example of the failure of transposition occurs in percussion instruments, where the fixed parts of the spectrum are essential to the timbre. Transposing a sample of a triangle or a tambourine produces instruments which merely sound wrong!

4.3.3 Storage

There are two forms of storage used in a sampler. The short-term internal storage is usually inside the sampler itself, whilst the longer-term storage is often external to the sampler and is frequently removable.

The storage which is used to hold the samples as they are made or replayed is normally fast read–write memory called RAM. RAM is an acronym for random-access memory, and the name refers to the ability to rapidly access any location in the memory device at random. In contrast, a tape recorder is much more restricted in its access: it either plays back the audio, or it requires to be wound or rewound to reach an alternative location on the tape. RAM storage does not have this problem: any location can be accessed as quickly as any other. RAM storage comes in two forms: static and dynamic. Static RAM chips will hold their contents for as long as they are powered up, which makes them ideal for short-term storage using battery backup. Dynamic RAM chips lose their contents if they are not continuously 'refreshed' by the host microprocessor chip. Dynamic RAM chips are considerably cheaper than the static version, and so low-cost samplers are more likely to have dynamic RAM which will require backing up to another more permanent type of storage before powering down the sampler.

Longer-term storage is often associated with magnetic or even optical media, although a variation of ROM technology called 'flash' memory allows long-term storage of samples in memory chips which do not require a backup battery. Flash memory can be internal to the sampler or on a plug-in memory card. Suitable magnetic and optical media include floppy disks, as well as hard disks and CD-ROMs, in either fixed or removable forms. Memory cards can be either RAM or flash-based, or may include a miniature hard disk drive, and will typically use one of the many flash memory card formats. Samplers in the 1980s and 1990s typically used the parallel-organised small computer system interface (SCSI) bus to interface to external memory devices, and this allowed additional protocols like the SCSI variant of MIDI (SMIDI) to be used to transfer samples at higher data rates. In the 21st century, USB, USB2 and IEEE488 or FireWire serial connectors provide fast external storage interconnections with lighter cabling and smaller connectors. Networking of samplers together over a local area network, or LAN, allows samplers to share common storage devices. The use of large amounts of online storage forces the use of detailed management of the storage to enable specific samples to be located, and then loaded samples into memory for editing and playback, with the edited versions then being catalogued and stored again.

Digital audio signals require large amounts of storage. For a 44.1-kHz sample rate, stereo 16-bit samples produce just over 1.4 megabits per second, or about 600 Mbytes per hour. This is easy to remember when you consider that an audio CD lasts for about an hour, and contains about 600 Mbytes. The 8-bit resolution samples halve these figures, but with a very significant loss in quality. Reducing the sample rate restricts the bandwidth, which is only useful with sounds which have limited bandwidths like some bass and drum sounds. Table 4.3.1 shows some examples of storage requirements for sampled audio.

Storage on hard disks has seen a halving of cost every 12 months or so for some years. In 2003, the 120-gigabyte external FireWire drive became a common sight, with a cost comparable to a mid-range hi-fi CD player. By the mid-2000s, half terabyte drives may well be replacing them.

The early 21st century has seen the rise in popularity of sophisticated audio compression schemes. MP3 (actually the audio encoding part of a moving pictures expert group (MPEG) video and audio encoding standard) was the first, with typical data rates of about 128 kilobits per second for stereo audio. Reducing the data rate to about a tenth of the CD uncompressed rate can only be achieved by removing redundancy first, and then by reducing quality. MP3 coding does this by hiding the deficiencies in parts of the spectrum where they are

Table 4.3.1 Storage requirements for sampled audio

Resolution (bits)	Sample rate (kHz)	Mono/stereo	Time (seconds)	Storage			Time
				Kilobits	Kbytes	Mbytes	
8	32	Mono	1	256	31.3		
8	32	Mono	10	2560	312.5	31.3	
8	32	Mono	60	15360	1875	1.8	1 minute
8	32	Mono	3600	921600	112500	109.9	1 hour
12	32	Mono	1	384	46.9		
12	32	Mono	10	3840	468.8	0.5	
12	32	Mono	60	23040	2812.5	2.7	1 minute
12	32	Mono	3600	1382400	168750	164.8	1 hour
16	44.1	Mono	1	705.6	86.1		
16	44.1	Mono	10	7056	861.3	0.8	
16	44.1	Mono	60	42336	5168	5	1 minute
16	44.1	Mono	3600	2540160	310078	302.8	1 hour
16	44.1	Stereo	1	705.6	86.1		
16	44.1	Stereo	10	7056	861.3	0.8	
16	44.1	Stereo	60	42336	5168	5	1 minute
16	44.1	Stereo	3600	2540160	310078	302.8	1 hour
24	48	Stereo	1	1152	140.6		
24	48	Stereo	10	11520	1406.3	1.4	
24	48	Stereo	60	69120	8437.5	8.2	1 minute
24	48	Stereo	3600	4147200	506250	494.4	1 hour

masked by other louder sounds. MP3 also exploits the wide acceptance of small light headphones, and it is quite instructional to listen to MP3 encoded audio on a hi-fi system. AAC and other coding schemes are reducing the data rate even lower. Whilst some gains can be made by increasing the processing, the ultimate loser is the quality, even if it is well hidden. Severe compression algorithms also have the side effect of making the audio very sensitive to errors: a single error can have very serious effects on the audio output.

In samplers, there are different criteria to meet. Unlike music tracks, samples do not have the same broad spectrum of sounds offering places to hide distortion. Compressing sounds which feature one instrument at one pitch is not straightforward, and can easily expose any weakness in the compression technique. But there are some genres of music which thrive on distortion, and some genres which use broad-spectrum sounds, and for these compression may be appropriate.

4.3.4   Transfer of samples

In the early 1990s, using an external SCSI hard drive, or using a computer as a storage and editing device, was the limit of most sample transfers. The MIDI Sample Dump Standard (SDS) was intended to allow sample data to be transferred between samplers, but this was slow because of the large size of 16-bit, 44.1-kHz sample rate samples and the slow transmission rate of MIDI. SCSI-MIDI, or SMIDI, was an attempt to use SCSI as the transport for samples, but it did not see wide acceptance. The music industry has lagged behind the computer industry slightly by continuing to use SCSI even as USB2 and FireWire have increased in popularity. The mid-2000s should see a change in rear panels to reflect the wide adoption of USB2 and FireWire, whilst front panels should increasingly see flash-based storage replacing floppy disks.

The 1990s saw an increasing dependence on the CD-ROM as the medium of sample exchange, especially as the cost of CD-writers, and then CD-rewriters, dropped to affordable prices: 600 Mbytes of samples on one CD-ROM is sufficient to store more than the complete sample RAM of many samplers. Removable hard drives have seen some acceptance, with the Iomega Zip drive of 100 Mbytes upwards being one of the longest surviving and most widespread examples. The low-cost, robustness and re-recordability of the CD-RW has made it very popular, whilst the one-time write CD-R has become very low cost and very quick to write: 52 times recorders produce a 600-Mbyte CD in just over a minute. Recordable digital versatile disk: DVD-R is still being developed, and there are a number of different formats with varying amounts of compatibility. As with CD-Rs before, DVD-Rs should converge on just one or two common formats by the mid-2000s.

SoundFonts and MIDI downloadable sounds (DLS) are formats which allow samples to be transferred over computer-to-computer connections, typically a LAN or the Internet. These are descended from the .MOD files which were first used in the 1980s to create music on computers from very simple sound generating resources. mLAN has yet to see wide adoption, but it, and the Ethernet variant of MIDI that is being developed by the IEEE, should ultimately make sample and audio file transfer an everyday part of electronic musical instrument interchange.

4.3.5 Types

There are three main types of digital sampler:

- stand-alone
- keyboard
- computer-based.

Stand-alone

Stand-alone samplers are normally designed to fit into a 19-inch rack-mount case. Control and editing functions are carried out using the MIDI protocol, although some samplers also have provision for an external monitor and keyboard to provide improved access to the editing functions. Samplers are often controlled from a master keyboard or synthesizer keyboard, but some samplers are designed to be controlled from the front panel; for example, for adding sound effects or replaying drum sounds.

Keyboard

Keyboard samplers are essentially a stand-alone sampler placed in a larger case and with an added keyboard. Whereas S&S instruments have seen considerable success with this format of instrument, keyboard samplers have been less successful commercially. S&S sample players where part of the ROM memory is replaced by RAM have been slightly more successful, although these are better described as user sample replayers, since they usually lack any way for the user to actually sample sounds.

Computer-based

Computer-based samplers were initially manufactured in the form of plug-in 'sound cards'. Some of the NuBus cards were very large and complex, although advances in electronics have meant that the more recent peripheral component interconnect (PCI) bus equivalents are considerably smaller in many cases. Some computer-based samplers have taken advantage of the built-in audio capabilities of some computers and are then merely software, but their performance is very much dependent on the computer's audio circuitry. For computers where the audio system is not adequate, the cards are merely converters where the audio storage uses the computer's own RAM. Other cards may provide special-purpose processors to carry out digital signal processing (DSP) functions: these are sometimes referred to as DSP 'farms'. It is also possible to find all of these separate parts on a single card.

The conversion from analogue audio signals to and from digital data is sometimes carried out in a separate box outside the computer in order to optimise the conversion accuracy – the interior of a computer case is not an ideal location for a sensitive conversion system. In some systems, the external box is merely used to provide a convenient way to house all the connection sockets: plug-in cards for computers normally provide only a very small area of panel in which to locate input and output sockets.

Direct-to-disk recording can be thought of as a variation on computer-based sampling, although it has a different set of design goals. Whereas a sampler will normally record into RAM memory, and this sets a time limit on the length of the sample which can be recorded, a direct-to-disk recording unit will store the converted audio data on the hard disk directly, which means that the length of the recorded audio is limited only by the available hard disk size. This process places considerable demands on the computer and the hard disk, and in fact, the number of tracks which can be recorded and/or replayed simultaneously is determined by the computer's processing power and the rate at which data can be transferred to the hard disk storage.

4.3.6   Sample sound-sets

The open nature of samplers means that they are considerably more customisable than S&S ROM-based sample-replay instruments. Although it is possible to populate a sampler's RAM with the single-cycle waveforms that might be found in an S&S sample set, it is more common to use longer samples. The demands of the computer industry for plug-in RAM in the form of single and dual inline memory modules (SIMMs and DIMMs, respectively), as well as ever larger and faster hard drives, have meant that samplers have acquired very long total sample times, and large libraries of samples on hard disk to fill the RAM.

Apart from the extremely detailed pianos, violins and other orchestral instruments, sample sound-sets are available for a very wide range of vintage and ethnic instruments – a much wider range than is found in S&S sample-replay units – even where specialist sound-sets are available in ROM cards. But a huge number of sample sound-sets are available which are intended as pads and special effects sounds. Complex evolving textures and ambient soundscapes can make further processing inside the sampler almost unnecessary.

Samplers are also widely used in a field where S&S sample replay has very limited facilities: loops. Loops are 1 or more bars of rhythmic patterns made up from drums, bass and other accompaniment instruments, sometimes with melodies as well. They are intended to be the raw material from which pieces can be constructed in much the same way that a groove box or phrase sequencer works (see Chapter 8) Longer loops exist, but these are often intended for audio-visual presentations where background music is required.

Samplers often allow loops, and particularly drum loops, to be separated out into short samples, often on a 'per beat' basis. This means that a single drum beat in a loop can be extracted for use independently, or can be moved in time inside the loop. Some samplers allow shuffling of these loop fragments to be carried out with varying degrees of randomness. The loop is one of the evolving areas of sampler technology, especially in the live performance context.

4.3.7 Using samplers

Samplers can be used as pure replay instruments in much the same way as S&S synthesizers. But this is not exploiting the capability of the sampler to change its complete sound-set, and especially to reproduce sounds which are not purely instrumental.

Because a sampler can be loaded with a specific sample set made up of a number of sounds, even the sounds produced by synthesizers, then it can be used as a way of producing the sounds from a large variety of instruments from one piece of hardware or software. It is also possible to record samples of several synthesizers played together and mixed into one complex sound, and then to use the sampler to replay the sound. A complete rack of hardware synthesizer modules can be replaced with one or more sample sets, and so the combination of a single keyboard and a sampler can be used to replace complete racks of synthesizers.

Samplers are also good when either a complete backing track or loop is required without needing a sequencer, several synthesizers, a mixer and some outboard effects units. Vocal performances can also be sampled and used as backing vocals, sometimes even as solo lead vocals. Special effects are another type of sounds where samplers can be very useful for playing back a range of sounds chosen from a larger library.

The flexibility of samplers requires considerable investment in time and sampled sounds if the capability is to be exploited properly. Auditioning sample sounds from CDs or CD-ROMs is a slow task, and turning these selected sounds into sample sets for specific songs takes careful planning and consistency of assignment of sounds to MIDI channels or sequencer tracks. Making backups of any sample set definitions is also essential with many samplers, although some samplers are now incorporating flash memory, which reduces the need to take backups but does not remove it completely.

4.4 Convergence of sampling with S&S synthesis

The fundamental differences between an S&S synthesizer and a sampler are often described as being related to the sample memory and the sample processing.

- **Sample memory**: There is a popular misconception that S&S synthesizers have permanent ROM memory whilst samplers have volatile RAM memory. This view ignores the way that both types of instruments have evolved. S&S synthesizers have acquired user sample RAM, whilst CD-ROM drives in samplers virtually relegate them to replay-only status.

- **Sample processing**: S&S synthesizers normally have a restricted set of controls for the replay of samples, but this is usually compensated for by the provision of a sophisticated synthesis section with a resonant filter and voltage controlled amplifier (VCA). Samplers often concentrate more on the sample-replay controls, with multi-sampling, looping, sample stretching and interpolation between one sample and another, although their subsequent processing is often just as capable as many S&S instruments.

The differences are thus less apparent than is often supposed. There is an ongoing convergence of functionality in both instruments. S&S instruments can have user sample RAM memory, and external sampling units can provide samples, although CD-ROMs are more frequently used to provide additional 'off-the-shelf' sounds, much as with a sampler. Samplers now use CD-ROMs and hard drives to provide rapid access to raw sounds in much the same way that S&S instruments provide sample replay from ROM or RAM memory.

Samplers are sometimes used merely as replay devices. This wastes the creative potential of the synthesis sections which can be used to give great effect in processing the samples and providing new sounds. There is some evidence of a stigma being associated with samplers because of this 'replay-only' reputation, with some people preferring to use S&S instruments with user sample RAM memory instead of a 'sampler'. The convergence between S&S and sampling should soon produce instruments which are so difficult to categorise into either of the two types that the bias against samplers may change.

4.5 Example equipment

Ensoniq Mirage (1985)

The Ensoniq Mirage was the first affordable commercial sampler (Figure 4.5.1). Although only monophonic and with a grainy 8 bits of sample resolution, with 8-note polyphony and a very restricted memory (2 seconds total sample time at 15 kHz), this instrument changed synthesis and ushered in S&S instruments and samplers. In contrast to the basic sample replay that might be expected from a first instrument from a new company, the Mirage has LFO modulation and separate filter/VCA envelopes, with a velocity sensitive keyboard. The user interface is minimalistic, with a two-digit light-emitting diode (LED) display and yet there are plenty of features. Up to 16 multi-samples can be assigned across the keyboard, the low-pass filters have a resonance control and keyboard tracking, and samples can be looped. There is also a simple sequencer too!

Akai S900 (1986)

The Akai S900 (Figure 4.5.2) was probably the first serious rackmount professional-quality sampler. Although only 12-bit and 8-note polyphonic had the facilities (like eight individual outputs) and software

Figure 4.5.1 Ensoniq Mirage (1985).

Figure 4.5.2 Akai S900 (1986). LCD: liquid crystal display.

to make it almost a *'de facto'* standard for sampling for several years, with the floppy disk format used for the samples also acquiring the status of a common exchange medium. The S900 had a maximum sample rate of 40 kHz giving 12 seconds of high quality monophonic sampling. Up to 32 samples could be assigned across the keyboard, and it provided facilities such as velocity switching/cross-fading, and cross-fade looping. Complete setups of the sampler can be saved onto disk, with 32 of these available at any one time.

Akai S1000 (1988)

The Akai S1000 (Figure 4.5.3) introduced 16-bit stereo sampling with 16-note polyphony, eight separate outputs, increased memory size, additional controls on the front panel, and sample compatibility with the S900, as well as an optional SCSI hard disk interface. Sampling time on the standard model was almost 50 seconds at 22 kHz for mono-phonic samples. The modifier section looks much like an analogue synthesizer, with LFO modulation, and separate filter VCA envelopes. The sample playback control includes all the cross-fading/velocity switching, looping points and forwards/backwards/alternating loop modes (and more) that you would expect on a second-generation pro-fessional sampler.

217

Figure 4.5.3 Akai S1000 (1988).

Akai CD3000 (1993)

The Akai CD3000 was a 16-bit stereo sampler, which allowed sampling from the built-in CD/CD-ROM drive. This was a reflection of the growth in popularity of sample CDs.

Sonic Foundry ACID (1998)

ACID is a sequencer that plays samples. It comes complete with sample library and has seen continuous development since its release: rather like a hardware sequencer plus hardware sampler, but without the hardware.

Roland VP-9000 (2000)

The Roland Variphrase VP-9000 provided almost complete independence of time and pitch, with very sophisticated (and patented) signal processing giving huge control and flexibility of working with samples.

4.6   Questions

1. What are the differences between a sampler and an S&S synthesizer?
2. What is the relationship between the speed of a tape and the playback pitch? What happens when the tape is played backwards?
3. What three electronic devices form the basis of a sampler?
4. What do anti-aliasing filters do?
5. What do reconstruction filters do?
6. What criteria need to be considered when looping a sample?
7. What editing functions would you expect to find in a sampler?
8. Outline the convergence of samplers with S&S synthesis.
9. Describe the limitations of analogue sampling techniques using optical and magnetic storage. Then show how these limitations were then overcome by the use of digital techniques.
10. What electronic devices use analogue sampling? What audio applications have they been used for?

Time line

Date	Name	Event	Notes
1876	Alexander Graham Bell	Invented the telephone	Start of the marriage between electronics and audio
1877	Thomas Edison	Thomas Alva Edison invented the cylinder audio recorder, the 'Phonograph'. Playing time was a couple of minutes!	Cylinder was brass with a tin foil surface, replaced with metal cylinder coated with wax for commercial release
1878	David Hughes	Invented moving coil microphone	
1888	Emile Berliner	First demonstration of a disc-based recording system – the 'Gramophone'	Disc was made of zinc, and the groove was recorded by removing fat from the surface, and then acid etching the zinc
1898	Valdemar Poulsen	Invented the Telegraphone, which recorded telephone audio onto iron piano wire (also known as the 'Dynamophone')	30 seconds recording time, and poor audio quality
1903	Double-sided record	The Odeon label released the first double-sided record	Two single-sided records stuck together?
1920–1950	Musique Concrète	Musique Concrète	Tape manipulation
1920s	Harry Nyquist	Developed the theoretical basis behind sampling theory	Nyquist frequency named after him
1920s	Microphone recordings	First major electrical recordings made using microphones	Previously, many recordings were 'acoustic', using large horns to capture the sound of the performers
1920	Louis Blattner	The first magnetic tape recorder	Blattner was a US film producer
1930s	Record groove direction	Some dictation machines record from the centre out instead of edge in	This pre-empts the CD 'centre out' philosophy
1930s	Run-in grooves	Run-in grooves on records invented	Previously, you put the needle into the 'silence' at the beginning of the track …
1935	AEG, Berlin	AEG in Germany used iron oxide-backed plastic tapes produced by BASF to record and replay audio	Previously, wire recorders had used wire instead of tape
1937	Tape recorder	Magnetophon magnetic tape recorder developed in Germany	The first true tape recorder
1940s	Wire and ribbon recorders	Major audio recording technology uses either steel wire or ribbon	High speed, heavy and bulky, and dangerous if the wire or ribbon breaks!
1948	Pierre Schaeffer	Coined the term 'Musique concrète'	
1949	C.E. Shannon	Published book *The Mathematical Theory of Communications*, which is basis for subject of information theory	Shannon's sampling theorem is the basis of sampling theory
1950s	Tape recorder	Magnetic tape recorders gradually replaced wire and ribbon recorders	There were even domestic wire recorders in the 1950s!

Date	Name	Event	Notes
1951	Herbert Eimert	North-west German Radio NWDR in Cologne starts experimenting with sound using studio test gear	Used oscillators and tape recorders to make electronic sounds
1958	Charlie Watkins	Charlie Watkins produces the CopyCat tape echo device	The CopyCat is still in production
1958	RCA	Radio Corporation of America (RCA) announces the first 'cassette' tape, a reel of tape in an enclosure	Not a success
1960s	Mellotron	The Mellotron, which used tape to reproduce real sounds	Tape-based sample playback machine
1963	Philips	Philips in Holland announces the 'compact cassette': two reels plus tape in a single case	A success well beyond the original expectations!
1974	George McRae	'Rock Your Baby' was the first record to completely replace the drummer with a drum machine	
1974	Kraftwerk	'Autobahn' album was of huge success. A mix of music concrete technique and synthetic sounds	
1978	Philips	Philips announced the compact disk (CD)	This was the announcement getting the technology right took a little longer …
1979	Fairlight	Fairlight Computer Musical Instrument (CMI) announced. Sophisticated sampler and synthesizer	The start of the dominance of computers in popular music
1979	First Digital LPs	First LPs produced from digital recordings made in Vienna	A mix of analogue playback and digital recording technology
1979	Roland	Roland Space-Echo launched: used long tape loop and had built-in spring line reverb and chorus	A classic device, used as the basis of several specialist guitar performance techniques (for example, Robert Fripp)
1980	E-mu	Emulator, the first dedicated sampler	
1980	E-mu	E-mu emulator released. This was the first widely available digital sampler	8 bit, unusual in appearance and with limited facilities
1981	Roger Linn	The Linn LM-1: world's first programmable digital drum machine	Replays samples held in EPROMs
1982	Philips/Sony	Sony launched CDs in Japan	First domestic digital audio playback device
1983	Philips/Sony	Philips launched CDs in Europe	Limited catalogue of CDs rapidly expanded
1985	Akai	The S612 was the first affordable rackmount sampler, and the first in Akai's range	12-bit, quick-disk storage and only 6-note polyphonic
1985	Ensoniq	Introduced the 'Mirage', an affordable 8-bit sample recording and replay instrument	Related to the Ensoniq ESQ-1 hybrid sample-replay synthesizer

Date	Name	Event	Notes
1985	E-mu	EII (Emulator II) released. Low-pass analogue filters and 1/2 MByte of sample RAM memory	Companded 27.7 kHz, 12-bit samples stored as 8 bits
1986	Yamaha	Launched Electone HX series organ	Mixture of frequency modulation (FM) and advanced wave modulation (AWM) (sampling)
1987	DAT	DAT was launched. The first digital audio recording system intended for domestic use	Worries over piracy severely prevented its mass marketing, and it became a professional format
1987	Yamaha	TX16W 12-bit, 16-note polyphony sampler	The operating system was completely overhauled by a Swedish programming team and released as 'Typhoon'
1988	Akai	S1000. One of the first professional-quality stereo 16-bit samplers	S1100 was expanded and enhanced version
1988	Yamaha	TX16W sampler released. Rackmount, stereo, 12-bit model with sophisticated filters	Lots of editing facilities. Now supported by third party software
1989	Akai	XR10 drum machine launched	A digital drum machine using sampled drum sounds
1990	Yamaha	SY77, a digital FM/AWM hybrid synthesizer, mixed FM and sample replay technology	Followed in 1991 by the larger and more powerful SY99 with MIDI SDS sample dump capability
1991	Sony	MiniDisc released. A small rewritable magneto-optical (MO) disk enclosed in a shell; a mixture of CD-R and floppy disk thinking	
1992	Akai	S01 released. Low-cost 16-bit, monophonic sampler	A move away from the top-of-the-range pro-Akai samplers
1992	Kurzweil	The K2000 was launched. A complex S&S instrument, which mixes sampling technology with powerful synthesis capability	
1993	Akai	CD3000. 16-bit stereo sampler with sampling from the built-in CD/CD-ROM drive	A reflection of the growth in popularity of sample CDs
1993	Roland	S760 sampler released. A low-cost 16-bit stereo sampler with optional external television monitor display	Very popular in some composing circles
1994	E-mu	SoundFont 1.0 specification released with the AWE32 PC sound card	The first in the SoundFont series of wave file sound-card sample formats
1995	E-mu	ESI32 sampler released. 32-note polyphonic, 16-bit stereo sampling at a low price	A contrast to E-mu's top-end samplers
1997	Yamaha	A3000 Sampler and effects unit	Budget sampler with lots of DSP-based effects processing. Followed by the even more effects and processing-equipped A4000 and A5000

Date	Name	Event	Notes
1998	E-mu	ESi4000: sampler with effects processor	Samplers go low-cost ...
1998	Sonic Foundry	ACID, loop-based sample sequencing software	A low-cost alternative to hardware samplers ...
1998	NemeSys	GigiSampler, very capable software-based sample-replay software	Uses patented kernel-level software and so is available on the Windows PC platform only
2000	Roland	VP-9000 Variphrase processor	Effectively a sampler with astonishing control over the pitch and time parts of samples
2001	Tascam	Buys NemeSys, creators of the GigaSampler software sample-replay product	When a hardware tape and mixer company buys a software company, you know that the market has changed
2003	Sony	Sony Pictures Digital bought Sonic Foundry music software, including ACID	The market continued to change ...

5 Digital synthesis

As digital audio transmission formats like S/PDIF/AES/EBU and mLAN become more widely adopted as the outputs of synthesizers, and the inputs of mixers, fully digital instruments may eventually appear where there is no DAC at all. Software synthesizers which are used in computers are already purely digital.

Digital synthesis of sound is the name given to any method which uses predominantly digital techniques for creating, manipulating and reproducing the sounds. Often the only 'analogue' part of a 'digital' instrument will be the audio signal that is produced by the digital-to-analogue converter (DAC) chip at the output of the instrument.

Most digital synthesis techniques are based very strongly on mathematics: even methods like digital samples and synthesis (S&S) which often attempt to mimic, in software, the analogue filters found in subtractive synthesizers. The precision with which digital synthesizers operate has both good and bad aspects. Repeatability and consistency might seem to be a major advantage over the uncertainty which often occurs in analogue synthesizers, but this precision can also be a disadvantage. For example, frequency modulation (FM) synthesis in an analogue synthesizer is difficult to control adequately because of the slight non-linearities of the FM inputs of many oscillators, in a digital synthesizer the precision of the calculations can mean that 'unwanted' effects like the cancellation of harmonics in a spectrum can happen. In an analogue synthesizer, the minor variations in tuning and phase would prevent this from happening: in a digital system, these may need to be artificially introduced.

This illustrates a very important point about digital sound synthesis. The degree of control which is possible is often seen as an advantage. But it also requires a considerable investment of time in order to be able to take advantage of the possibilities offered by the depth of detail which may be required, especially when there are potential problems if one does not fully understand the way that the synthesis works. This is very important in techniques like Fonctions d'Onde Formantique (FOF), where forgetting to set some of the phase parameters can result in major changes to the sound which is produced.

In summary:

- **Analogue synthesizers** offer the rapid and often intuitive production of sounds, but they have intrinsic non-linearities, distortions and inconsistencies which can contribute to their characteristic 'sound'. If the speed of use and the available sounds are suitable, then the limitations may not matter.
- **Digital synthesizers** can provide a wider range of techniques, some of which are very powerful at the cost of complexity and difficulty of understanding. But they do not suffer from the built-in imperfections of analogue circuitry, and so these may need to be

simulated, which adds to the task of controlling the synthesis and makes them less intuitive. The creative possibilities offered by digital synthesis are obtained at the expense of the detail required in setting up and controlling them.

Digital sounds

The synthesis techniques covered in this chapter are:

FM
waveshaping
modelling
granular
FOF
resynthesis
hybrids.

It has been said that digital sounds 'clean', whilst analogue sounds 'natural'. As a vague generalisation, this is almost acceptable. It is possible to make very crisp, clear timbres using digital technology, but this is by no means the only tone colour which is available. In fact, digital technology often introduces its own distinctive 'dirt', 'grunge' and distortion into the signal.

Two of the commonest artefacts of using digits instead of analogue signals are quantisation noise and aliasing.

- **Quantisation noise** is the grainy, roughness that is typically found on the decay and release of pianos or reverbs. It happens because of the limited resolution of the numbers that digital systems use to represent audio signals, as the numbers get too small then errors get introduced and this appears as extra noise.
- **Aliasing** is a side effect of the process of sampling: it is caused by a combination of imperfect filtering and 'just good enough' sampling rates. Aliasing sounds like ring modulation, and is often heard as harmonically unrelated frequencies towards the top of the frequency spectrum.

Notice that both of these 'distortions' are due to imperfections in the way that digital works and as such, are very similar to the limitations that are found in 'real' instruments, or even 'analogue' synthesizers. So the gross distortion which can be produced by overloading a filter in an analogue synthesizer is fundamentally no different to the aliasing in a digital sampler, or to the 'wolf' tones that can be obtained by careful blowing into wind instruments.

The important thing is that applying descriptions like 'natural' or 'clean' to a sound is a very personal and subjective thing. Some very 'natural-sounding' flutes and harps can be entirely synthetic in origin, whilst some 'clean' sounding clavinets might have high levels of distortion. Digital does not have any better claim on 'clean' sounds than any other method, nor does 'analogue' have a special reason to sound 'natural'. For example, there is no way that an analogue resonant filter-based 'Moog bass' sounds like anything in nature, because most real instruments do not have resonances that change in frequency quite to that extent!

5.1 FM

FM is an acronym for frequency modulation, an old idea which although possible on analogue synthesizers, was not really practical. Analogue synthesizer voltage controlled oscillators (VCOs) are subject to frequency drift, variation with temperature, non-linearities, high frequency mis-tuning and other effects which lead to unrepeatable results when you try to use FM for generating melodic timbres. In fact analogue-based FM is very good for producing a variety of 'non-analogue' sounding special effects: sirens, bells, metallic chimes, ceramic sparkles and more. It was not until the advent of digital technology that FM really became possible as a way of producing playable sounds rather than special effects.

FM essentially means taking the output of an oscillator and using it to control (modulate) the frequency of another oscillator. If you try this with two VCOs in a modular synthesizer then you are almost guaranteed to get some bell-like timbres at the output of the second, the 'modulated' VCO, especially if the only control input to the VCO is the exponential control input (FM should really use the linear control input – often marked FM!). In synthesizers, FM is used as a synonym for 'audio FM', where both the oscillator frequencies are approximately in the audio frequency range: about 20 Hz to 20 kHz. FM radio uses an audio signal to modulate a very much higher frequency which is then used to carry the audio waveform as radio waves. This use of the word 'carrier' persists even in audio FM, where radio transmission has nothing to do with the sound that is produced.

FM synthesis is not really like any of the major synthesis techniques described so far, although it was briefly described in the context of a modulation method in Sections 2.3 and 2.4. It is not a subtractive or additive method, and it does not fit easily into the 'source and modifier' model either. FM has its roots in mathematics, and is concerned with producing waveforms with complicated spectra from much simpler waveforms by a process which can be likened to multiplying. The simplest waveforms are sine waves, and FM is easiest to understand if sine waves are used for the initial explanations. In fact, unlike analogue synthesizers, where the waveshapes are often the main focus of the user controls, FM is much more concerned with harmonics, partials and the spectrum of the sound (Figure 5.1.1).

5.1.1   Vibrato

So, if an oscillator is set to produce a 1-kHz sine wave, and another oscillator is used to change the frequency with a 20-Hz sine wave, then the 1-kHz tone will have a vibrato effect. The oscillator producing the 1-kHz tone is called the *carrier*, whilst the oscillator which is producing the modulation waveform is called the *modulator*. Although it is technically only correct to call the two oscillators the 'carrier' oscillator and the 'modulator' oscillator, it is more usual to call them the carrier and modulator for brevity.

If the modulator output is increased then the depth of the vibrato will increase, which means that the carrier is sweeping through a wider range of frequencies. If the modulator output is decreased, then the carrier will be swept through a smaller range of frequencies around the original 1-kHz unmodulated frequency. The difference between the highest and lowest frequency which the carrier reaches is called the *deviation*: so an unmodulated carrier has no frequency deviation.

Figure 5.1.1 The terminology of audio FM is different from analogue subtractive synthesizers, although many of the component parts are the same. In this example, the basic FM is produced by two identical 'modules' which can be thought of as consisting of a sine wave DCO, digital VCA and EG.

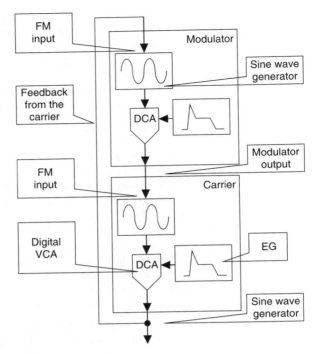

If the speed of the vibrato is increased, then above about 30 Hz, the result will stop sounding like vibrato, and above 60 Hz it will be perceived as several sine-wave tones of different frequencies all mixed together which is the 'characteristic 'bell-like', clangorous timbre often associated with FM.

5.1.2 Audio FM

Audio FM replaces the low frequency modulator with another audio frequency. Suppose that the modulator level is initially zero. The output of the carrier oscillator will thus be a sine wave. As the level of the modulator is increased, then the sine wave will gradually change shape as extra partials appear. Initially two sidebands (partial frequencies on either side of the carrier) appear, but as the depth of modulation increases, so does the modulation index, and thus more partials will appear. The timbre becomes brighter as more partials appear, although unlike opening up a low-pass filter, partials appear at higher and lower frequencies. The lower frequencies can alter the perception of what the pitch of the sound is: if a 1-kHz sine wave acquires an additional sine wave (it is a partial, not a harmonic, because it need not be related to the fundamental by an integer frequency ratio) at 500 Hz, then it can sound like a 500-Hz tone with a partial at 1 kHz.

As described in Chapter 2, the output consists of the carrier frequency and sidebands made up from the sum and difference frequencies of the carrier and multiples of the modulator frequency.

The number of sidebands depends on the modulation index, and a rough approximation is that there are two more than the modulation index. The modulating frequency is not directly present in the output. The amplitudes of the sideband frequencies are determined by a set of functions called Bessel functions (Chowning and Bristow, 1986 is unfortunately now out of print). So, as the level of the modulator signal increases, the output gradually changes into a much more complex timbre. The transition is a smooth addition of frequencies much as you would expect with a low-pass filter gradually opening up, but with the added complication of extra frequencies appearing at lower frequencies too.

5.1.3 Bessel functions

Mention of Bessel functions normally means that mathematics takes over and the next few pages should be filled with formulae. FM is often presented as being inaccessible because of its complexity, so here I will attempt to try and describe how FM works in as simple and non-mathematical a way as possible.

We will start by taking the filter analogy a little further. Imagine an additive synthesizer (Section 2.3) which has individual envelopes for each of a number of frequencies. If we want to simulate a low-pass filter opening up, then we need some envelopes which allow first the low frequencies to appear, then middle frequencies, and finally the higher frequencies. The envelopes would look like a series of delayed attack and sustain segments, where the delay in the start of the attack was related to the frequency which was being controlled – the higher the frequency the longer the delay time. Triggering the envelopes would cause the lower frequencies to appear, then the middle, and finally the high frequencies. A similar set of envelopes could be used to produce frequencies which were lower than the fundamental (Figure 5.1.2).

The shapes of these envelopes control the harmonic/partial structure of the sound produced by the synthesizer. By changing the shapes of the envelopes, we can change the way that the frequencies will be added as time passes. Actually we do not need to use envelopes: we could use any controller which can map one input to lots of outputs whose behaviour we can control. It just happens that an envelope is one way of using time as the controller. If we used a control voltage and lots of voltage modifiers, it would be possible to control the frequencies from the additive synthesizer in just the same way, and the same envelope shapes could be used to describe what would happen: the only difference would be that the envelopes are now curves which show how the frequencies change with the input voltage instead of with time (Figure 5.1.3).

Bessel functions are the name given to the curves which relate how the frequencies are controlled in FM. Although they are smooth curves instead of the angular envelopes, the principle is exactly the same. In much the same way as the filter envelopes have time delays

Figure 5.1.2 These envelopes can produce an output equivalent to a filter frequency being swept upwards. Each envelope processes one frequency component.

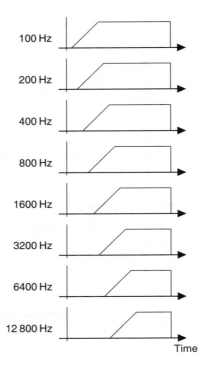

Figure 5.1.3 Bessel functions describe how the amplitude of the sidebands in FM vary with the modulation index. In this example, an FM output is produced using a modulation index of 10. Only the first four Bessel functions are shown.

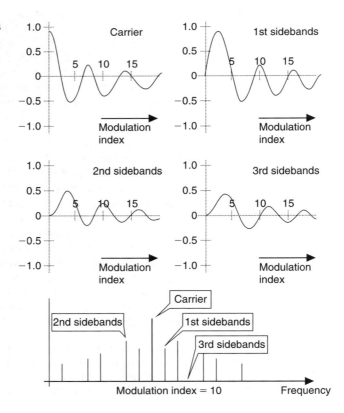

built into them, the Bessel functions vary in a similar way – the further away from the carrier frequency, the higher the value of the control needs to be to have an effect. Instead of time or a control voltage, the control is the modulation index. So as the modulation index increases (when the modulator level increases, or the modulator frequency decreases) the number of partials increases. And that is really all there is to Bessel functions: they merely describe how the level of the partials changes with the modulation index. There are one or two complications in reality, and you should look at the references if you need more details.

The only other thing which needs considering for FM is how the frequencies are controlled. In FM the spacing between the partials is related to the carrier and modulator frequencies, but there are really only three basic relationships:

- **Integer**: Integer relationships between the carrier frequency and the modulator frequency produce timbres which have harmonic structures which are similar to those of the square, sawtooth and pulse waveforms – harmonics at multiples of the fundamental. The only complication is that the fundamental is not always at the carrier frequency because of the extra frequencies which can appear below the carrier.
- **Slightly detuned from integer**: Slightly detuned carrier and modulator frequencies produce the same sort of 'multiples of the carrier' harmonic structures, but with all the harmonics detuned from each other too. This can produce complex beating effects, although the amount of detuning needs to be carefully controlled to avoid too rapid beating effects.
- **Non-integer**: Non-integer relationships between the carrier and the modulator frequencies produce the bell-like, clangorous timbres for which FM is famous. If either the carrier or modulator is fixed in frequency (i.e. it does not track the keyboard pitch), then the relationship will change with the pitch of the note being produced.

In all of these cases, the basic timbre produced is set by the relationship between the carrier and modulator frequencies, whilst the number of harmonics and partials which are produced is controlled by the modulation index, using the values in the Bessel functions. There are some additional complexities caused when the modulation index is so large that the spreading out of the partials causes some frequencies to go below the zero-frequency point and get 'reflected' back, which can cause some additional cancellation effects.

That is really all there is to FM: you choose the timbre, and then control it usually dynamically. FM may be different from subtractive or additive synthesis, but the controls and the way that they work are relatively straightforward: once you understand what is happening. Almost all of the other functions, like low frequency oscillators (LFOs), portamento, envelope shapes and effects, should be very similar to the same functions in other synthesizers.

Yamaha's TX81z was their first FM synthesizer to provide non-sine waveforms. These are in some ways equivalent to extra operators, because a single pair of non-sine wave operators can produce sounds which are like those from three or more sine wave operators.

FM is normally produced using oscillators which are made available as general-purpose building blocks called *operators*. These consist of an oscillator, an envelope generator (EG) and a voltage controlled amplifier (VCA), all in digital form, of course. The oscillator is a variable speed playback of a wavetable, whilst the VCA is a multiplier connected to the EG. In the first audio FM implementations, the wavetable held a sine wave, but later versions of FM had additional waveforms.

On a larger scale, FM may use more than one pair of operators (carrier and modulator), several modulators onto one carrier, or even a stack of operators all modulating each other. It is also possible to take the output of a carrier operator and feed it back to the input of a modulator, which can be used to produce noise-like timbres, although it is still the modulation index and the frequency relationships which determine much of the timbre. Learning how to make the most of FM involves analysing the FM sounds produced by other people, and programming sounds yourself, but the model described here should provide the basic concept of how FM works.

FM is good for producing sounds with complicated time evolutions and detailed harmonic/partial structures, but it can be difficult to program: the explanation above has been simplified, and there can be quite a lot of parameters to cope with. It is also possible to produce FM with non-sine wave oscillators, although all that happens is that each sine wave, which is present in the waveform, acts as its own FM system, and so you get lots of FM happening in parallel, which can lead to very noise-like timbres because of the large numbers of frequencies which are produced. FM is especially suited for 'metallic' sounds like guitars, electric pianos and harpsichords (Figure 5.1.4).

FM really only requires the following parameters to define a timbre:

- carrier frequency
- modulator frequency
- modulator level.

But most FM sounds change the modulator level dynamically by using the modulator envelope, and the carrier also has an envelope, but even

Figure 5.1.4 This overview of a simple FM synthesis system shows how the individual component parts contribute to the final sound output.

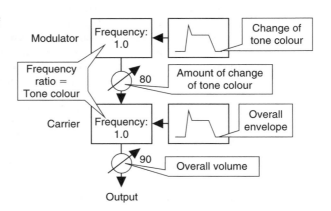

then the number of parameters required to specify a given sound is less than 20 parameters. In comparison to subtractive or additive synthesis, this is a much smaller number of parameters to deal with. For this reason, FM has been investigated as the synthesis part of an analysis–synthesis resynthesis system, but there are problems in extracting the FM parameters from sounds. In particular, it is not easy to take a specified waveshape or spectrum and calculate the required FM parameters, especially if there are any partials: non-harmonic frequency components. See Section 7 of this chapter for more information on resynthesis.

Filters did not appear until comparatively late in audio FM history. The SY77 in 1990 had twin 12-dB/octave digital high- and low-pass resonant filters, whilst the DX200 in 2001 had a modelled voltage controlled filter (VCF) that had high-pass, low-pass, band-pass and notch modes with up to 24-dB/octave cut-off slopes.

Having a small number of important parameters also enables FM to be a very powerful synthesis method when using real-time control. By changing the carrier or modulator frequency, or the modulator depth with specialised musical instrument digital interface (MIDI) commands (or front panel controls), FM can be used to produce sounds which can change rapidly and radically. On an analogue subtractive synthesizer the only comparable parameter change is the filter cut-off; and this is much more restricted in the timbral changes which it can produce.

5.1.4   *Realisation*

The actual details of the way that FM is realised differ on different platforms. Computer-based software will probably use a different approach from the hardware-oriented custom digital signal processing (DSP) solutions used in synthesizers. But the basic elements are much the same in all cases, although the terminology may be very different. The descriptions which follow are mostly based on Yamaha's FM, mainly because it has been the most widely accepted and most commercially successful of any of the digital FM implementations.

Oscillators

Initially FM was produced using only sine waves. The mathematics behind FM are easiest to understand if sine waves are used, and early FM work was at an academic level, where both the understanding of the sound production process and the aesthetics of the resulting sounds are important. The first commercial FM synthesizers also used just sine waves, probably for reasons of price: the Yamaha DX7 was introduced at a time when digital consumer electronics was virtually unknown – the compact disk (CD) player was not introduced for another year. The high ratio between the features and price of the DX7 was partially due to the use of digital technology, but also to a careful minimisation of functionality: after all, Yamaha were testing the market with a very different type of synthesizer. The lack of front panel control knobs shows that they were prepared to take radical design decisions in both the synthesis and user interface areas.

Implementing a digital FM sine wave generator has been covered in Section 3.3 on digitally controlled oscillators (DCOs). With only one waveform, the size of read-only memory (ROM) required can be quite small, especially if the symmetry of the sine wave is used to reduce the storage requirement – you can produce a complete sine wave cycle from just one quarter of a cycle of sine wave waveform points. The Yamaha design multiplexes the oscillator: it is used to provide the waveforms for all of the oscillators, with storage of the successive outputs used to give the equivalent of four or six separate oscillators. Further multiplexing is used to provide the DX7's 16-note polyphony, which was about twice the normally expected polyphony of polyphonic synthesizers in 1983.

Later, more advanced FM synthesizers like the TX81z used more complex waveforms. But this was often only achieved by deriving additional waveforms from the same sine wave ROM memory, by changing the way that the quarter cycle is reassembled to form a waveshape. Quite minor changes to the waveshape can have significant effects on the spectrum, and using waveforms which contains additional frequencies can produce FM sounds which are very rich in harmonic and partial content – even to the extreme of becoming noise-like. The second generation of commercial FM synthesizers from Yamaha (the SY77 and SY99) also added the ability to use samples as part of the FM synthesis, but this did not prove to be very popular with users, and subsequent models in the 'SY' series concentrated on sample-replay technology rather than FM. Since then, Yamaha have only released two further devices that use FM synthesis: a rackmount expander module, the FS1R; and a desktop synthesizer plus step sequencer, the DX200.

Most FM oscillators can be used in two modes. The usual mode is to allow the oscillator to track the keyboard pitch, although this need not be the normal keyboard scaling. The second mode is usually called 'fixed frequency', and here the oscillator frequency does not change. Fixed oscillators can be used in several ways. At low frequencies they can be used as carriers or modulators to produce vibrato-type cyclic timbral change effects, whilst at higher frequencies they can be used to partially emulate a very resonant system. Fixed frequencies of a few hundred hertz are often used to produce vocal sounds, since the resulting sound has many of the qualities of the formants which determine human vocal sounds. A fixed oscillator within an FM algorithm produces an output spectrum with frequency components which are related to the fixed oscillator frequency, or harmonics of it, and this can sound somewhat like a resonant tube. This was exploited and extended in the FS1R in 1998.

In each of these cases – sine wave, sine-derived waveforms and samples – the technology used to produce the FM waveform is very similar to the advanced DCO designs described in Chapter 3. The output of a period of time (not necessarily a single cycle) for each oscillator is stored, and then used as the basis for the modulation of the next oscillator, that is, ending with the carrier oscillator. The high precision of the sine wave, the frequency resolution and the linearity of the FM, all enable FM synthesis to be achieved in a precise, repeatable and controllable way – a big contrast to producing FM using analogue technology.

Envelopes and VCA

Although DCOs were found in hybrid synthesizers, FM required digital control over the amplitude of the oscillator outputs, and for this they used the multi-segment rate/level type of envelope. Rate/level envelopes provide comprehensive control over the shape of an envelope by using function generator controls to set the characteristics of each segment. As the name implies, two parameters are used to control each segment: a rate and a level. The rate specifies how long the segment lasts, whilst the level sets the final level which that segment reaches. The initial level is normally the same as the final level, although some later instruments do not have this restriction.

Yamaha had identified that conventional attack decay sustain release (ADSR)-type envelopes were not suitable for envelopes which had complex attack stages – especially where the start of the sound did not rise at a constant rate. The multi-segment envelopes which they used in the DX7 had three segments to cover the 'key on' part of the envelope, plus one segment for the 'key off' or 'release' portion of the envelope. Because the final level of the third segment is held whilst the key is held down, it effectively produces a separate 'sustain' segment, which is the fixed level which the attack segments end at. This produces a categorisation of the segments according to their function within the envelope. The first three segments are used to control the 'attack' portion of the envelope, the level of the last attack segment sets the sustain level, whilst the final segment controls the release behaviour (Figure 5.1.5).

The envelope used in DX-series Yamaha six-operator FM synthesizers is a five-segment rate/level. There are three 'attack' segments, one 'sustain' segment and one 'release' segment. EG forced damping is

Figure 5.1.5 The EGs used by Yamaha in their FM synthesizers provide great flexibility because of their structure. The four levels can be anywhere in the permitted range, which allows a wide variety of envelope shapes, including inverted envelopes and pseudo-exponential attack segments.

Some possible envelope shapes

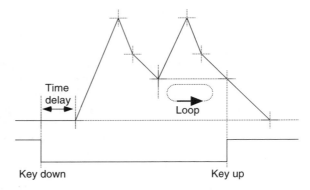

Figure 5.1.6 Loopable envelopes allow previous segments to be continuously looped whilst the envelope is in the sustain segment. In this example, the first three segments are looped. The use of delayed attack segments enables the production of echo-like and arpeggio effects.

found in the Mark II DX7, and forces the envelope to restart when a note is reassigned because of note stealing at the limits of the polyphony.

SY-series Yamaha six-operator FM synthesizers use eight-segment envelopes which add a separate initial level, delay time, two release segments and the ability to loop the envelope whilst the key is held down. Also, the final level is not necessarily the same as the first level. The two release segments enable additional control over the end of the note, whilst the delay time is used for special effects like arpeggiated operators or allocating operators to specific parts of the final sound – using one set of operators to generate the initial portion of a sound, whilst delayed envelopes produce the remainder of the sound. This effectively increases the apparent number of envelope segments at the expense of using operators for only part of the duration of the sound. The looping enables the sustain segment to be less static: simpler envelopes just reach a sustain level and stay there as long as the key is held down (Figure 5.1.6).

Low-cost FM implementations with four operators have simplified ADBDRR envelopes which had only six controls: five rates and two levels (break-points for the decay and release segments).

Not all multi-segment EGs use the word 'rate'. Time and slope are sometimes used as synonyms. There is also no standardisation on how the parameters relate to the duration of the segment: some manufacturers use small numbers to mean short times, whilst the converse is used by others.

The digital VCA in FM is almost always treated as part of the EG.

Operators

The combination of an oscillator, envelope and VCA is such a fundamental building block of FM synthesis that it is often treated as a single module. Because the first major commercial success of utilising FM in a digital synthesizer was from Yamaha in the early 1980s, the terminology that they used has become widely adopted. Yamaha used the word 'operator' for the block formed by an oscillator, envelope and digital VCA (Figure 5.1.7).

The initial FM prototype instruments were the Yamaha GS1 with eight operators, and the GS2 with four operators, whilst the initial DX

Figure 5.1.7 A Yamaha FM operator typically consists of a sine wave DCO, digital VCA and EG. Operators can be connected together in arrangements called *algorithms*, where the description of the operator as a carrier or a modulator is determined only by its position in the algorithm.

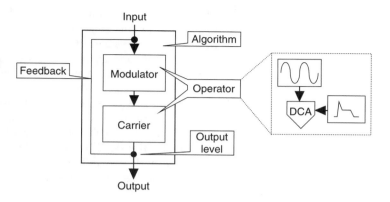

synthesizers had six operators. (The DX9 had two operators deliberately disabled to give it reduced functionality.) Lower-cost FM implementations followed using four operators with restricted frequency control and limited internal calculation precision and the chips for these were made available to other manufacturers – these became a '*de facto*' standard for the basic implementation of a personal computer (PC) sound card. Prototype FM synthesizers like the V80 were produced by Yamaha with eight operators, although these never progressed beyond the development laboratory. Some of the Yamaha HX-series organs were released with eight operators, but these did not have the depth of user-programmability of the synthesizer products. Some FM implementations use multiple 'pairs' of operators, which do not provide the same flexibility as being able to arbitrarily connect more than six operators together.

In 1998, Yamaha released the FS1R rackmount expander module, which had eight operators with built-in band-pass filtered 'formant' noise generators in place of the simple feedback or noise generators of previous implementations. The FS1R harked back to previous products based on speech synthesis concepts like the SFG-05 FM-plug-in module for the CX-5M MSX computer, which had Japanese speech synthesis software. In the FS1R, the combination of two types of operator: voiced and unvoiced, reflects speech synthesis terminology. The 'voiced' operators were standard FM operators with pitched or fixed frequency modes: vowels in terms of speech, whilst the 'unvoiced' operators provide the 'f' and 's' sounds, and combinations of the two can produce consonants like 'b' and 't'. Of course, the FS1R could not only use these facilities for speaking, it could also be used to produce singing (note the release of the 'Vocaloid' software a few years later), and instrumental and percussive sounds too. As with other implementations of FM, the FS1R requires an understanding of the principles behind the design in order to make the most use of the available facilities, and programming requires skill and time.

In 2001, Yamaha released the DX200, a tabletop synthesizer with a built-in 16-step sequencer. The DX200's FM had six operators and is DX7 voice compatible. But the DX200 introduced a new feature derived from the AN1X modelled analogue synthesizer: interpolation between two

235

sounds: by using a front panel controls, the parameters defining one sound can be smoothly changed to the parameters for another sound. The result is not always perfect: it can sound like a 'morph' between the two instrumental sounds, or a blend from an instrumental sound to noise and metallic klanging and then to another instrumental sound. The morphing can be very effective, whilst the blend can be useful for adding just an edge to a sound by only moving slightly towards the noisy, metallic sound. The DX200 also attempts to provide an alternative user interface to FM, taking concepts from analogue synthesizers and the Korg DS-8 and 707, to give a set of front panel controls that are intended to provide live 'interactive' control over the sounds, and for some algorithms, this approach is very effective.

Algorithms

Yamaha use the word 'algorithm' for the arrangement and interconnection of operators. Although there are many ways of arranging the topology of four or six operators, there are only a few important types:

- additive
- pairs
- stacks
- multiple carriers
- multiple modulators
- feedback
- combinations.

Additive

Although not actually FM, parallel operators can be used as a simple additive synthesizer producing several frequencies simultaneously. Unlike many additive synthesizers, the frequencies need not be harmonically related, and so slightly detuned oscillators can be used to provide chorused 'additive' sounds. Each operator provides a single frequency component, or partial, or 'formant', with the EG controlling just that frequency.

Pairs

The simplest FM algorithm (apart from a single operator, which can only produce sine waves, of course) is a pair of operators: one carrier which is modulated by one modulator. The carrier EG and level control give control over the overall volume of the sound which is produced, whilst the level control and the envelope of the modulator control the modulation index of the FM. The timbral controls are thus the two operator frequencies, plus the modulator envelope and level controls.

Stacks

By taking a second modulator, and connecting this so that it modulates the modulator in an FM pair, then a stack of three operators can be produced. Additional modulators can be added, although a stack of four operators is normally the most that is required. Since the pair formed by the two modulators produces an FM sound, the carrier which is modulated by this sound (and not by the sine wave which would be produced by a single modulator) is much more complex because each frequency in the modulating signal creates FM with the carrier operator. Stacks are often used for pad sounds, where lots of slightly detuned harmonics and partials are used for producing rich, chorused sounds.

Multiple carriers

By connecting one modulator to more than one carrier, the same modulator can be used to control two carriers. By having different frequencies for the two carriers (or different envelopes), the output is two FM sounds which are related but separate. If the modulators have slightly detuned frequencies, then two similar but detuned sets of harmonics and partials are produced.

Multiple modulators

If several modulators are connected to one carrier, then each modulator can be used to produce part of the final sound, which can simplify the development of sounds. Having only one carrier operator means that controlling the output envelope is easier, but it also restricts the timbral possibilities because there is less flexibility in choosing the ratio between the carrier and modulator frequencies (Figure 5.1.8).

Figure 5.1.8 FM algorithms' summary. In these diagrams, C indicates a carrier operator, whilst M indicates a modulator operator. There are six basic arrangements of operators, plus a seventh consisting of combinations of parts of these. The examples shown here are for six operators, but the same topological arrangements apply to other numbers of available operators.

Feedback

By connecting the output of an operator back to its frequency control input, the resulting feedback signal affects the output signal of the operator. In the simplest case, a single operator with a feedback loop will produce additional frequencies with a large amount of feedback, and this can sound similar to a sawtooth or pulse type of waveform. Feedback around several operators can be used to produce very complex sounds. If too much feedback is applied, then noise-like sounds can be produced.

Feedback has always been one of the more interesting and less well-understood aspects of FM. The basic idea is that you take the output of the operator and connect it to the frequency control input (in some algorithms this is the same operator, in others you get a loop of two or three. On an SY-series FM synthesizer you can patch several operators together and apply feedback between them. The 'feedback level' is a control over the level, and so controls the modulation index: 0 is no feedback, whilst 7 is an index of about 13 on a DX7 or SY. A modulation index of 13 is going to produce quite a lot of frequency deviation, and so the original sine wave is deviated well away from its basic frequency, but at a rate which is tied to itself. This produces lots of extra harmonics and perhaps even partials (the spectrum for a single operator on a DX7 with a feedback value of 7 has 23 harmonics) and a very contorted waveform. In fact, with a feedback value of greater than 5, the underlying precision of the FM synthesis implementation used by Yamaha begins to become significant and the output begins to get noise-like: although the operator output level also affects the feedback, since the two level controls are in series! Below 5, the sound produced is merely richer in harmonics and partials.

Although the sounds produced by the feedback are described as 'noise-like', this does not usually mean that they are like the 'white' or 'pink' noise found in analogue synthesizers. With two operators, things rapidly get out of control once you start connecting a harmonic-rich waveform from a feedbacked operator as the modulator of another operator. Aliasing and the finite precision of the FM synthesis 'engine' combine to produce a plethora of noise-like sounds, with not-so-flat spectra and lots of harmonics and partials – especially if you use non-integer ratios for the carrier and modulator frequencies. Careful use of feedback level and operator output levels can keep things non-noise-like, and still in the realm of complex but interesting timbres. Because of the effects of aliasing, and the way that FM folds harmonics or partials when the modulation index is large, the resulting spectra may be rich in frequency content, but they are rarely flat: the noise is not white, nor is it really coloured: various shades of grey, perhaps!

Because the output of a 'fedback' operator is a spectrum relatively full of harmonics and partials, changing the carrier or modulator frequencies merely changes some of the harmonic and partial amplitudes and the aliasing components. The only audible effect is often a change in the timbre or 'colour' of the noise. Only with low modulator indexes and low feedback values will the carrier and modulator frequencies make any significant difference.

In the SY synths, these problems with producing 'white' noise were solved by providing a noise generator. This produces white noise, and this side-steps any need to use feedback to try and get a flat noise spectrum. Feedback noise sounds tend to be slightly too structured or grainy to fool the ear, whereas a simple maximal-length pseudo-random sequence is probably used by Yamaha's noise generator to provide white noise on tap. Using feedback creatively with all the flexibility offered by the SY synths is worthy of further exploration.

In the FS1R, the noise generation is extended further by adding filtering, and the result is called an 'unvoiced' operator, referring to speech synthesis terminology. As a general rule, the pitched, harmonic or 'voiced' parts of FM have remained relatively the same throughout FM synthesis development, whilst the 'noise-like', inharmonic or 'unvoiced' part has seen the most development to try and extend and enhance the capabilities of FM synthesis. The FS1R might even be classified as being a combination of FM synthesis with formant synthesis.

There is some basic information on DX-series FM feedback (Figure 5.1.9) in Dave Bristow and John Chowning's now out-of-print book: *FM Theory and Applications for Musicians* (1986) pages 133–136.

Combinations

Most FM sounds are made up from combinations of the simple algorithms. Two parallel stacks of three operators are often used because they enable two separate sounds to be combined, whilst each stack of three operators is a versatile FM sound source. Multiple modulators can produce complex sounds where each modulator contributes a distinct element to the sound, and they can each be controlled separately. The development process for FM sounds tends to be iterative, with operators being turned on and off to determine their effect on the sound each time, as their parameters are changed. This technique is especially important where groups of operators are used to provide different parts of the sound. Unlike many methods of synthesis, FM allows the programmer to investigate the effect of minor changes both in isolation and in context.

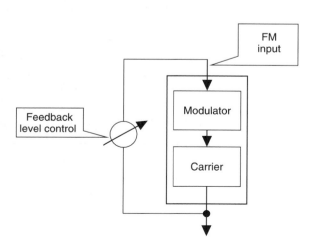

Figure 5.1.9 By feeding back the output of an FM algorithm to the FM input, it is possible to generate noise-like outputs. Some implementations have added specialised noise generators to supplement this method of generating noise-like sounds.

DX-series FM synthesizers provide fixed algorithms where the topology can be selected from a number of presets. The presets provided typically include all of the possible arrangements of operators, and many of the additional possibilities provided by adding one feedback loop. SY-series and subsequent FM synthesizers provide user control of the interconnections and multiple feedback connections, as well as preset topologies. Choosing a specific algorithm is largely a matter of experience. But in many cases, starting from a simple pair of parallel stacks is a good idea, because extra modulators (or carriers) can then be added as required. Familiarity with the timbral possibilities of a simple pair or stack of operators can be very useful in helping to produce FM sounds. Examining pre-programmed sounds can also help to reveal some useful techniques.

5.1.5 History

John M. Chowning's paper: 'The synthesis of complex audio spectra by means of frequency modulation', in the *Journal of the Audio Engineering Society* in September 1973, was the first serious description of the practical use of digital technology to implement audio FM as a way of synthesising timbres. This is very much a 'landmark' in digital synthesis; unlike additive synthesis, where the large number of required parameters made a digital realisation unwieldy, FM showed that digital synthesis could be powerful and yet require only a relatively small number of controls.

5.1.6 Implementations

There are three strands of FM development from Yamaha: four, six and eight operators. Four-operator FM tends to be used in the lower-cost and computer-oriented areas, whilst six- and eight-operator FM is used in 'professional' instruments. From 1982, Yamaha have continued to release FM instruments through to the early 21st century, although from the mid-1990s onwards their main focus has been towards S&S synthesis (Table 5.1.1).

Korg's DS-8 and 707 synthesizers from 1987 used FM technology as a result of a temporary pooling of research facilities by Korg and Yamaha

Table 5.1.1
FM implementations

	Sine waves						Sine variants		Samples					
	1982	1983	1984	1985	1986	1987	1988	1989	1990	1991	1992	1998	2001	2002
4-Operator	GS2	DX9	CX5M	DX21	DX100 FB01	TX81z 707	DX11	V50 TQ5	SY22 TG33	EVS1	SY35		GS2	DX200
6-Operator		DX7	DX1 TX7	DX5 TX816		DX7II DX7S			SY77	SY99				FM7
8-Operator	GS1					HX...		V80				FS1R		

in the middle 1980s. Many PC soundcards of the 1980s and 1990s used a Yamaha FM chip set to produce musical sounds.

Until the end of the 1980s, the FM chips and DACs used in FM implementations had limited resolution, and the resulting sounds have some background quantisation noise. From the 1990s onwards, higher resolutions were used, and the FM has less of these digital artefacts. As with many 'retro' music fashions, the older sound has been subject to cyclic peaks of popularity, although adding in suitable emulated noise is also possible with more modern implementations.

The SY77 and SY99 from the early 1990s add resonant filtering and sample replay to enhanced FM synthesis. The FS1R from 1998 added additional filters in the form of formants (see Section 5.5.1) to eight-operator FM. The DX200 added interpolation between two sets of sound parameter settings in 2001.

Although Yamaha had acquired the patent rights to the commercial exploitation of FM in the early 1980s, there were several variants on FM which differ enough to be usable without actually infringing the patent. For example, phase modulation changes the phase of the carrier instead of the frequency, which produces very similar sounds (and spectra) to FM.

Casio's CZ series of synthesizers used waveshaping, but their later VZ series of synthesizers used an FM-like phase modulation method, calling it phase distortion (commonly abbreviated as PD). Eight operators were available, with eight-stage envelopes.

Peavey has also used the term 'phase distortion' to describe synthesizers which appear to be S&S instruments, and not a variant on FM.

FM produced on computers using software initially offered enhanced sophistication at the price of non-realtime operation, although faster processors and DSP technology now makes FM more accessible and immediate. In 2002, the FM7 plug-in from Native Instruments offered real-time six-operator FM which was DX7 sound compatible, and added a number of enhancements like more sophisticated SY-series-style noise generation and additonal resonant filters. Some purists complained that the background quantisation noise inherent in the early FM implementations was missing.

5.2 Waveshaping

Waveshaping is a way of introducing controlled amounts of distortion onto a waveform. This differs from the 'fuzz box' type of distortion which is used by guitarists, because it is used on the 'monophonic' outputs of the oscillators, and so it merely changes the shape of the waveform without adding in all the intermodulation distortion which happens when more than one note is passed through a waveshaper or fuzz box.

Waveshapers are non-linear amplifiers. This means that they provide control over the way that the amplifier processes incoming signals to produce an output. For an amplifier with a fixed gain of two, you expect to get an output which is twice the input. If a graph is plotted of input against output for a linear amplifier, it would be a straight line through the origin (zero) of the graph. This line is called the 'transfer function' of the amplifier – it shows the way that the input is 'transferred' to the output. In fact, the straightness of this line can be used as a measure of the quality of an amplifier, since a perfect amplifier would have a perfectly straight-line transfer function. If the amplifier did not have a gain of two for high levels of input signal, then the transfer function graph would be curved at high input levels, which means that an audio waveform which is passed through the amplifier will change shape. Changing the shape of the transfer function changes the shape of the waveform.

It is the convention that transfer function graphs always have the input plotted horizontally, and the output vertically. The scaling is also arranged so that the input and output ranges are from −1 to +1, with the zero point of both axes being in the centre of the graph. The input sine wave moves completely across the horizontal axis once per cycle: from a value of 0 to +1, then back through the zero position to −1, and then back to zero again. The output waveform is dependent on the transfer function, although the maximum and minimum outputs are normally +1 and −1, respectively (Figure 5.2.1).

Although this sounds like an easy way to produce extra waveshapes, it actually does rather more than that. Distorting the shape of a waveform changes the harmonic content of the waveform: in fact, in

Figure 5.2.1 A transfer function is a graph which relates an input to an output. In this example, a straight-line transfer function allows a sine wave to pass unconverted, whilst a transfer function which has a steeper slope and two flat zones converts a sine wave into a trapezoidal waveform.

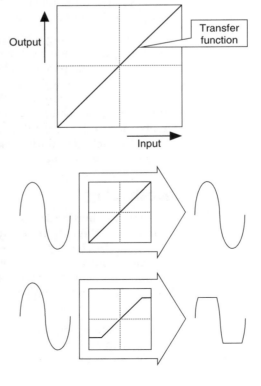

most cases, it adds harmonics rather than subtracts them. If the transfer function is symmetrical about the horizontal axis or has a rotational symmetry, then the harmonics which are added will be the odd harmonics, whilst if the transfer function is symmetrical about the vertical axis or shows a mirror symmetry, then only even harmonics will be produced. So with a sine wave and a waveshaper, it is possible to use different shapes of transfer functions curve to produce outputs which have a wide variety of harmonic contents. Now using a sine wave and producing extra harmonics from it sounds like FM, and in fact, with the right type of transfer function curve, waveshaping can produce sounds which are very FM-like in character. But it can also produce sounds which do not have 'FM-like' characteristics. When FM was at its peak of popularity in the mid-1980s, Casio used a waveshaping-based synthesis technique in their CZ-series of synthesizers, but called it *phase distortion*.

Using a sine wave has advantages and disadvantages. It is possible to calculate a transfer function which will produce any given spectrum (but not waveshape) from a sine wave input. The technique involves the use of Chebyshev polynomials and enables the input sine wave to be changed in frequency. The resulting output frequency is multiplied by the order of the Chebyshev function: so a fourth-order function would produce an output sine wave which is four times the input frequency. By adding together several Chebyshev functions it is possible to produce a composite transfer function which will then produce any required spectrum. The calculations of the relevant Chebyshev polynomials are simplified if the input waveform is a sine wave. Because the sine wave shape has two different times when the same value occurs, then some waveshapes cannot be produced, but this restriction is often not a problem, since the harmonic content is normally more important for a specific timbre.

The input waveform to a waveshaper need not be a sine wave. If a sawtooth wave is used, then the waveshaper is little more than a look-up table for output values, and so resembles a wavetable oscillator. The positive and negative half cycles of the sawtooth wave just map onto images of the output waveform and thus the two half cycles can be different. Effectively, the two halves can be thought of as two separate transfer functions, although they normally share a common point at the origin. But for a sine wave, the symmetrical nature of the waveshape means that there is more redundancy, which in turn means that there is scope for more independence of transfer functions. A sine wave can only be converted into other waveforms of a particular class of shapes by using a single non-linear transfer function – basically only those where the first quarter of the cycle is the same as the next quarter cycle, but where the second cycle is time-reversed. As the sine wave input moves up and back down the transfer function horizontal axis, the symmetry is inevitable. The same applies to the third and fourth quarters of the cycle.

But by providing different transfer functions for each of these quarter cycles, the waveshaper can be used to convert a sine wave into waveforms which do not have this first and second quarter-cycle mirroring. This means that a single transfer function graph can have two separate halves for each half cycle of the sine wave, and the symmetry produces waveforms which have a large content of even harmonics. In contrast, if there are two separate transfer functions, with each quarter cycle having its own graph, then any waveshape can be produced (Figure 5.2.2). This type of quarter-cycle waveshaping is used in the second generation of Yamaha FM synthesizers to produce additional waveforms from a wavetable ROM containing just a single high precision sine wave (see also Figure 3.3.4).

Although waveshaping can be used as a general-purpose tool for changing the shape of a waveform, it can be arranged so that the audible behaviour of the waveshaper is similar to the VCF found in an analogue synthesizer. Using digital technology to emulate familiar 'analogue' characteristics is a continuing theme of most digital synthesis methods. In the case of an analogue low-pass filter, harmonics are successively added as the filter cut-off frequency is increased; hence the output waveform is initially a sine wave at the fundamental frequency. As harmonics are added the shape of the output waveform will change until with the filter fully 'open', then all frequencies will pass through and the output waveform should have the same shape (and frequency spectrum) as the input.

For a basic waveshaper implementation, this 'filter emulation' behaviour would seem to imply that the transfer function is changing dynamically, and it is possible to produce transfer functions which

Figure 5.2.2 By using separate transfer functions for each half or quarter cycle of a waveform, it is possible to produce almost any required output waveform from an input sine wave. In this example, each quarter cycle of the input sine wave has its own transfer function. The output waveshape is the concatenation of the four output quarter cycles.

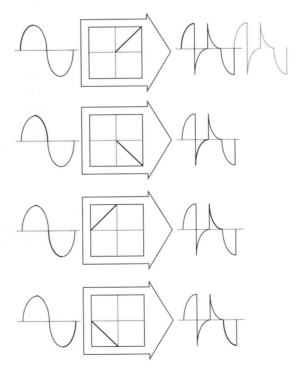

scale in size to produce this effect. However, by designing the transfer functions carefully, and by ensuring that the transfer function curve passes through the origin (zero) of the graph, it is possible to produce simple waveshapers with just one fixed transfer functions which can be used with inputs which are smaller than the +1 and −1 maximum and minimum levels.

As a simple example, consider a transfer function which is a straight line as it passes through the zero points of the input and output axes, but which gradually curves away from a straight line as it moves away from the zero. At low amplitudes of input sine wave, the output will also be a sine wave, because the linear portion of the transfer function will be used. But as the input level is increased, the non-linear parts of the transfer function will be used, and the waveform will be distorted. As the level increases to the maximum, then the largest waveform distortion will be produced. This tends to produce an output signal which starts out as a sine wave, but which gradually acquires additional harmonics as the amplitude increases, in much the same way as an analogue VCF. By arranging an amplifier to correct for the amplitude changes, it is possible to produce an output which does not change in level as the 'filtering' action takes place (Figure 5.2.3).

The audible result of this 'waveshaping' process is a smooth transition from a sine wave to one containing a number of harmonics. But unlike an analogue VCF, the evolution of the waveform is dependent on the way that the transfer function changes with input amplitude. This means that the harmonics do not need to be added in a progressive sequence comparable to a low-pass VCF, but can change in other ways

Figure 5.2.3 Dynamic waveshaping alters the input level and then scales the output to compensate. In this example a sine wave is passed through an asymmetric transfer function which is linear for positive inputs, but a complex function for negative inputs. The outputs for different levels are shown; it can be seen that the output waveform changes as the input level is increased in much the same way as opening a VCF does on an analogue subtractive synthesizer.

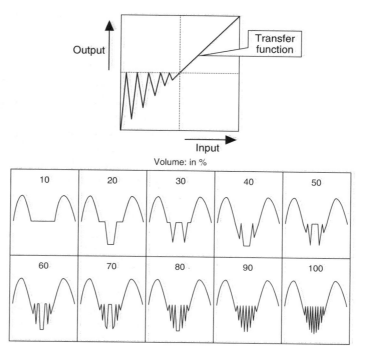

which can be more interesting to the ear. These complex changes of harmonic content are also found in FM, although the evolution of FM waveforms is fixed by the Bessel functions. For a waveshaper-based synthesizer the transfer function is not fixed and so can produce more sophisticated and varied harmonic changes: at the price of an increased need for mathematical understanding on the part of the designer of the transfer function. Unlike FM, the additional frequencies which are produced by waveshaping are always harmonically related to the input frequency since the waveshaping is based around the shape of one cycle of the waveform.

Some manufacturers have used waveshaping in a much more limited sense. For example, the Korg 01 series S&S synthesizers implement waveshaping, but it is a very limited form of single non-linear transfer function waveshaping. It is used to process the outputs of the oscillators and is really limited to just adding in a few extra harmonics to the raw samples. Casio-style dynamic waveshaping is a much more powerful technique: if Korg had moved the waveshaper to after the VCF or VCA, or made the transfer curve controllable or dynamic, then the possibilities for timbral change would have been much greater.

5.3 Modelling

Whereas other digital methods of sound synthesis tend to try and emulate the terminology of functions of analogue synthesis, mathematical modelling breaks away from these conventions. There are no samples, no function generators and much less use of envelopes and filtering, and yet despite throwing away almost everything with which the synthesizer user may be familiar, instruments which use modelling techniques can produce sounds which feel so much like real instruments that it is hard to think of them as electronically produced.

There are many variations on the basic idea of using mathematical models to produce sounds. In this section, just three will be examined:

- '**Source-filter synthesis**' is a simplified modelling technique that concentrates on the interactions between the two major component parts that produce an instrument's sound.
- '**Physical modelling**' attempts to describe the complete instrument with a complex and sophisticated model.
- '**Analogue modelling**' describes analogue synthesizer circuitry.

5.3.1 Source-filter synthesis

Instead of trying to describe how a complete instrument works in terms of equations, source-filter synthesis looks for a way that the important elements can be encapsulated in a form which provides control, but is easy to use. It turns out that there is a way, and it comes from research into speech. When you speak, your vocal cords are vibrated by the air which rushes past them, and this raw sound is then modified by the complex set of tubes and spaces formed by your throat, nose, mouth, teeth, lips and tongue. A physical model of this

Figure 5.3.1 The driver produces a raw sample sound which has had the effect of any resonance removed artificially. This is then coupled to a resonator section through a coupler section, which allows control by the performer.

would need to consider the velocity of air, pressure, tension in the vocal cords, the space between them, their elasticity, etc.; and trying to work out the exact mechanisms for how they vibrate could be difficult and time-consuming. The more pragmatic approach of source-filter synthesis asks: what does the raw sound produced by the vocal cords sound like, what sort of filter do the throat, mouth and nose form, and how do these two parts interact with each other?

Source-filter synthesis assumes that musical instruments can be split into three parts (Figure 5.3.1):

- **Drivers**, which produce the raw sound. Examples are the hammer hitting a piano string, or the pick plucking a guitar string, or the reed vibrating in an oboe.
- **Resonators**, which colour the sound from the driver. Most musical instruments exhibit some sort of resonance: often the whole of the instrument vibrates along with the sound to some extent, and the way that it vibrates affects which frequencies are emphasised and which are suppressed.
- **Coupling between the driver and the resonators**, which determines how the two interact with each other.

In a real instrument, the drivers and resonators are very closely connected. They interact with each other: the hammer hitting a piano string causes the string to vibrate, but the vibration of the string is affected by the fact that the hammer is touching the string, has probably stretched the string slightly when it moved the string, and has added in a low frequency thump. The act of setting the string vibrating depends on the hammer – you cannot have the sound without it, but the hammer affects the sound. The two are inextricably interconnected. In source-filter synthesis, the two are separated, but the same interactions can be produced by controlling the way that the driver and resonator are connected together.

The basis of this technique is to separate the driver and resonator, and then couple them together so that they can interact. Instead of trying to model the driver, the technique assumes that the raw driver is more or less fixed, whilst the coupling to the resonator is the important aspect. This means that a driver 'sample' can be used to provide the stimulus for a resonator model via a coupling device – there is no need to try and create a model for the driver at all. Modelling resonators is much easier, since they are just filters, and filter theory is well understood. This means that it is easy to produce a number of driver 'samples', and resonator specifications, and couple them together. This approach means that a large number of possibilities are opened up without any need for careful research into musical instruments.

247

The coupling part of source-filter synthesis deals with the interconnection and interaction between the driver and the resonator. This is probably the major part of the technique to use the same approach as 'physical modelling'. A bowed string is a good analogy for the process. The player of a stringed instrument can control parameters like the position of the bow on the string, and how hard the bow is pressed onto the string. The resonator can be changed as well: for example, it may be a fixed resonance or one that changes with the playing pressure. The combination of a simple model for the coupling, plus the fixed driver 'sample' and the variable resonance, produces a versatile synthesis 'engine'.

The driver output is not a conventional audio sample. Because this is the raw driving force without any modification by a resonator, it is not possible to actually place a microphone and sample it directly. One approach to determining what it would sound like is to take the final sound of the instrument, and then remove the effect of the resonances. If you listen to a raw driver signal, then it will sound very bright with an emphasised initial transient, almost like high-pass filtering. But since most resonators act as band-pass or low-pass filters, coupling this driver signal to a resonator transforms it into a sound which suddenly takes on a more normal sound. In fact, it sounds much like the sample that you would actually hear in a recording (which is what a sample is, of course: the result of a driver coupled to a resonator). The difference is that by separating out the driver and the resonator, and by changing the parameters which control the resonator, you can change the timbre. This is not possible with a conventional sample at all. It is easy to design resonators which behave like strings, tubes, cones, flared tubes, drums and even customised ones. Most will have a combination of band-pass or low-pass response, combined with one or more narrow peaks or notches.

The Technics WSA1 keyboard, released in 1995, used source-filter synthesis to produce its sounds. Although widely praised for its sound, there were no follow-up instruments using the same technique.

Although this may sound like the S&S 'pre-packaged' sample concept, in fact, the combination of driver 'sample', coupling and resonator produces sounds which can change their harmonic content much more than any S&S sample which can be merely filtered. Remember that this is not a physical modelling instrument, although it is similar in some respects: especially the coupling section. What you lose is the transition between notes and the behaviour outside of the basic sound generation: so whereas an instrument based on physical modelling will move from one note to another in much the same way as a real instrument, one using source-filter synthesis will merely play two notes, one after the other. This is most noticeable for brass sounds, where a physically modelled instrument like the Yamaha VL1 will exhibit the characteristically 'overblown' brassy natural series of notes when the pitch is changed with the pitch-bend control, whilst a source-filter synthesis instrument will merely bend the note.

It remains to be seen whether source-filter synthesis is going to reappear in the future as a major method of synthesis or it will remain as part of the hybrid digital synthesis methods (see Section 5.7) or even just part of the tools used to produce the inharmonics and transient samples used in S&S synthesizers to augment the basic instrument sounds.

5.3.2 *Physical modelling*

The 'physical modelling' technique uses DSP chips to create a mathematical model of how some real musical instruments work. Instead of the conventional 'source and modifier' approach used by many S&S instruments, where a basic sample sound is modified by a filter and envelopes to produce a finished sound, a physical modelling instrument uses its internal model of an instrument to create the sound whole in one operation. Because the model covers the entire instrument, it behaves like the actual thing, and so it also produces realistic transitions between notes, not just the notes themselves. It can produce sounds which emulate the behaviour of the real thing, often with astonishing realism. But the depth of detail which is required is formidable: you need to know a huge amount about the physics of musical instrument, acoustics and mathematics and then you need to convert this into software and electronics.

The techniques and algorithms for modelling musical instruments did not reach the level of sophistication where they could be done in real time without the aid of rooms full of supercomputers until the mid-1990s, and the number of types of instrument which can be adequately described is still quite small. The future may produce additional instrument descriptions, and physical modelling will be able to utilise these, but physical modelling has so far been only a limited success. In particular, it tends to be used for minor variations on existing instruments rather than in producing new synthetic sounds. Paradoxically, it may be the very precision and detail that is required to produce a physical model that prevents it from being a user-programmable synthesis tool.

Mathematical models

Using mathematics to make models of real-world objects is common in engineering, but it is more unusual to find it used in musical applications. The underlying concept is the same for any model: you look at the inputs, outputs, their interconnections and dependencies, and then determine the equations which connect them all together. Imagine a tap and a bucket with a hole in it. Suppose that the tap can provide anything up to 10 litres of water per minute, the bucket holds 20 litres, and that the hole leaks at the rate of 1 litre per minute. Ignoring the leak, the fastest time taken to 'fill' the bucket by the tap (when full on) is the time it takes for the tap to provide 20 litres of water, which would be 2 minutes (that is 20 litres at 10 litres per minute = 2 minutes) (Figure 5.3.2).

When the effect of the hole is taken into account, the figures change correspondingly. In the 1st minute, 1 litre of water will escape out of the hole, and so only 9 out of the 10 litres supplied by the tap will be in the bucket at the end of the 1st minute. During the 2nd minute,

Figure 5.3.2 This 'bucket' diagram shows the power of a mathematical model in predicting the behaviour of a real-world system. Physical modelling uses much more complex models of musical instruments to produce sounds.

10 litres per minute

Bucket capacity = 20 litres

1 litre per minute

another litre of water leaks away and so there will only be 18 litres in the bucket, thus it will obviously take slightly longer than the original estimate of 2 minutes because the tap will still need to provide just over 2 more litres of water …

By using this simple 'tap and bucket with hole' model, it is possible to make several other deductions based on how the system works. For example, if the tap supplies less than 1 litre per minute, then the bucket will never fill up because the hole leaks at 1 litre per minute. When there are 20 litres in the bucket, then it will begin to overflow and if you subtract the 1 litre per minute leak, then the overflow rate is the tap supply rate (from just over 1–10 litres per minute) minus the leak rate; so for the tap fully on, the bucket will overflow after just over 2 minutes have passed, and the overflow rate will be 9 litres per minute.

As you can see, with just simple calculations we can make some quite complex predictions about the way that the real-world works. The models of how musical instruments work which are used in physical modelling are obviously more complex than this example, but it is based on the same principles: you measure what happens, produce a description of what is happening, and then you use this information to work out what will happen.

Model types

Physical modelling synthesis falls into two distinct areas: continuous and impulsive:

- **Continuous models** deal with blown or bowed instruments, where there is a continuous transfer of energy into the instrument from the air flow, or the bow. The sound which is produced thus carries on as long as the energy is transferred. Typical examples include a trumpet and a violin.
- **Impulsive models** are for plucked or struck instruments, where a sudden 'impulse' of energy is transferred to the instrument, which then produces a sound as it responds to this input. The sound decays away naturally since energy is lost as friction, sound and movement once the initial input is taken away. Typical examples include a piano and a snare drum.

Apparently continuous events which are actually discrete are more common than many people expect. A narrow stream of water from a tap may appear to be continuous, but high-speed cameras show that many are formed from many individual droplets of water.

In some modelling terminology, the drivers are referred to as the excitation signal.

Sometimes, the distinction between a continuous and an impulsive model is not immediately obvious. In the case of a violin, the bow scraping on the string transfers energy to the string because it is rough, and the string catches on the rough surface of the bow and is pulled away from its rest position, then is released when the tension in the string exceeds the friction, and the string then jumps back to its original resting position. Each of these tiny movements of the string is an impulse, but they happen quickly enough to have much the same effect as a continuous transfer of energy.

For continuous models, the two major parts of most blown/bowed musical instruments are: the bit that you blow or move; and the part that vibrates. In a reed instrument, air is blown into a mouthpiece, whilst for a trumpet the lips move and control the flow of air. For a stringed instrument, the bow scrapes across the string. In all of these, the player is forcing the instrument to make a sound; hence these are called *drivers*, just as in source-filter synthesis. In contrast, the air inside a saxophone or trumpet vibrates inside a tube and so makes a sound; or the string vibrates and moves the air around to make a sound, and these make up the resonator part of the model. Whereas in a real instrument there are normally fixed combinations of drivers and their corresponding resonators, with physical modelling a reed type of driver feeding into a string-type resonator is entirely possible, even though a real-world equivalent would be difficult to construct.

The drivers in continuous models transfer energy into the resonator, but in order for this to be converted into a sound, the energy needs to be converted from a steady stream into a repeated cycle of variations in air currents to produce a sound. In the case of a violin, the bow rubbing against the string produces vibrations in the string, and the resonator formed by the string and the body of the violin then reinforces some vibrations, and dampen others. For a stream of air in an oboe, the opening and closing of the reed produces a stream of air which varies in pressure, and the resonator formed by the tube and holes in the oboe reinforces some of the variations, and dampen others. The driver specification thus needs to take into account how these initial vibrations are produced, and how they are coupled to the resonator.

The 'Karplus–Strong' (Karplus and Strong, 1983) plucked string algorithm is just one example of many impulsive models. This algorithm uses a damped resonator and a step input of energy to simulate what happens when a string or bar is plucked or struck. The resonator produces a note at its resonant frequency, with additional harmonics caused by its other resonances, and the decay of the sound occurs because the resonator has no source of power except for the initial input of energy. So as the energy leaks away, the sound decays. The way that the resonator loses energy, and the way that it produces the sound output are critical to the harmonic content and the way that it changes with time. The damping depends on the way that the string or bar is mounted or supported, whilst the mass of the string or bar, the tension in the string and the dimensions of the bar can all affect how the sound changes with time.

The Karplus–Strong algorithm is simulated by using a time delay to model the movement of waves along the string or bar. The reflections at the end of the bar or string are set so that some energy is removed from the wave, and so the reflected wave is reduced in amplitude. The initial step input can be just a sudden change in level, but it can also be a brief pulse of noise. More complex models may also take into account more details of the initial input of energy, which need not be a sudden step input of energy, but may have an 'envelope' and other characteristics which affect the way that the energy is transferred to the resonator.

The hammer of a piano is one example of the complexity of characteristics that needs to be considered in an impulsive model: the hammer is accelerated by the piano action and hits the string. It then moves the string away from its rest position, but this is cushioned by the felt, so the transfer of energy does not happen instantaneously. Whilst the felt is being compressed by being pressed against the string, the string itself is starting to vibrate. The hammer continues to move the string away from the rest position until the tension in the string is equal to the force expended by the hammer, and the string then moves back towards its rest position, and the hammer bounces off it. This is not a simple 'step change' transfer of energy to a resonator, but a coupled system where the string is part of the driver and the resonator; and the felt acts to smooth the transfer of energy to the string both when the hammer hits the string and bounces away from the string.

Practicalities

The complexity of the mathematical models which have been used so far in physical modelling synthesis have been such that the manufacturers of commercial units have usually chosen to present a number of fixed preset instrumental sounds. The user cannot program these sounds, other than changing their response to performance controllers and changing some modifiers. Although this is very different to most previous synthesizers, it is exactly how real instruments are treated – you do not take a drill to a saxophone and try making holes in the metalwork! Instead, you use the mouthpiece to control the sound through a combination of air pressure, lip pressure, throat resonance, vocal cords and your tongue.

The ordinary household bath can be used to illustrate how a digital waveguide works. Having filled the bath to about half capacity, a hand is used to cyclically move the water back and forth by a few centimetres at a frequency of approximately 1 Hz at one end of the bath. Some experimentation on the frequency of movement will be needed, but when the correct frequency is reached, then the

The models which have been used in the initial physical modelling instruments are complex enough to provide exactly the sort of subtle and expressive control over timbre and pitch that you would expect from a real instrument. And there appears to be quite a lot of scope for modelling a wide range of instruments, but several academic papers have commented that there are only good models for a limited number of real instruments, and that much more research still needs to be carried out. Digital waveguides crop up several times in the research literature of physical modelling, and are a very computationally efficient way of simulating a resonator pipe or string by using DSPs, and can be used in continuous and impulse models. A digital waveguide is essentially a delay line that has one or more time taps for feedback ripples or waves in the water will travel along the bath, bounce back from the far end, and return to the end where the hand is still moving the water. At the right frequency, the returning ripples or waves will reinforce the ripples generated by the hand, and the size of the ripples or waves will increase. The movement of the hand should be stopped before the waves are large enough to go over the side of the bath.

from the output to the input, and where the input is not a conventional audio signal, but a driver signal consisting of a series of shaped pulses.

Digital waveguides are used in different way to produce different types of resonator. Simple tube-based instruments can be modelled with a simple waveguide for the tube, but often require complex driver models. Stringed instruments can be modelled with two waveguides: one for each side of the point where the string is plucked or bowed. Brass instruments can be modelled with several linked waveguides for the exponential horn.

Physical modelling can require a large amount of data to specify a specific timbre. For example, the Yamaha VL1 duophonic 'virtual acoustic' synthesizer uses 387 Kbytes to store 128 patches, which is roughly 3000 bytes per patch. For comparison, a DX7 FM patch uses only 155 bytes, and only 128 bytes in the compressed form! Even so, the size of the VL1 file is still tiny compared to the size of a sample in an S&S synthesizer, where about 88 Kbytes of storage are required for each second's-worth of sample.

Controlling the instruments provided by a physical modelling synthesizer can be difficult because of the large number of parameters which may need to be manipulated. Keyboard control is useful for pitch and velocity control, but it is not as natural and interface as a wind controller. Keyboards have the disadvantage of a naturally polyphonic keyboard, whilst a blown instrument is normally monophonic. Unfortunately, despite the advantages of a wind-instrument-like controller, the keyboard has still appeared on the first generation of commercial physical modelling instruments. Blowing can be easily simulated by using a breath controller, but lip or bow pressure, muting or string damping are less obvious, and foot controllers, velocity and after-touch can be used, although it requires practice for a keyboard player to become familiar with the use of additional controllers.

Experimentation

It is not necessary to have sophisticated digital workstations to experiment with physical modelling synthesis. Using conventional recording studio equipment, it is possible to try out the underlying principles for real. All that is needed is an audio delay line (Figure 5.3.3) with a

Figure 5.3.3 A delay line can be used as the basis for experimentation into physical modelling using analogue audio equipment.

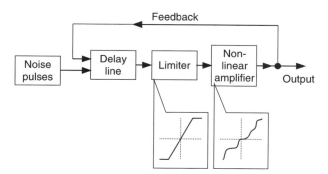

few milliseconds of delay (almost any effects processor with an echo or delay setting will do), a limiter or compressor/limiter, a noise generator (or synthesizer with noise generator) and a non-linear amplifier (or a dynamics processor or a fuzz box). The non-linear amplifier is an analogue equivalent of the waveshaper described earlier in this chapter – almost any operational amplifier (op-amp) can be used to provide this function (Clayton, 1975).

The basic idea is to connect the output of the delay line to the limiter, the output of the limiter to the non-linear amplifier, and then the output of the amplifier back to the input of the delay line. The noise generator should be mixed into the input of the delay line as well. The output of the delay line also serves as the output of the system (*Sound on Sound*, February 1996). By adjusting the feedback and injecting pulses of noise into the system, it should be possible to get percussive sounds which decay away as per Karplus–Strong synthesis, whilst with higher levels of feedback, sustained continuous tones should be produced, whose timbre can be changed by adjusting the non-linear amplifier settings. By sampling the results into a sampler, some of the more interesting or useful timbres can be stored for future use.

Notice that the amount of delay is inversely proportional to the pitch. The minimum delay time thus determines the highest pitch which can be produced. Also notice that the delay time needs to be very precisely controllable to produce specific pitches. For example, a 440-Hz note requires a delay of 2.2727 recurring milliseconds. Table 5.3.1 shows the relationship between time delays and frequency for this experiment.

Summary

Physical modelling is just one of many possible methods of digital synthesis based on sophisticated software rather than just DSP hardware. It can produce expressive, astonishingly 'real' feeling instrument sounds, and this can apply even to the impossible synthetic ones extrapolated from the models. In common with other synthesised sounds, these are not a replacement for real instruments, more a whole new set of them.

Physical modelling technology began to appear in a range of products in the mid-1990s. Technics produced a source-filter-based physical modelling synthesizer in 1995, whilst Yamaha and Korg produced several physical modelling products, and MediaVision produced a PC card using physical modelling techniques. These were the first examples of physical modelling in commercial instruments, and whilst successful, they were limited in the instruments that they could model, and the lack of user control meant that they were seen in many ways as being the equivalent of samplers that could only replay the sounds of a few instruments. Whilst this replay was very good, and in many cases better than a sampler in terms of performance accuracy, the limitations were not appealing. When the first physical modelling instruments appeared, they were expensive and monophonic or duophonic, whereas the first source-filter synthesis instruments which appeared were polyphonic for about the same

Table 5.3.1 The relationship between time delays and frequency

Delay time (milliseconds)	Frequency (Hz)	Delay time (milliseconds)	Frequency (Hz)	Delay time (milliseconds)	Frequency (Hz)	Delay time (milliseconds)	Frequency (Hz)
		3	333.33	6	166.66	9	111.11
0.1	10000	3.1	322.58	6.1	163.93	9.1	109.89
0.2	5000	3.2	312.5	6.2	161.29	9.2	108.69
0.3	3333.33	3.3	303.03	6.3	158.73	9.3	107.52
0.4	2500	3.4	294.11	6.4	156.25	9.4	106.38
0.5	2000	3.5	285.71	6.5	153.84	9.5	105.26
0.6	1666.66	3.6	277.77	6.6	151.51	9.6	104.16
0.7	1428.57	3.7	270.27	6.7	149.25	9.7	103.09
0.8	1250	3.8	263.15	6.8	147.05	9.8	102.04
0.9	1111.11	3.9	256.41	6.9	144.92	9.9	101.01
1	1000	4	250	7	142.85	10	100
1.1	909.09	4.1	243.9	7.1	140.84	10.1	99
1.2	833.33	4.2	238.09	7.2	138.88	10.2	98.03
1.3	769.23	4.3	232.55	7.3	136.98	10.3	97.08
1.4	714.28	4.4	227.27	7.4	135.13	10.4	96.15
1.5	666.66	4.5	222.22	7.5	133.33	10.5	95.23
1.6	625	4.6	217.39	7.6	131.57	10.6	94.33
1.7	588.23	4.7	212.76	7.7	129.87	10.7	93.45
1.8	555.55	4.8	208.33	7.8	128.2	10.8	92.59
1.9	526.31	4.9	204.08	7.9	126.58	10.9	91.74
2	500	5	200	8	125	11	90.9
2.1	476.19	5.1	196.07	8.1	123.45	11.1	90.09
2.2	454.54	5.2	192.3	8.2	121.95	11.2	89.28
2.3	434.78	5.3	188.67	8.3	120.48	11.3	88.49
2.4	416.66	5.4	185.18	8.4	119.04	11.4	87.71
2.5	400	5.5	181.81	8.5	117.64	11.5	86.95
2.6	384.61	5.6	178.57	8.6	116.27	11.6	86.2
2.7	370.37	5.7	175.43	8.7	114.94	11.7	85.47
2.8	357.14	5.8	172.41	8.8	113.63	11.8	84.74
2.9	344.82	5.9	169.49	8.9	112.35	11.9	84.03

price. Unfortunately, source-filter instruments did not seem to be a huge advance on S&S with a simple audition, and S&S had the advantage of a simple and familiar control metaphor. Physical modelling had limited polyphony, and either preset sounds, or sounds with very restricted ranges of variation.

5.3.3 Analogue modelling

Low-cost analogue modelling reflects the low entry cost and excellent support that now exists for programming DSP chips like the Motorola 56000 series. One analogue modelled synthesizer recently cost less to purchase than a mid-range digital versatile disk (DVD) player.

Analogue synthesizers are a mixture of the mathematics (waveforms) with electronic engineering (filters), and underneath, both are just numbers turned into voltages and circuitry. So if physical modelling is complex, then analogue modelling (also known as virtual analogue) is merely a matter of converting analogue circuits into software. And after a slow start, with the Clavia Nord Lead taking the early lead, just about everyone else seems to have played catch-up and succeeded. The last years of the 20th century saw analogue modelling gradually gaining popularity, and the 21st century has seen analogue modelling become very widely implemented, with some examples at very low cost indeed.

Simple 'two-oscillator, low-pass VCF, twin envelope with VCA and LFO modulating everything' type analogue synthesizers are available in 2003 as synthesizers, as tabletop units, as modules, on small plug-in cards, and in software to run on general-purpose computers as plug-ins. In order to find a differentiator, the manufacturers have explored morphing between sounds, adding FM, feedback around the signal path, sample playback, complex modulation routings and controllers, subtle distortion and noise to mimic the limitations of the original analogue circuitry, and more. There is considerable attention to the details of implementation.

In 1995, the Clavia Nord Lead provided a very standard analogue mono-synth type of synthesizer, but in four-note polyphony, and with a distinctive red case. Korg's Prophecy added a number of additional physical models, and let the programmer mix analogue modelling with FM with S&S with physical modelling simultaneously, but in a mono-synth. Two years later, Korg's Z1 provided the same type of sound generation as the Prophecy, but in a 12-note polyphonic synthesizer. The Z1's architecture allows you to combine sound modules to produce the final sound. The modules include: a two VCO, VCF, VCA analogue synthesizer; a comb filter; variable phase modulation, also known as FM; ring modulation; oscillator sync; a resonant filter bank; additive synthesis; an electric piano physical impulsive model; a reed physical continuous model; a plucked string physical impulsive model; and a bowed string physical continuous model. Of the major synthesizer manufacturers, Korg seem to have the broadest range of modelling capability in production instruments, and this is probably due to their investment in their Open Architecture SYnthesis System (OASYS) development system, which is the basis for the development of their modelling technologies.

The sounds of analogue modelled instruments are close emulations of analogue synthesizers. The controls are the same, and whilst the early implementations had noticeable stepping or quantisation as some of the control knobs altered the modelling values, the 2003 models behave like an analogue. Where things are different, it is in the additions made possible by digital modelling. FM or cross-modulation of analogue VCOs exposes every slight non-linearity or lack of tuning or scaling match, whilst on a modelled synthesizer the results are predictable and consistent.

There are two very different types of oscillator which are used:

- **Waveform playback**, where a sample of the analogue waveform is replayed.
- **Oscillator modelling**, where the oscillator itself is modelled mathematically.

The waveform playback is simpler to implement, but suffers from a number of problems: the sample itself is not perfect, and so any unwanted noise or frequencies will be pitch-shifted as the waveform is played back at different pitches, which gives a characteristic 'pitched buzz and noise' effect. Oscillator modelling requires more careful study of the source oscillator's fine detail in terms of how it performs when outputting various pitches, but produces more consistent results at different pitches.

Modelled filters have a similar division into 'perfect' mathematical filters that behave as the theory suggests, and modelled filters which reproduce the behaviour of real-world filter circuits. A hybrid technique also exists where a 'perfect' filter is deliberately degraded by adding noise to the cut-off frequency control, the feedback circuitry or to the resonance control so that the stability of the filter is compromised. Most analogue synthesizers had a low-pass filter, with either a 12- or 24-dB/octave cut-off slope. In 2003, modelled synthesizers have the capability to model different types of analogue filters from many of the manufacturers of the 1970s: that is, 30 years of progress in a selection from a menu.

It is now clear that modelling represents the same sort of technological leap that the GS1/DX7 did in the early 1980s, when analogue synthesizers were replaced by digital FM-based ones almost at a stroke. But it is not physical modelling that has changed things. Modelling of analogue synthesizers has been the dominant growth area in the early years of the 21st century, with true analogue (also sometimes known as 'pure' analogue) now seen as an expensive luxury, and physical modelling seen as a very specific solution for producing real-sounding instruments.

The adoption of analogue modelling is reflected in the terminology used in commercial synthesizer adverts. Modelling has come to mean the modelling of analogue synthesizers, where 'analogue' is implied, whilst 'physical modelling' is a specific and different type of modelling. Physical modelling's role could almost be seen as showing that it was possible to use DSP chips to create musical sounds with modelling techniques, and this then opened the way for the modelling of analogue synthesizers.

5.4   Granular synthesis

Granular synthesis is used to be regarded as an unusual technique. Unlike many of the other methods of synthesis described so far, it has not been used in commercial synthesizers, although it has been used by some composers working in the academic and research fields. It does not fit into the source and modifier model, but instead approaches the production of sound from a bottom-up point of view, which is very different to most other methods of sound synthesis. But software synthesis has opened up new opportunities for otherwise obscure techniques for making sounds, and granular synthesis is now available as software for use on computers within commercial sound creation programs. Reason, from Propellerhead software in Sweden, is one example.

Granular synthesis builds up sounds from short segments of sounds called 'grains'. In much the same way that many pictures in colour magazines are made up from lots of dots, granular synthesis uses the tiny sound fragments to produce sounds. The grains are of very short duration: 20 or 30 milliseconds, which is close to the 10–50-millisecond timing 'resolution' of the human hearing system; audio events which occur closer together than this tend to be heard as one event instead of two. The controls are relatively straightforward: the number of grains in a given time period, their frequency content and their amplitude are the major parameters. The difficulty lies in controlling these parameters: rather like the large number of parameters in additive synthesis, manipulating a large number of grains requires envelopes, function generators and other controllers, and can become a very large overhead.

Grains are normally enveloped so that they start and finish at zero amplitude, so that sudden discontinuities are avoided; any sharp change in the resulting waveshape would create lots of additional unwanted harmonics and the result would sound like a series of clicks. Grains may contain single frequencies with specific waveforms, or band-pass filtered noise, and each grain can be different. In some ways, granular synthesis can be considered as the limiting case of wavetable synthesis, where the table of waveforms is swept very rapidly to give a constantly changing waveshape; but few wavetable synthesizers have the control of wavetable selection and the zero-crossing smoothly enveloped grains that are found in granular synthesis. In fact, granular synthesis is normally produced by software, and so the grains can be produced using a number of techniques from additive sine waves to filtered noise or even processed samples of real sounds. Some experimenters have worked on coupling granular synthesis with mathematical systems like chaos theory, John Conway's 'life', and fractals (Figure 5.4.1).

Granular synthesis seems to be somewhat analogous to the way that film projectors work. By presenting a series of slightly different still images at a rate which is just about the limit of the eye's response to changes, the impression is one of a smooth continuous movement. In granular synthesis, the rapid succession of tiny fragments of spectra combines into an apparently continuously changing spectrum. This constant change of grains is reflected in the timbres which are produced by granular synthesis: words like 'glistening' or 'shimmering' are often used to describe the complex and busy sounds which can result, although the technique is also capable of producing more subtle, detailed sounds too. As digital synthesizers have become increasingly

Figure 5.4.1 Granular synthesis uses small 'grains': short segments of audio which are arranged in groups. The contents can be waveforms, noise or samples. The major controls include the number of grains, their lengths and their repetition rate.

software-based, granular synthesis has become one of a number of synthesis techniques that are offered in commercial software-based plug-ins, and maybe the future will see it appearing in real instruments. Despite several attempts to produce a musically and commercially acceptable computer with a music keyboard for stage use, there is still a gap between what can be achieved on a computer and on stage.

5.5   FOF and other techniques

One example of 21st century programmability is the Cameleon from Spanish company Soundart. This is a rackmounting DSP engine which can be configured via MIDI system exclusive dumps. It is a general-purpose audio box, and it is completely programmable: it can be an effects unit, a poly-synth, a mono-synth, amplifier emulation and more. The manufacturer provides extensive support for developers via the Internet, including lots of documentation, including some examples from Motorola on how to program 56000 series DSP chips as sine wave generators, or as 10-band stereo graphic equalisers. There is even a Soundart tutorial on programming a complete mono-synth.

Mass-market digital synthesis technology first appeared with the Yamaha DX7 in 1983. After a pause whilst the other manufacturers looked around for other viable methods of digital synthesis, the additive and S&S instruments began to appear. Over the next 10 years, S&S gradually took over until by the early 1990s, it was virtually the only digital method of synthesis. After such a slow and steady development over 10 years, the mid-1990s marked a sudden change when a number of sophisticated instruments were released which could utilise combinations of additive, subtractive and FM synthesis, and these were soon joined by instruments based on physical modelling techniques. It is strange that commercial S&S instruments have not been joined by the large number of techniques which are still used in academic research. Since digital techniques are making it increasingly easy to implement these alternatives, then maybe the problem is the metaphor used for the representation. Analogue modelling has been very successful, perhaps because it has presented exactly the same user interface and programming model as that of the analogue synthesizers of 30 years ago.

This section looks at some of the synthesis techniques which may well be incorporated into the digital synthesizers of the near future. They all have a common theme, which is derived from a combination of research into musical sounds, acoustics and human speech and singing. Many are the result of a fusion of the worlds of telecommunications, computing and music.

5.5.1   Formants

All of these methods are focused around the sounds which are produced by strong resonances, wherever you get a fixed set of 'formant' frequencies (see also Section 2.4.4). The human voice is one example of this sort of system – the mouth, nose and throat can be thought of as a complicated tube-like arrangement where particular frequencies are emphasised whilst others are suppressed, so the resulting frequency response is a series of peaks. The vocal cords produce a spiky pulse-like waveform which has lots of harmonics in it, and this is then processed by the vocal tract (the mouth, nose and throat) which acts as a filtering mechanism. The result of the filtering is to produce an output which contains predominantly those frequencies, from the original pulse sound, which match the resonant peaks of the filter. Since you can only make minor changes to the physical shape of the tubing formed by the mouth, nose and throat (e.g. changing the size and shape of your mouth cavity with your tongue), then the peaks are mostly fixed, and so what comes out is a set of harmonics which have peaks which are fixed by the formant frequencies, regardless of the pitch of the note being sung! The only things that do change are the fundamental and the underlying harmonics (Figure 5.5.1).

This can be regarded as another type of 'source and modifier' model, where the source is the vocal cords, and the modifier is the filter or resonator formed by the mouth, noise and throat. The vocal cords can be emulated by using a short burst of sound whose frequency is fixed, and then by triggering this at the rate of the fundamental frequency that you want to produce. The pulse repeats, producing the harmonics associated with the fixed resonances of the formants that it represents, whilst the pitch that you hear is the repetition rate. The modifier part can be emulated by combining several band-pass and notch filters; although since changes of the shape of the 'tube' can happen, these filters need to be dynamically changeable in real time. In fact, the human ear is very sensitive to exactly these changes in formant structure.

Instruments exhibit the same sort of formant structures: the analogy between the human vocal apparatus and some of the woodwind and brass instruments is probably the strongest. The abstraction of a source of sound connected to a 'resonant set of formants' acting as a modifier

Figure 5.5.1 Formants are peaks in the frequency spectrum of a sound. This example shows two large peaks in the output spectrum, regardless of the spectrum or frequency of the input.

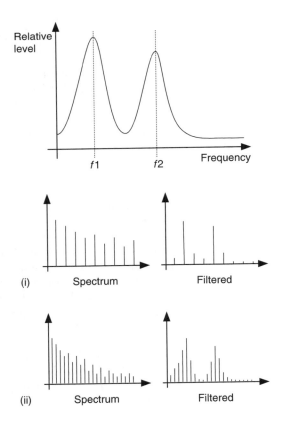

can be applied to almost any instrument. For string instruments the formants are determined by the string characteristics, its mountings and the structure of the body of the instrument. For some instruments other external factors can be very important: an electric guitar is designed to provide a rigid support for the vibrating string, and the heavy wooden body is not a very strong resonant system. But the combination of the guitar string, amplifier, speaker, speaker cabinet and feedback between the acoustic output and the guitar pickups forms a very complex resonant system which is often exploited to great effect in live performance. In contrast, synthesizers and most other amplified musical instruments tend to be used as self-contained systems, and the amplification is merely used to make them louder.

5.5.2 *Vocoder*

Finding more efficient ways to transmit human speech along wires has been one of the major activities of telecommunications research for many years. Most of the raw information content of speech can be found between 300 and 3400 Hz, and so telephone systems are designed with a bandwidth of about 3 kHz. Frequencies outside of this range add to the clarity and personality of the voice, which is why it is difficult to distinguish between an 's' and an 'f' on the telephone, or why people may sound very different in real life to hearing them over the telephone.

Research at the Bell Telephone Laboratories in New Jersey, USA, in the early 1930s, was looking at how different parts of this 3-kHz bandwidth were used by speech signals. By using band-pass filters, the speech could be split into several separate 'bands' of frequencies, and the contribution of each band to the speech could then be determined. By using an envelope follower, the envelope of the contents of each frequency band could be determined. Once split into these bands, the audio signal could be mixed back together again in different proportions, and even have new envelopes applied to each band. As basic research into the properties of speech, the results were interesting (you need all of the 3-kHz bandwidth: removing bands alters the timbre of the speech too radically to be useful for telephony), but they had no practical application at the time. It was not until digital processing techniques became available in the 1960s and 1970s that vocoders were to find reuse in telecommunications.

The band-pass filters are similar to the graphic equalisers that are found in applications as diverse as recording studios and car radios.

But the vocoder proved to be a powerful tool for processing audio signals. By splitting an audio signal into separate bands, analysing the contents and then allowing separate processing of these bands, it allows sophisticated control over the timbre of the sound. More importantly, by separating the analysis and processing functions of the vocoder, it is also able to extract the spectral characteristics of one sound and apply them to another (Figure 5.5.2). The fidelity with which this can happen depends on both the number of bands and the characteristics of the envelope followers. As the bandwidth of the bands decreases, more filters are required to cover the audio spectrum. For 'octave' bands, each covering a doubling of frequency, only

261

Figure 5.5.2 A vocoder is made up of two parts: analysis and synthesis. The analysis section converts the incoming audio signal into frequency bands and produces a control voltage (CV) proportional to the envelope of the contents of that frequency band. The synthesizer section has identical band-pass filtering, but this time it acts on a different audio signal. Each band is controlled by a VCA driven from the analysis section. The characteristics of the analysed signal are thus superimposed on the synthesised signal. Although this diagram shows analogue blocks, implementing a vocoder is now easier in digital circuitry or on a DSP chip.

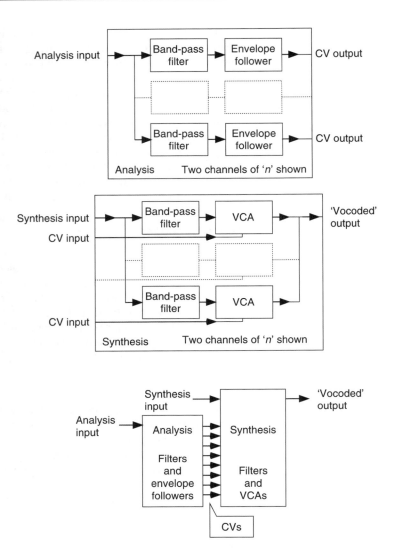

eight filters are required: six band-pass, one low-pass and one high-pass. This produces only a coarse indication of the spectral content of the audio signal which is being analysed, and correspondingly coarse changes to the signal which is being processed. For 'third-octave' bands, 30 or 31 filters are required, and the resulting finer resolution significantly improves the processing quality. The envelope followers determine how quickly the spectrum can be imposed on the processed signal: if the time constant of the envelope follower is too long then the bands will not accurately follow the changes in the signal which is being analysed, whilst if the time constant is too short then the controlling of the amplitude of the bands can become noticeable.

Vocoders began to be used to process musical sounds in the 1950s. The basic vocoder structure had some features which were specific to processing speech, most importantly the voiced/unvoiced detection.

This determines if the speech sound is produced by the vocal cords or by noise. Voiced sounds are produced by the vocal cords, and modified by the resonant filter formants in the mouth, nose and throat: 'ah', 'ee', 'mm' and 'oh' are examples of voiced sounds. Unvoiced sounds are modifications of noise produced by forcing air through gaps formed by the mouth, tongue, teeth and lips: 'sh' and 'f', 't' and 'puh' are examples of unvoiced sounds. Many vocal sounds are combinations of these two basic types: 'vee', 'kah' and 'bee' have a mixture of noise and voiced parts. The noise tends to be wideband, and so can be detected by looking for a simultaneous output in many bands of the analysis filters. In order to produce intelligible speech in the processing section, a noise signal needs to be substituted for the audio signal when an unvoiced sound is detected.

With this emphasis on speech, the first uses of the vocoder were to superimpose the spectrum of speech onto other sounds. The processing requires a harmonically rich source of sound in order to be able to produce good results: using a sine wave will give an output which occurs only when that band is activated by the analysis section, for any other bands there will be no output. Using the voiced/unvoiced detector can be used to substitute for noise which is present in the analysed signal, but this only affects unvoiced sounds, not voiced sounds. Some military communication systems use the minimalistic technique of providing either noise or fixed frequencies in the bands for the processing section. The only information that then needs to be transferred along a communications line is the parameters for the bands and the voiced/unvoiced detection. This results in a very robotic sound which has high intelligibility but almost no personality. Using a vocoder to superimpose the spectral changes of speech onto music instruments has a similar effect: the output has a robotic quality and sounds synthetic. This has been used for producing special effects like singing pianos, laughing brass instruments and even talking windstorms.

Implementing large numbers of filters in analogue circuitry is expensive, and so analogue vocoders tend to have restricted numbers of filters, whereas digital vocoders can have much finer resolution. Digital vocoders can also extract additional information about the audio signals in the bands, and the 'phase vocoder' is one example – it can work with narrow, high resolution bands and can output both amplitude and phase information, which improves the processing quality and enhances the creative possibilities for altering musical signals.

5.5.3 *VOSIM*

VOSIM is an abbreviation for VOice SIMulation, and uses a simple oscillator to produce a wide range of voice-like and instrumental timbres, although the original intention was to use it for speech synthesis. The original hardware was developed in the 1970s at the University of Utrecht, and has since been adapted for software-based digital generation. The oscillator produces asymmetrical waveforms which are made up of repetitions of a series of raised sine-squared

Figure 5.5.3 VOSIM produces pulse trains with controllable pulse width, repetition rate, amplitude decay and gap width. It is similar to FOF in many ways.

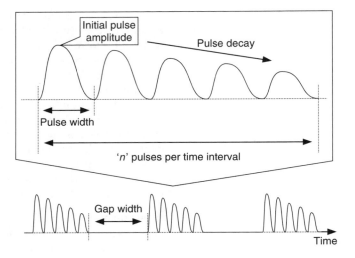

waveforms called a *'pulse train'*. The series of waveforms reduces in amplitude with time, and so only a small number of parameters are required: the width of the pulses, the decay rate of the amplitude, the number of pulses and the repetition rate of the pulse trains. Because the spectrum that is produced is dependent only on the parameters which control the pulse trains, and not on the repetition rate, the harmonic content is independent of the pitch. This is exactly the opposite of a sample playback system, and is useful for simulating the fixed formant frequencies which are found in vocal and instrumental sounds (Figure 5.5.3).

The simple controls, versatility and small number of parameters used in VOSIM are ideally suited to the real-time control requirements of a speech synthesis system. In many ways VOSIM has a similar 'minimal parameter' interface to FM, although whereas FM has been commercially successful in musical applications and has only seen limited use as a speech synthesis method, VOSIM is more suited to speech synthesis and has not been used for mass-market musical applications.

5.5.4   FOF

FOF was first developed by Xavier Rodet in Paris in the early 1980s. It is a French acronym for Fonctions d'Onde Formantique, which translates to something like formant-wave-function synthesis, and it is sometimes referred to as formant synthesis. It can be used to produce simulation of vocal-type sounds, and incorporates similar frequency splitting elements to vocoding, and the oscillators use a more complicated variation of VOSIM.

The basic idea is to generate each required formant separately, and then combine them to form the final output. Each formant 'oscillator' produces an output which deals with just one formant and instead of having an oscillator and resonant filter, it combines the effect of the filter on the oscillator output into the oscillator itself. The oscillator produces a series of pulses which are each the equivalent to what would

Figure 5.5.4 FOF produces pulses whose shape is determined by the impulse function of the sound which is required. The repetition rate determines the frequency of the sound, whilst the pulse contents determine the formants of the sound.

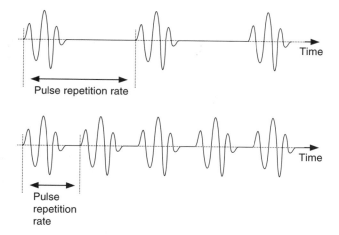

Pulse repetition rate

Pulse repetition rate

be output from the filter if a single rapid step signal was passed through it, called the impulse response of the filter. The pulse contents are thus derived from the impulse response of the filter, and if a series of these pulses is then output, the resulting sound is the same as if the filter was still processing the original step signal. More importantly, the rate of outputting these pulses can alter the frequency of the sound which is produced, but the filtering will remain the same, since it is the shape and contents of the pulse that determine the apparent 'filtering', not the repetition rate of the pulses (Figure 5.5.4).

The output from a typical FOF oscillator is a succession of smoothly enveloped (as in granular synthesis) audio bursts which happen at a repetition rate which is the same as the pitch of the required sound. Each burst of audio has a peak in its spectrum that is the same as the required formant frequency. If the repetition rate is above 25 Hz, then these bursts produce the effect of a single formant with spectral characteristics determined by the audio burst itself. For lower repetition rates, it provides a variant of granular synthesis. Digital implementations of FOF normally provide both FOF and granular modes, and this allows continuous transformations to be made between vocal imitations and granular textures. Each FOF oscillator produces a single formant, and the output of four or more of these can be combined to produce sounds which have a vocal-type quality.

FOF can be produced using conventional synthesizers by taking a sound which has a fast attack and decay time, with no sustain or release, and then triggering it repeatedly so that it produces a rapid series of short bursts of audio. If the synthesizer produces these short audio bursts at 100 Hz, then the fundamental frequency of the output will be at 100 Hz, but the apparent filtering of the signal will be determined by the contents of the sound itself and so changing the repetition rate will change only the pitch: the formants (filtering) will remain the same because the sound which is being repeated is also staying the same.

In MIDI terms, this usually means choosing a single note and making a simple and very short enveloped sound which has the right harmonic content and then sending note on and off messages very rapidly for just that one note, where the repeat rate sets the fundamental frequency, and thus the pitch, of the resulting sound. This is easiest to do by creating lots of messages and then changing the tempo of playback! Unfortunately, MIDI is too slow to create high frequency note repetitions. This limits the maximum frequency which can be generated using this method to monophonic sounds at just under 800 Hz under ideal conditions (see also Table 5.3.1).

Producing suitable sounds for FOF involves throwing away some of the instinctive approaches that many sound programmers have. In fact, it is not necessary to use sounds which approximate to the impulse response of a filter; all you need is a quick burst of harmonics. For simulating real instruments and voices then you need to have something which sounds like a single click processed by whatever it is you want to sound like, whilst for synthetic tones almost anything will do.

5.5.5 Dynamic filtering

There are a large number of techniques that utilise the same model of the throat, mouth, nose and vocal cords as the other methods in this section, but which approach the design from the opposite viewpoint. Most were originally developed for use in telecommunications speech coding applications, but they can also often be used to synthesise formant filter-based sounds. One of the best known is LPC, which is an acronym for linear predictive coding. LPC techniques can also be used in resynthesis to help design suitable filters. Other techniques include CELP, PARCOR and the 'Z-plane' dynamic filters used by E-mu, initially in their Morpheus and UltraProteus products, and later in many other products including samplers.

To generalise the dynamic filtering method: a digital filter is used to approximate the formants, and this filter is used to process a source waveform into the desired output. This is very different to extracting the formants and synthesising them individually, since a single multi-formant filter can produce the equivalent of several separate FOF oscillators simultaneously. The filter shape is controlled by a number of parameters, and can usually be changed in real time to emulate the changes which can occur in a real-world resonant system like the mouth, nose and throat.

5.5.6 Software

For stand-alone instruments, digital synthesis is a combination of digital hardware and software, although strictly there is usually an analogue output stage and low-pass filter connected to the output of the DAC. But it is also possible to use digital synthesis to produce sounds using a general-purpose computer. In this case the software is normally independent of any hardware constraints – the use of specialised DSP chips to carry out the DSP is often only required to improve the calculation speed. The output of such software is in the form of 'sound files'. Some of the common formats are shown in Table 5.5.1.

Table 5.5.1 File formats for sound files

Suffix	Type	Format
.aif	Audio	AIFF
.aifc	Audio	AIFF
.aiff	Audio	AIFF
.au	Audio	μ-law
.au.gsm	Audio	GSM μ-law
.avi	Data	Intel Video
.gm	Data	MIDI
.gmf	Data	MIDI
.mid	Data	MIDI
.mov	Movie	QuickTime
.mp2	Audio	MPEG Audio
.qt	Movie	QuickTime
.ra	Audio	Real Audio
.sds	Audio	MIDI SDS
.smf	Data	MIDI
.snd	Audio	SND: System Resource
.voc	Data	SoundBlaster
.wav	Audio	WAV
.mod	Data	MOD specification
.mp3	Audio	MPEG Audio
.asf	Data	Streaming format
.dls	Data	MIDI DLS

The 1996 computer platforms were the Amiga, Atari ST, Macintosh, PC and Unix. 2003 has just the Macintosh, PC and Unix/Linux.

These sound files can be used as the basis for further processing, transferred to samplers for replay, or replayed using a computer sound card or built-in audio facilities. It should be noted that in the first edition of this book, in 1996, there were at least five different types of computer platform in general use for music, and that many file types were restricted to specific platforms. In 2003, there are only three major platforms, and the file formats are almost always usable on any platform.

This software-only synthesis comes in several forms. Commercial software tends to be either simple sample editing programs, or else sophisticated audio processing software. Freeware and Shareware software is much more varied: ranging from complete digital synthesis systems to sample processing programs, although there is less emphasis on the detailed audio editing that is found in the commercial software.

5.6 Analysis–synthesis

Analysis–synthesis techniques are the basis for the resynthesizer, which takes a sample of a sound, extracts a set of descriptive parameters, and then uses these parameters to recreate the sound using a suitable synthesis technique. There are two major problems in achieving this:

- converting the sample into meaningful parameters
- choosing a suitable synthesis method.

267

Figure 5.6.1 Resynthesis takes an existing sound sample, and analyses it to produce a set of parameters. These parameters can then be edited and used to control a synthesizer which produces an edited version of the original sample.

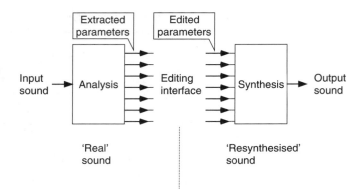

The conversion is between a sample of a sound and a set of parameters which describe that sound is not straightforward. There is also the issue of mapping those parameters to the chosen synthesis method (Figure 5.6.1).

5.6.1 Analysis

The first stage is to analyse the sample. Parameters which might be required to describe the sound adequately to allow subsequent synthesis include:

- pitch information
- pitch modulation: LFO and/or envelope
- harmonic structure
- formant structure
- envelope of complete sound
- envelopes of individual harmonics
- relative phase information for individual harmonics
- dynamic changes to any parameter in response to performance controls.

There are a number of techniques which can be employed to produce this information. Fast Fourier transforms (FFTs) are a way of transforming sample data into frequency data, and they are widely used for spectrum analysis. FFTs require considerable computation in order to convert from the time domain (a waveform) into the frequency domain (a spectrum). The detail which can be obtained from an FFT is inversely proportional to the length of the sample which is analysed. So short samples have only coarse frequency resolution, whilst long samples can have fine resolution: if a sample of 20 milliseconds is converted, then the resolution will be 50 Hz. If the harmonic content of the sample is changing quickly, then a compromise will need to be made between the length of sample which is analysed, and the required frequency resolution. Successive FFTs can move the sample 'window' in time, overlapping the previous sample, and so build up detailed spectrum information, even though the majority of the sample data is the same. An alternative approach is to use interpolation between the spectral 'snapshots' (Figure 5.6.2).

Figure 5.6.2 FFTs convert from the time domain to the frequency domain by processing blocks of samples. The larger the block of sample material, the better the resolution of the spectrum: provided that the sample material has a constant spectrum.

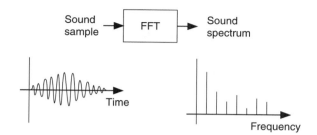

Linear predictive methods can be used for formant analysis, since they output the parameters which describe a filter which emulates those formants.

Pitch extraction employs a number of techniques in order to determine the pitch of a sampled sound. Because the perceived pitch of a sound is concerned more with the periodicity rather than the frequency of the fundamental, pitch extraction can be difficult. Methods include:

- **Zero-crossing**: The simplest is to count the number of zero-crossings, but this is prone to errors because of harmonics causing additional zero-crossings. Filtering the sample sound to remove harmonics and then counting the zero-crossings can be more successful, but a better technique is to use the peaks of the filtered sample since the harmonics have been removed and a simple sine-like waveform is all that should be left after the filtering. This method has problems when the fundamental frequency is weak, since filtering the harmonics still leaves a noisy, low-level signal.

- **Auto-correlation**: Auto-correlation is a technique which compares the waveform with a time-delayed version of itself, and looks for a match over several cycles. When a delay equal to the periodicity of the waveform is reached, then the two waveshapes will match. This assumes that the sample sound does not change rapidly and that there are no beat frequencies or large inharmonics.

- **Spectral interpretation**: Spectrum plots derived from FFTs can be used to determine the pitch. The spectrum is examined and the lowest common divisor for the harmonics shown is calculated. For example, if harmonics at 500, 600, 1000 and 1200 Hz were present, then the fundamental frequency would probably be 100 Hz. Again, beat frequencies and large inharmonics can produce significant errors with this technique, normally producing fundamental frequencies which are too low (a few or tens of hertz).

- **Cepstral analysis**: By further processing the spectrum, it is possible to produce plots which quite clearly show peaks for the fundamental. The process involves converting the amplitude axis of the spectrum into a decibel or logarithmic representation instead of the normal linear form, and then calculating the spectrum of this new shape, that is, using an FFT to treat the spectrum as if it is a waveform! The resulting 'cepstrum' (a reworking of the word 'spectrum') will show a peak in the upper part of the time or 'frequency' axis that indicates the fundamental frequency of the sound. The cepstrum merely indicates the underlying spacing of

269

Figure 5.6.3 Pitch extraction needs to be able to cope with a range of inputs: from simple sine waves (i) which can be processed by a zero-crossing method; through waveforms which change slightly from cycle to cycle (ii) where auto-correlation or cepstral analysis can produce useful pitch outputs; and finally noise (iii) where the pitch extractor should indicate that it is noise rather than a rapidly changing pitch. Although the human ear can readily achieve this, the process is less straightforward for electronics and computers.

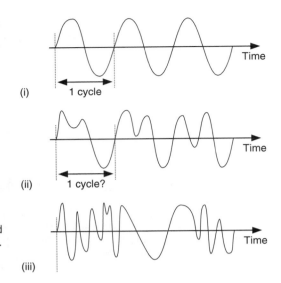

the harmonics shown in the spectrum, and so spectra with only odd harmonics, or very sparse harmonics (like a sine wave!) can be difficult to interpret because of processing artefacts that may obscure the important information (Figure 5.6.3).

Envelope following

Extracting the envelope from a sample sound is relatively straightforward in comparison to pitch extraction. The sample sound is low-pass filtered, and then a 'leaky' peak detector is used to produce a simple curve that approximates to the original volume envelope. The setting of the low-pass filtering and the peak detector decay time constant govern the effectiveness of the envelope detection. The low-pass filter should be set so that its cut-off frequency is lower than the lowest expected frequency in the input sample, but setting it too low can slow down the response of the envelope: resulting in slow attack, decay or release times.

Additional parameters

Pitch and formant analysis may also produce outputs which change with time, and so these may need to be converted into envelope format. Pitch modulation is likely to be in two parts: *cyclic modulation* (vibrato) and *time-varying* (pitch bending), and so further processing may need to be employed to separate these two parts.

Some sounds require interactions between notes to be taken into account. For example, the sympathetic vibrations that are set up in other strings on a piano when a note is played.

In order to produce a realistic sound from a resynthesizer, it is not sufficient to take a single sample of the instrument sound and analyse it. The characteristics of the sound which is being analysed may change under the influence of external parameters used in performance, or when different notes are played. There is thus a need to take into account any changes caused by performance controls and different playing pitches. One example is the change in timbre which happens when an instrument is played harder or more vigorously: hitting a piano key harder, or bowing a string with more pressure. Other examples include damping strings or muting a brass instrument. Several samples will be required in order to measure the dynamic changes to parameters which occur in response to these performance controls. Different pitches can be dealt with by making several samples of the instrument throughout its playable range. The outputs of these dynamic measurements can then be interpolated to give approximations for all notes and performance control settings.

5.6.2 Synthesis

Almost any synthesis technique could be a candidate for the synthesis 'engine' for a resynthesizer. The most important consideration is how the parameters of the technique map to the parameters that can be extracted from the sample. The mapping needs to be complete and unambiguous, but it also needs to produce a parameter set which can be manipulated by the end user of the resynthesizer.

Additive

Additive synthesis appears to offer perhaps the simplest approach to resynthesising sounds from parameters. The only parameters which are required are detailed pitch, amplitude and perhaps phase information for each of the harmonics which are present in the sample sound. Unfortunately, this is likely to be a large number of harmonics, each with complicated multi-stage envelopes for the changes in the pitch, amplitude and phase parameters with time and performance controller settings. So, although the extraction of the parameters is relatively straightforward, presenting them to the end user in a manageable form is more difficult.

Analysis–synthesis using sine waves is often abbreviated to A/S.

In 1999 Xaxier Rodet, at IRCAM in Paris, published a paper describing SINOLA, which uses a measure of the peaks in a complex spectrum as the analysis part, and combines additive synthesis with sine waves and wavetable synthesis for the synthesis part. Work at IRCAM on analysis–synthesis techniques continues in the 21st century.

FM

This modulation has a much smaller set of required parameters than additive synthesis. In this case, the problem is how to convert the extracted parameter information about pitch, amplitude and phase for each harmonic, into suitable parameters to control FM. There is no simple way to work backwards from a sound to calculate the FM parameters which produced it – a process called deconvolution. An iterative process which tests possible solutions against the given parameters might be successful, but it is likely to require considerable processing power as well as time.

Subtractive

Subtractive synthesis requires more parameters than FM, but it provides a smaller set of controls than additive synthesis. The major problems with using subtractive synthesis are the fundamental limitations of the technique: the filtering is often a simple resonant low-pass filter; and there is a limited set of source waveforms. The combination of these problems means that subtractive synthesis has a very limited set of possible sounds, and this seriously restricts the possibility of being able to resynthesise a given sound.

Formant

Formant synthesis techniques like FOF and VOSIM have small numbers of parameters, and the conceptual model is similar to subtractive synthesis. But unlike subtractive synthesis, formant synthesis techniques are not restricted to simple filtering, but can recreate complex and changing formant structures. Although the source waveforms may be simple to control, the dynamic formant filter presents a considerable problem to a user interface designer.

In fact, FOF is part of a complete software package called CHANT, written at IRCAM in Paris by Xavier Rodet and others in the early 1980s. CHANT can be used to analyse a sampled sound and extract the harmonic peaks, and then use these formants as the basis of an FOF resynthesis of the sound.

Physical modelling

Physical modelling can be considered to be a type of analysis–synthesis technique, although the analysis process is more sophisticated since it involves a study of the physics of the instrument and its sound, and then the building up of a physical model of that instrument. The synthesis part is then relatively simple – just run the model to simulate the instrument's behaviour. At the moment, the process of analysing a real instrument is a time-consuming one, although the commercial development of physical modelling may facilitate the development of software tools for this task.

5.6.3   Resynthesis

Any resynthesis technique requires a compromise between the depth of required detail to describe the original sound, and the ability of the user to make meaningful changes to the sound. There are two types of editing methods which can be used to control the resynthesis of a sound:

- **Extracted parameters**: Editing the transforms which are used to map the extracted parameters to the synthesizer parameters. This requires a good knowledge of the analysis technique.
- **Synthesizer parameters**: Editing the synthesizer parameters. This only requires knowledge of how the synthesizer produces sounds.

Because analysis–synthesis techniques tend to produce information on the spectrum of the input sound during specific time windows, then the conversion of the extracted parameters into continuous controls for the synthesizer tends to be iterative. The process requires knowledge 'analogue' feel to some of its sounds. It provides a conventional 'analogue' synthesis emulation; FM, physical modelling of brass, reed and plucked instruments; and three variations on sync/cross-modulation and ring modulation analogue emulations. To control these methods, it has a wide range of performance controllers.

- Korg's Z1 extended the Prophecy's mix of synthesis to a polyphonic version, and is available as a plug-in card for Korg's S&S instruments.
- Technics' WSA1 mixed bits of 'physical modelling' with S&S to give a simplified 'driver and resonator', source-filter synthesis instrument which had the advantage of being polyphonic at a time when other physical modelling instruments are monophonic or duophonic. It was not followed by any further models.
- Kurzweil's variable architecture synthesis technique (VAST) provides many resources; but they are more like a modular approach to an S&S synthesizer than any combination of separate synthesis techniques.
- Roland have mixed sophisticated S&S technology with sampling in the Fantom-S workstation.
- Propellerhead's Reason is a combination of a sequencer, synthesizer modules, drum machine and effects units, but implemented in software. The synthesizers include wavetable and analogue modelling, plus a granular synthesizer.
- Native Instruments' Reaktor is a software S&S synthesizer, sampler, granular resynthesizer, effects and more. Running on Mac or PC, it provides powerful soft synthesis capability, and there are hundreds of instrument definitions (ensembles) available to download.

The future seems to lie with a combination of techniques, since none of the available methods offers a complete solution. As hardware becomes more powerful, the software functionality increases and also becomes more flexible. The limits are more likely to be the user interface and the processing power, rather than the synthesis methods.

Future synthesizers are likely to be general-purpose synthesis engines which can be configured to produce a number of different techniques, although it is unlikely that any standardised way of controlling these techniques will emerge in the near future. This means that even though the synthesis methods will converge, the user interfaces and sound storage formats will not. The commercial model for producing this type of general-purpose synthesis instrument is not clear, and it may be that the internal construction is common, whilst the external appearance may be very different.

In 2003, there are a number of examples of generic DSP-based audio engines in rackmount units: Soundart's Chameleon; Creamware's Noah; Manifold Lab's Plugzilla, and Symbolic Sound Corporation's Kyma.

Hybrid instruments are thus similar to the pre-MIDI analogue instruments – 'closed' systems where interconnecting synthesizers were not possible without sophisticated hardware. With complex software-based synthesis, the possibilities for interfacing become more remote which is very useful for commercial synthesizer manufacturers, but not as good for users.

5.8   Example instruments

Casio CZ-101 – waveshaping (1985)

The Casio CZ-series of synthesizers are one example of a commercial use of a full waveshaping implementation to produce sounds. Although it is called *'phase distortion'*, it uses waveshaping, but presents it in a way which is intended to present it in a way that emulates the operation of an analogue synthesizer. Two DCO oscillators provide the raw pitched sound source, and two parallel sets of modifiers follow. Each DCO has a separate EG for controlling its pitch, although vibrato is provided by a single LFO. The DCO output passes through the digitally controlled waveshaper (DCW), again with an associated EG, and finally through a digital VCA or digitally controlled amplifier (DCA). Ring modulation and noise can also be added. By using just one of the two sets of DCO and modifiers, the polyphony is doubled.

The DCW or waveshaper behaves and sounds much the same as the VCF found in an analogue synthesizer. As the control value increases, harmonics are gradually added to the sine wave, so that it changes into one of the eight waveforms, and this can be controlled by an EG as well as tracking the keyboard note. This implies that the transfer function is changing dynamically, which would suggest that a great deal of complex processing is being carried out. However, by working backwards from the waveform, it is possible to work out what is really happening. Because waveshapers tend to add harmonics, not take them away, then the only way that a sine wave can be produced is if the basic waveform at the input to the waveshaper is a sine wave. The waveshape selection is thus used to change the transfer function of the waveshaper, not the waveform produced by the DCO. The waveshapes shown represent the final output of the waveshaper when the full range of the transfer function is being used. The waveshapes which are provided reinforce this: the sawtooth, square and pulse shapes are joined by a 'double sine', 'saw pulse' and three 'resonant' waveshapes (Figure 5.8.1).

Figure 5.8.1 The Casio CZ-101 uses waveshaping, but presented to the user as a DCO followed by a DCW (a digitally controlled waveshaper). The waveshapes provided include the sawtooth, square and pulse waves of conventional synthesis, plus five more unusual ones.

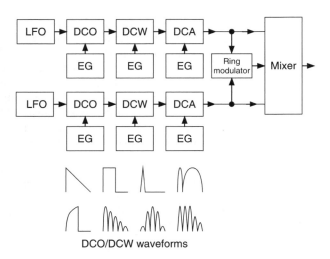

DCO/DCW waveforms

Roland JD-800 (1991)

The JD-800 is a 24-note polyphonic S&S synthesizer with CD-quality samples, and an intriguing user interface: nearly 60 sliders and nearly 60 buttons, lots of light-emitting diodes (LEDs), two liquid crystal displays (LCDs) and one LED display. Each slider is dedicated to a single function: reminiscent of early analogue synthesizers.

Yamaha SY99 and SY77 (1991, 1990)

The real-time convolution modulation (RCM) synthesizers incorporate advanced versions of Yamaha's FM and Sampling (AWM) technologies, as well as a way of using samples inside FM called RCM. Both methods incorporate detailed control over the source and modifiers: resonant filters can be used to process the samples and the FM, which makes the FM synthesis more powerful since dynamic timbral changes are not only controlled by the modulator envelopes. The built-in effects sections provide a wide range of chorus, reverb, EQ and echo effects. The SY99 also provides user random-access memory (RAM) for storing samples which can be loaded from disk or via the MIDI Sample Dump Standard (SDS) (Figure 5.8.2).

Note that the sample processing capabilities had advanced considerably since the early S&S instrument like the Roland D50 (see Figure 3.6.2).

Yamaha VL1 (1994)

Subsequent VL-series instrument, notably the VL-70m, allowed editing of the parameters via computer. But this was non-intuitive, and arguably harder to understand than FM.

The VL1 is designed as a performance instrument and provides duophonic sounds. It uses preset models of instruments (both real and imaginary) and allows them to be controlled via instrument controls; no user editing of the models is allowed. Although it uses a conventional keyboard with velocity and pressure sensing, as well as pitch-bend, dual modulation wheels, pedals and breath controller inputs,

Figure 5.8.2 The Yamaha SY99 gives a comprehensively equipped set of FM and sample-replay synthesizers, but allows the samples to be reprocessed through the FM.

277

these can be mapped to a large number of instrument controls, including:

- pressure (or bow speed)
- embouchure (tightness of lips or bow pressure on the string)
- pitch (the length of the tube or string)
- vibrato (affects pitch or embouchure via an LFO)
- tonguing (simulates half-tonguing damping of saxophone reed)
- amplitude (controls volume without changing the timbre)
- scream (drives the whole instrument into chaotic oscillation)
- breath noise (adds widely variable breath sound)
- growl (affects pressure via an LFO)
- throat formant (simulates the players lungs, throat and mouth)
- dynamic filter (controls the cut-off frequency of the modifier filter)
- harmonic enhancer (changes the harmonic structure of the sound)
- damping (simulates air friction in the tube or on the string)
- absorption (simulates high frequency loss at the end of the tube or string).

As you can see, most of the controllers are very specific to real instruments, although the parameter which is being controlled may be one that does not and cannot exist! There are individual scaling curves and offsets for each controlled parameter, so you can adjust the effect of a controller like breath control to do exactly what you want with great precision.

The outputs of two separate instrument models can be combined and then processed through the user-programmable modifiers:

- harmonic enhancer
- resonant dynamic filter (low-pass, high-pass, band-pass and notch)
- five-band parametric equaliser
- impulse expander
- resonator.

Whereas the VL1's self-oscillating virtual acoustic synthesis (S/VA) physical modelling synthesis is designed to synthesise real monophonic instruments, the companion VP1 was intended to produce polyphonic

Figure 5.8.3 The Yamaha VL1 uses a fixed physical model to produce the sounds, but the user model via MIDI controllers is very sophisticated.

synthetic timbres, and uses a different variation of physical modelling called free-oscillation virtual acoustics (F/VA) and this was probably based on something like the Karplus–Strong algorithm for producing plucked and struck sounds. The VP1 never saw a commercial release (Figure 5.8.3).

Technics WSA1 (1995)

The WSA1 takes the best parts of physical modelling and combines them with the familiar parts of S&S. Rather than model a complete instrument, it takes the driver and resonator split, provides preset driver 'samples', connects them to a programmable resonator and includes physical model-like interaction control of the resonator. The output of the resonator passes through a conventional digitally con-trolled filter (DCF) and amplifier (DCA) synthesis section.

The driver 'samples' in the WSA1 are not really equivalent to the samples you find in S&S synthesizers. They do not have the same emphasis on length/time and multi-sampling that conventional sam-ples have because of the use of resonators to modify the sound of the driver 'sample' in a way that would normally require multi-sample. The driver/resonator model works extremely well – the bass and snare drums are an excellent example where although the basic driver sounds usable on its own, putting it through a resonator suddenly makes it more 'drum-like' (Figure 5.8.4).

Korg Z1 (1997)

Korg's Z1 is a 12-note polyphonic modelling synthesizer derived from Korg's OASYS development computer. The oscillator section provides 13 different types of combinable modelling module includ-ing analogue modelling, FM and physical modelling. The dual multi-mode filters provide sophisticated control over the timbre, and are controlled by dedicated control knobs. Also have digital multi-effects

Figure 5.8.4 The Technics WSA1 uses source-filter synthesis to provide 'physical model' like capabilities. Two of the four synthesis sections are shown, and the resonators in each of these can be coupled to provide more complex resonators. The two front panel real-time controllers provide 'live' user control over the timbre: a conventional joystick and tracker ball.

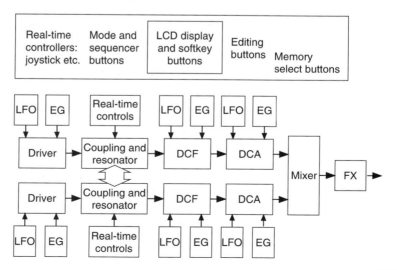

Figure 5.8.5 Korg's Z1 is a 12-note polyphonic modelling synthesizer.

and a polyphonic arpeggiator. Storage via a PCMCIA (PC card) flash memory card. The display has five assignable knobs, and there is an X–Y controller pad above the compact Prophecy-style pitch-bend and modulation wheels (Figure 5.8.5).

5.9 Questions

1. Compare and contrast the major features of analogue and digital methods of synthesis.
2. What are the two common artefacts which can result from digital synthesis?
3. What happens to the output of the carrier oscillator as the level of the modulator oscillator is increased in an FM synthesis system?
4. What are the three basic parameters which define a static FM timbre?
5. What is the difference between waveshaping and a guitar 'distortion' pedal?
6. What is important about the relationship between the input frequency and the additional frequencies which are produced by a waveshaper?
7. Why is an audio signal always sampled at a rate of at least twice the highest frequency component?
8. Multi-sampling each note on an instrument is one approach to obtaining maximum realism from a sample-replay instrument. Suggest an alternative way of using several samples to enable accurate reproduction of the dynamics of an instrument.
9. Describe one way of splitting a musical instrument into separate parts that may be useful in producing a physical model of the instrument.
10. What is the connection between FOF, VOSIM and human speech?

Date	Name	Event	Notes
1990	Technos	French-Canadian company, Technos announced the Axcel the first resynthesizer	There was no follow-up to the announcement
1990	Yamaha	SY77, a digital FM/AWM hybrid synthesizer/ workstation, mixed FM and sampling technology	Followed in 1991 by the larger and more powerful SY99
1991	Roland	JD-800, a polyphonic digital S&S synthesizer	Notable for its front panel – controls for everything!
1992	Kurzweil	The K2000 was launched. A complex S&S instrument, which mixes sampling technology with powerful synthesis capability	
1993	E-mu	Morpheus synthesizer module was launched. Uses real-time interpolating filter morphs to change sounds	Sophisticated DSP
1995	Yamaha	VL1, world's first physical modelling instrument was launched	Duophonic, and very expensive
1995	Roland	VG8 Virtual Guitar System, not a guitar synth and not a guitar controller	A physical modelling guitar sound processor ...
1995	Clavia	Nord Lead, a programmable digital analogue modelling synthesizer with a 'subtractive synthesis' metaphor	DSPs were used to emulate the sound of an analogue synthesizer in 4-note polyphony
1995	Korg	Prophecy, programmable monophonic analogue modelling, FM and physical modelling synthesizer	Features the 'log', a combined modulator wheel and ribbon controller, and was derived from OASYS (see Z1)
1995	Korg	Trinity S&S workstation with 32-note polyphony and a large touch screen	The effects are multi-timbral, so you get the same sound regardless of how other sounds use effects
1996	Native Instruments	Generator: hundreds of synthesizers on a PC	A modular, polyphonic, real-time software synthesizer
1997	Korg	Z1, 12-note polyphonic analogue modelling synthesizer	Derived from Korg's OASYS development computer
1997	Roland	JP8000, 8-note polyphonic analoguvese modelling synthesizer	Not a digital Jupiter-8, and only a four-octave keyboard, but covered with knobs, sliders and switches
1997	Yamaha	AN1X, polyphonic analogue modelling synthesizer	Allows morphing between two sounds (actually between the parameters determining the sounds)
1998	Ensoniq	Fizmo, a digital wavetable synthesizer	A descendant of the VFX ...
1999	Korg	Triton workstation, 64-note poly-S&S with sampling, multi-effects, and a large touch screen	Knobs come back into fashion. Also has user installable expansion boards (a multi-timbral Z1, etc.) and small computer system interface (SCSI)
1999	Native Instruments	Reaktor 2.0: oscillator-based synthesis, sampling, granular resynthesis and effects on a PC or Mac	Version 4 is a powerful and sophisticated sound generation facility. Hundreds of instruments are available for download

Date	Name	Event	Notes
2000	Korg	MS2000, four-note polyphonic analogue modelling synthesizer, vocoder, arpeggiator and sequencer	Reminiscent of the MS20 from 1978
2001	Yamaha	AN200, tabletop analogue modelling synthesizer with 16-step sequencer	Allows morphing between two sounds (actually between the parameters determining the sounds)
2001	Yamaha	DX200, tabletop six-operator FM synthesizer with 16-step sequencer	Allows morphing between two sounds (actually between the parameters determining the sounds)
2001	Roland	XV5080, S&S synth rackmount expander module with user sample replay as part of the S&S	Sounds from the JD, JV, XP and VX series, plus sample replay from WAV, AIFF, Akai …
2003	Roland	Fantom-S, S&S workstation with sampling	Has 16 pads to play the samples, a Smart Media slot instead of a floppy drive, and D-Beam Doppler controller
2003	Soundart	Chameleon, programmable rackmount DSP module	Completely programmable: can be an effects unit, a poly-synth, a mono-synth, amplifier emulation, etc.

Applications

6 Using synthesis

The chapters so far have concentrated on the theory behind the methods of synthesis. This chapter deals with the use of synthesis to make music and other sounds, whilst Chapter 7 looks at how synthesizers are controlled, and Chapter 8 covers performance.

6.1 Arranging

At the simplest level, the user interacts with the synthesizer to produce sounds. This often means directly controlling the synthesizer directly by using a keyboard or other controller. The synthesist is thus carrying out a role which is analogous to that of a player in an orchestra.

Note that whilst MIDI control is described in this chapter, the techniques also apply to control voltage (CV) and gates in analogue synthesizers.

But when the control is indirect, as in remotely controlling a synthesizer by using musical instrument digital interface (MIDI) or other computer-based means, then the role is much closer to the conductor of an orchestra but with the option of simultaneously being able to act as an individual performer as well. This dual role actually provides the synthesist with something much closer to the detailed control and freedom which is available to an arranger, rather than the decreasing degrees of freedom which are available to the conductor and the performer.

Synthesis thus provides the ability to control three layers of a performance:

- choice and control over the use of timbres (arranger)
- precise control of the timing and dynamics of individual sounds in context, as they are produced (conductor)
- very low level detailed control over the production of timbre (player).

For an arranger, working with a conductor and an orchestra requires a transfer of instruction via the score, demonstration or the spoken word. But for a synthesist, changes to the performance or the conducting require only a change of role. (This can make the task of a synthesist working as part of a larger orchestra a difficult set of compromises regarding control over the synthesizer.) Many multi-timbral synthesizers provide this 'three-layered' level of control within one instrument, since they can produce several different independent timbres simultaneously. The skills of arranging are thus applicable from single synthesizers up to large 'synthesizer orchestras', or even a combination of real and synthetic instrumentation. In this section, the emphasis will often be focused onto individual sounds, but the same principles equally apply to larger structures.

The sales literature for many synthesizers concentrates on the production of sounds, and not the wider aspects of arranging. This chapter is not intended to be a guide to arranging, since that is a complete topic of study, but instead to show some of the tools and techniques which are available to the synthesist. There are three major divisions:

- **Arranging**: This covers topics like stacking, layering and hocketing.
- **Timbres**: This covers topics like multi-timbrality and polyphony, general MIDI (GM) and the use of effects.
- **Control**: This covers topics like the use of performance controllers and editing.

Many of the terms which are frequently used in connection with synthesizers are only loosely defined or even redefined for this specialist usage. Stacking and layering are often confused and given different or overlapping definitions, and they are often used interchangeably in manufacturer literature. In this book, stacking will refer to a single composite sound produced from two or more timbres, whilst layering implies a composite sound which changes or evolves with time, and thus exhibits dynamic changes of separate 'layers' of sounds (Figure 6.1.1).

There are two different methods which can be used to produce stacking or layering: several separate synthesizers can be used, or alternatively, a single multi-timbral synthesizer can be used. These two ways

Figure 6.1.1 Stack versus layer. Stacking is a single composite sound produced from two or more timbres. Layering is a composite sound which changes or evolves with time.

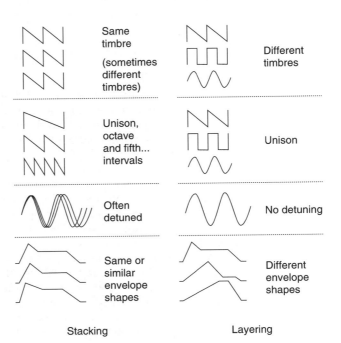

| Stacking | Layering |

Figure 6.3.1 Layering is concerned with changes in envelopes over time.

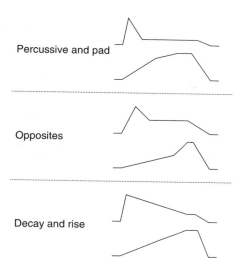

Percussive and pad

Opposites

Decay and rise

Layering involves a time separation of the sounds. This conventionally means that one sound is used to provide an initial attack sound, whilst the other is used for the sustain portion of the sound. But this is just one approach to using two sounds which are independent of each other in time. Some other possibilities include:

- **Percussive and pad**: The combination of a fast attack/rapid decay sound with a slow attack, slow release pad sound can be very useful for providing accompaniment timbres from a single performance keyboard. By altering the playing from staccato to legato, or by using the sustain pedal, the level of the pad sound can be controlled.

- **Opposites**: By providing the two sounds with opposite envelopes, the result is a complex sound which dynamically changes between the two timbres. This is particularly effective with slowly evolving sounds, or sounds which have some common elements and some contrasting elements.

- **Echo/reverb and dry**: By layering two sounds which have different acoustic 'spaces', it is possible to dynamically change the apparent 'position' of a sound. Dry sounds tend to be perceived as being close to the listener, whilst echoed or reverberant sounds are interpreted as being further away. Two sounds which have different timbres and different echo timings can be very useful for creating poly-rhythmic textures.

- **Pan position**: Layering sounds which have contrasting pan positions can be used to produce composite sounds which change their stereo position dynamically.

- **Slow decay and slow rise**: By using a sound which decays slowly and another sound which rises slowly whilst the first is decaying, the composite sound can have a 'sustain' sound which is not static, but which changes as the relative balance between the two layers changes.

293

Figure 6.3.2 By splitting a keyboard and sending the keyboard CV to two 'layers' of a sound, it is possible to use the separate gates to control the second layer independently.

Loops

Repeating or looped sounds can be played against either pads, percussive sounds, or even other loops. This provides continuous variation in the overall sound when it is held for a long period.

6.3.1 Splits

Although layering can be achieved by assigning both sounds to the same key, many synthesizers provide the ability to 'split' the keyboard, which allows different sounds to be used in different areas of the keyboard range. A typical application for jazz use might place a piano sound on the upper half of the keyboard, with a string bass sound on the lower half, perhaps with an additional brush sound. Less typically, the parts of a split can be used to produce different layers of a sound instead of different notes – this is easy to set up on an analogue modular synthesizer, although it can also be produced using MIDI equipment with specialised software. This allows control over which layers of a composite sound are produced on a note-by-note basis, which is more akin to the sort of control available to an orchestra and not a keyboard-based synthesist (Figure 6.3.2).

6.4 Hocketing

Hocketing is the name given to the technique of sending successive notes from the same musical part to different instruments, instead of to the same instrument (Figure 6.4.1). With the use of different (complementary and/or contrasting) timbres, or different pan positions, contrasting effects settings, or even slightly different amounts of detune, this can produce very complex sounding arpeggios and accompaniments, and combined with doubling and layering, can give the impression of a very detailed arrangement, when in fact all that has happened is that a few notes have been moved from one track to another, and a few tracks have been copied and pasted.

The general process to produce hocketing in a sequencer is to:

- listen to the track with one sound source
- choose a hocketing criteria (see below)

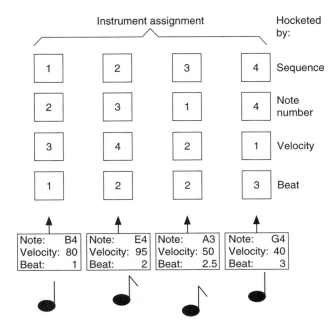

Figure 6.4.1 Hocketing involves changing the instrument that plays a note based on criteria like the MIDI note number, or the MIDI velocity, or the position of the note in the bar or in time. In this example, the four notes are hocketed to two instruments using the note number.

- select the notes to be hocketed
- filter the notes according to order, number, velocity, beat, time …
- copy the filtered notes to another track
- assign the hocketed notes to a different sound source
- set the two sound sources to contrasting or complementary timbres.

Hocketing can be achieved using several criteria:

- by note sequence order
- by note number
- by velocity
- by beat
- by time.

By note sequence order

Hocketing by note sequence order is the obvious one: you put the first note of the arpeggio or melody to the first instrument, the next note to a second instrument, and so on, with perhaps the third or the fourth note being allocated to the first instrument again. With similar timbres for each of the hocketed instruments, the effect is quite subtle, and by using slight detuning of the instruments, can produce very 'non-synthetic' ensemble effects. Using different pan positions for the hocketed instruments can be used to provide movement in a sound which would otherwise be static. With contrasting timbres, the effect of the hocketing is more like splitting the notes into separate parts.

By note number

By using the note number to hocket, there are a number of effects which can be produced. Hocketing with a fixed note number as a split-point will send all the notes above the split to one instrument, with those below the split going to the other instrument. Odd and even note number hocketing can break up chords into what sound like pseudo-inversions or complex arrangements (especially if the two instruments are transposed by one or more octaves relative to each other). Pairs of numbers can be used if the odd/even hocket results in too much spread of the notes in a chord. Chord splitting involves taking each chord and assigning hockets split around the middle note, with notes above the middle going to one instrument, and notes below going to the other. This can also be very effective if the assignment alternates so that the notes above go to one instrument for the first chord, then to the other instrument for the second chord.

By velocity

Using velocity to determine which instrument plays the sound involves selecting only those notes which have a velocity value above a specific value (half-way, 64, is a good starting point) and then allocating those notes to a different instrument. This is like the velocity splitting available on some samplers and S&S instruments, although the split-point can be edited as part of the score or sequencer information instead of as part of the sound. Hocketing by velocity can be particularly effective on sounds which move between an accompaniment and melody role. By using alternative sound sources for the velocity-hocketed notes, it is possible to have different snare sounds selected from different drum machines or samplers merely by editing the velocity. Reducing the velocity sensitivity of the sounds allows the velocity to be used as a mixing control rather than a dynamics control.

By beat or time

It is also possible to use the position in the bar to determine which instrument a note is hocketed to, or even the absolute time, in which case the hocketing is not related to the bar position at all. The hocketed instruments can have different timbres, in which case the timbre is reflected by the location in the bar, or by pan position, in which case the sounds can be made to move around in the stereo field in time with the music. This is much more effective and controllable than the 'random pan position' option which is available on some commercial synthesizers.

With a little practice, all of these 'hocketing' edits are relatively quick and easy to make on a computer sequencer, and yet they can greatly improve the detail and quality of the finished music.

6.5 Multi-timbrality and polyphony

Multi-timbrality is the name given to the ability of a single synthesizer to produce several different timbres at once. This has the effect of turning a single physical synthesizer or expander module into several 'virtual' ones, although some of the functions remain common: normally the effects, the overall control and the management of the sounds.

Multi-timbrality is an extension of the concept of 'stacking'. If an instrument can produce two different timbres at once when a key is played, then it is logical to extend this capability so that the two different timbres can be played independently. The history of multi-timbrality is closely connected to the development of polyphonic synthesizers and the differences between analogue and digital synthesis.

Many sounds and instruments are naturally monophonic, that is, one sound at once. Most people can only sing one note at once, and many acoustic instruments will only produce one note: flute, tuba and triangle are some examples of naturally monophonic instruments. The availability, price and complexity of early modular analogue synthesizers meant that they tended to be used as monophonic instruments too. Recording onto tape allowed the single notes to be combined into complete 'polyphonic' performances. Even when the synthesis resources were available, trying to keep two or more notes in tune and with the same timbre could be harder than using tape.

The first true polyphonic synthesizers were based around simplified analogue monophonic synthesizer voltage controlled oscillator/filter/amplifier (VCO/VCF/VCA) circuits to provide the notes. These cards are often called 'voice' cards, since each card provides a single 'voice', rather like independent singers in a choir. These synthesizers could be operated in two modes: either with common control of timbre and independent control of pitch, or with independent control of timbre and pitch. The mode with common control of timbre allowed true polyphonic operation, where several notes with the same or similar timbre could be played simultaneously. The mode which provided the independent control of timbre allowed stacking and layering of sounds, which reduced the number of different notes which could be produced simultaneously. Because of the problems of tuning and controlling voice cards, there was a practical limit of about 8-note polyphony. Even with eight voice cards, the minor variations in tuning and timbre can produce a distinctive open 'feel' to the timbres produced, rather like the ensemble playing of a violin section in an orchestra. Eight-note polyphony is very restricting for producing multi-timbral music, and so analogue polyphonic synthesizers were used more for their polyphony than their multi-timbrality.

Polyphonic digital synthesizers like the Yamaha DX7 started out as mono-timbral in 1983, although digital technology provided larger polyphony than many of the analogue mono- or multi-timbral polyphonic synthesizers: 16-note instead of the 4-, 5-, 6- or 8-note polyphony found in the analogue instruments of the time. But the precise tuning and identical timbres gave a very different 'feel' to these early digital instruments, and it was not until the early 1990s that the technology for emulating the imperfections of analogue polyphonic synthesizers began to be implemented in digital synthesizers. But the availability of larger polyphony meant that utilising the multi-timbrality was easier and more effective. The introduction of 8-part multi-timbrality in 16-note polyphonic synthesizers was quickly followed by 16-part multi-timbrality, which has formed the basis of most synthesizer specifications ever since.

By the beginning of the 21st century, 64- or even 128-note polyphony had become more common, particularly in S&S synthesizers, and this allowed a single synthesizer to produce many polyphonic parts simultaneously. The need for more than one rackmounted synthesizer expander module was increasingly driven by sounds rather than polyphony. The fundamental limit of 16 channels on one MIDI port has been overcome by some manufacturers by providing more than one MIDI-in socket. Some synthesizers and workstations arrange for the built-in keyboard and sequencer to access one set of 16-part multi-timbrality, whilst one or more MIDI-in sockets provide access to the remaining sets of 16 parts of multi-timbrality.

6.5.1   Definitions

- **Polyphony** is the total number of different pitches or sounds that an instrument can make at any one time. For a single sound, the polyphony is the number of different pitched notes that the instrument can play simultaneously (Figure 6.5.1).
- **Multi-timbrality** is concerned with the number of different sounds or timbres that can all happen at once, and played by at least one note each.

The word *part* is frequently used to indicate a separate timbre, and although it can be used in the same sense as a musical 'part', the two terms are not necessarily synonymous.

The word *voice* is also sometimes used to mean a single sound generating section: a monophonic part.

Thus, a 16-note polyphonic instrument can play a sound using up to 16 different pitches at any one time, whilst a 16-'part' multi-timbral instrument can play up to 16 different sounds or timbres at any one time.

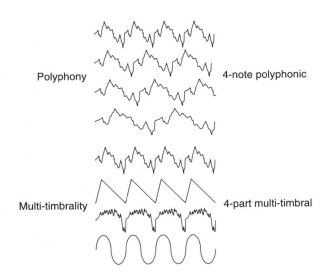

Figure 6.5.1 Polyphony is the total number of different pitches or sounds that an instrument can play simultaneously. Multi-timbrality is the number of different sounds or timbres that can happen simultaneously.

Polyphony   4-note polyphonic

Multi-timbrality   4-part multi-timbral

The two different numbers are independent, although the specifications for synthesizers frequently quote the same number for both the polyphony and the multi-timbrality. This has continued even when the multi-timbral parts have exceeded the 16 parts which can be carried by a single MIDI cable. Instruments with 32-note polyphony are frequently quoted as having 'up to' 32-part multi-timbrality, even though using all 32 parts requires the use of the keyboard and on-board sequencer as well as one MIDI-in socket, or two separate MIDI-in sockets and a multi-port MIDI interface on a sequencer or computer.

As a general rule, the polyphony should always equal or exceed the multi-timbrality, so a 16-note polyphonic, 16-part multi-timbral instrument is feasible, whilst a 2-note polyphonic, 16-part multi-timbral instrument is not possible: it can only ever make the sounds for two parts. Actually utilising the 16 independent monophonic parts from a 16-note, 16-part multi-timbral instrument can be harder than it might initially appear because of the need for true monophonic sequences of notes for each part: any overlap can increase the polyphony, which automatically reduces the available multi-timbrality.

- **Maximum polyphony** is a term used where the polyphony is dependent on the synthesis resources inside the synthesizer. In such a case, the use of simple sounds will result in the greatest polyphony, whilst complex sounds will reduce the effective polyphony. The allocation of synthesis resources is very dependent on the synthesis technique used, but a simple example based on oscillators will illustrate the general principles. If the synthesizer has 16 'oscillators', then the maximum polyphony is 16 notes, provided that each note only requires only one oscillator. The 'maximum' polyphony is therefore an exceptional case where only single oscillator sounds are used for each note. If the sound for a note is made up of two or more oscillators, then the polyphony reduces accordingly: 8-note polyphony when two oscillators are required per note, and 4-note polyphony when four oscillators are required for each note. The 16-note polyphony is thus a 'best case' rather than a typical value, and this should be taken into account when reading the specifications of synthesizers. In general, the best sounds will tend to use the full capabilities of the instrument, and so from the example above, this would be four oscillators per note, giving a 'minimum' polyphony of 4 notes. Typically, not all sounds will utilise the full synthesis capabilities and so the 'typical' polyphony would be between 4 and 8 notes.

The important element of the definition of multi-timbrality is the simultaneity. With a monophonic synthesizer it is possible to make a large number of different timbres, but it is always mono-timbral, that is, one timbre at once. If a synthesizer has two separate sound generating sections, then it has two-part multi-timbrality, thus it is two-part multi-timbral. But by adding in a third sound generating section, each separate part that can happen simultaneously with other parts only adds one to the multi-timbrality. The number of available timbres does not affect the multi-timbrality; there may be a large number of ways of combining two timbres: but the multi-timbrality is fixed by the polyphony and the number of simultaneous parts, and not by the timbres.

6.5.2 Notes per part

Using multi-timbrality can require a surprisingly large polyphony: even a 64-note polyphonic instrument can only play four simultaneous pitches on 16 multi-timbral parts. Trying to produce 16 multi-timbral parts using only 16-note polyphony can result in only one note per part if everything plays at once, and many synthesizers have limitations as they approach their polyphony or part limits. Producing music using only one note per part also has problems; although orchestral arrangements use monophonic melody lines played on 'one note at once' instruments like oboes, flutes and trombones, they use several players to get the required polyphony (and volume, of course). And an orchestra also has fixed limits to polyphony: if there are four violin players, then you can only have four separate violin parts.

The transitions between the notes are the cause of many of the problems when a synthesizer reaches its polyphony or part limits. This is easiest to understand by considering the case of making music using only one note per part on a polyphonic keyboard. *Staccato playing*, with gaps between the notes, gives controlled use of polyphony. But *legato playing*, tied notes, or overlapping notes can all cause short overlaps. Each overlap uses up 2 notes of the available polyphony, and because of the way that music tends to concentrate events around bar or beat intervals, the overlaps 'cluster' across parts. So it is likely that an overlap between 2 notes within one part will happen at the same time as overlaps in other parts. The required polyphony can thus be much higher than the apparent polyphony of the music: up to double in exceptional cases. A piece of music written for a quartet of four monophonic instruments can have a peak polyphony requirement of eight notes.

The release time of sounds can be a major contributing factor in causing problems with overlapping notes. Although the start of a note is when the key is pressed (or the MIDI note on message is received), the end of a note can last after the note has been released (or the MIDI 'note off' message has been received), because the synthesis resources are still being used to produce the sound as it fades away during the release segment of the envelope. 'Staccato' playing thus has a subtle but significant difference when playing electronic keyboards: whilst the notes written on the score may be separated by rests, the actual synthesizer may actually be playing legato notes because of long release segment times. The result is that an apparently low polyphony piece of music may well require up to double the polyphony to avoid problems (Figure 6.5.2).

Here, 'old' means notes which are already making sounds, whilst 'new' means notes which have been pressed and which are about to make sounds.

When the polyphony limit is reached, notes are inevitably lost. If there are not enough resources to play all the required notes at once, then some that are already playing will have to stop to enable the new notes to be heard. This produces the effect known as 'note stealing', where 'old' notes make way for 'new' ones. The behaviour of

Figure 6.5.2 The peak polyphony depends on the overlaps between notes. Even if there appears to be a gap between the 'gate' signals, the release of one sound can carry on past the start of the attack of the next note.

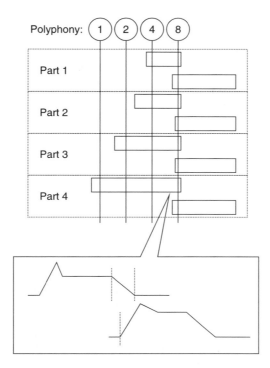

synthesizers when note stealing is taking place varies: it depends on the method used to assign the notes to the available sound-making resources inside the instrument, in an analogue synthesizer these would be the voice cards, and so these resources are often called 'voices'. A 'voice' is thus an imaginary part of a synthesizer which is capable of playing only one note, rather like a virtual polyphonic synthesizer voice card. In digital instruments, the distinction of physical hardware does not usually exist, and so 'voices' are more flexible (see Section 6.5.1).

6.5.3 Note allocation

The way that the required notes are assigned to the available 'voices' in a synthesizer is called *note allocation* or *voice assignment*. There are a number of techniques for doing this, and they all extend the basic idea of reusing the voice resources on demand. This underlying process is called *cyclic assignment*, and it assigns the incoming notes to voices in order: the first note is assigned to the first voice, the second note to the second voice, etc. The notes which are played could come from either the instrument's keyboard, from an internal sequencer, or via MIDI from an external keyboard, instrument or sequencer (Figure 6.5.3).

An 8-note polyphonic synthesizer would contain eight of these 'voice' resources, and so after all eight voices had been assigned, then the next available voice would be the first one again. The 9th note

Figure 6.5.3 Cyclic assignment assigns incoming notes to available voices in sequence. In this example, there are four voices available, and the first four notes are assigned to these. The fifth note will replace the first note on voice 1.

would thus cause the first voice to stop playing the first note, and this voice would be used to play the 9th note. Exactly the same 'stealing' would occur with the 17th note – the first voice would again be used.

Cyclic assignment does not take into account the status of the voices. If the first note is sustained, whilst the next seven are short staccato notes, then seven of the eight voices will be 'available' for the 9th note since they are not actually producing a sound. A pure cyclic note assignment strategy would ignore this and assign the 9th note to the first voice, thus stopping the sustained note. This approach thus wastes the voices, since the assignment is arbitrary and does not use the available resources efficiently.

There are many ways to improve the assignment strategy by making it responsive to the incoming required notes and the availability of voice resources. Some of these 'dynamic voice allocation (DVA)' approaches include:

- note reserving
- part priority
- voice status
- envelope status
- volume status
- repeated note detection
- sustain pedal detection.

Note reserving

Sometimes a particular part needs to always have a specific polyphony. For example, a note stolen from a sustained melody line, or a solo passage will be very noticeable to the listener whilst a note stolen from an accompaniment pad sound will be less apparent. 'Note reserving' allows a fixed allocation of polyphony to timbre where a specific part will always have a given polyphony. So a melody part might be allocated a polyphony of 2 notes so that any overlaps will not cause assignment problems. The disadvantage is that when the melody part is not being used then the voices are still not available, and so the utilisation of voices is not very efficient.

By using fixed polyphony, it is possible to make 8-note polyphony sound like 12 notes. Two contrasting timbres are allocated to two parts, and assigned to the same pitch control channel (or MIDI channel). Note reserving is then used to allocate the voice resources asymmetrically: 6 and 2 notes, for example. The result is like a 6-note polyphonic version of the two-timbre stack, because the listener will tend to hear the last two notes played, which are always played with the two timbres, and not the remaining 4 notes which are played on only one timbre. This is very effective when the two timbres are detuned relative to each other, since the result sounds like 12 detuned notes when only 8 are actually being used.

Part priority

Assigning priorities to parts allows a melody or a drum part to be relatively immune from note stealing, without the need for the permanent allocation of note reserving. By assigning a priority number to parts, it is possible to force notes to be stolen from the 'less important' voices or parts which have lower priorities. The highest priority parts are only stolen from when all other resources are in use. Time priority also needs to be considered: a higher priority needs to be given to the most recent notes. Despite the complexity of allocating priorities, this method gives a more efficient use of the available polyphony than fixed reservation.

Voice status

By monitoring the status of the voices, it is possible to alter the cyclic assignment so that incoming notes are only allocated to voices which are not playing sounds. This means that some mechanism is needed to keep track of the 'playing/not playing' status of each voice. Note stealing only needs to occur when there are no 'not playing' voices available.

Envelope status

Some portions of the envelope of a sound are less important than others. Stealing a note from a sound which is in the release segment is probably less audible than stealing from a sound which is in the attack segment. By monitoring the envelope status of each voice, it is possible to arrange to only steal sounds which are in the appropriate segments: normally the release or sustain segments.

Volume status

Volume, either controlled by the note velocity, overall volume or envelope position, can also be used to determine if a voice is a suitable candidate for stealing. Lower volumes or audio levels will be less audible if they are stolen by a voice assigner.

Repeated note detection

If the same note is allocated using a cyclic assignment, then the allocation will run through each of the available voices in turn, even though the same pitch is being played each time. By detecting the repeated note, and reassigning it to the same voice, perhaps with a retriggering of the envelope, the remaining voices will not be disturbed.

Sustain pedal detection

The sustain pedal on a synthesizer or MIDI device causes the notes which are being played when it is activated, or are played whilst it is active, to be held at the sustain segment of their envelope. If the number of incoming notes exceeds the available polyphony, then notes will be stolen, and because of the time-based nature of the sustain pedal event, the note stealing decision should probably be on a cyclic 'oldest-note' basis.

In order to allow the reservation, priority, status and other parameters to be tracked, the synthesizer needs to maintain running lists of all the parameters which might affect the note allocation. This changes the 'note assigner' circuitry, from the simple counter that is required for a cyclic assignment technique to one which has to maintain detailed records for several separate voices. Such an 'intelligent' assigner can add a considerable software overhead, which may mean slower response times from the synthesizer to incoming notes. There is thus a compromise between the complexity of the note allocation scheme and the response of the synthesizer.

6.5.4 Using polyphony

Multi-timbrality and polyphony tend to encourage the use of the layering, splitting and hocketing techniques described in this chapter. Polyphony can be either overused or underused: too much doubling or detuning can over-thicken a sound and instead make it dense and obscure the harmonic structure; too little available polyphony can be improved by reserving or prioritising parts, and perhaps even thinning out the chords used for accompaniment or reducing the number of simultaneous drum events.

There are a wealth of low-cost synthesizer expander modules which are available, both new and second-hand which can be used to increase the available synthesis polyphony. As has been mentioned several times, a hybrid or mix of several instrumental sounds using different synthesis techniques can be very powerful.

6.6 GM

General MIDI, abbreviated as GM (White, 1993) was introduced as an extension to the MIDI specification in 1991, and is a shared set of specifications and guidelines agreed by the MIDI Manufacturer's Association and the Japan MIDI Standards Committee.

GM uses the idea that most music can be reproduced using a small set of commonly used instrumental timbres. In effect, the concept provides a minimalistic 'electronic' orchestra. Many of the sounds chosen are orchestral or 'popular' in origin, although some additional synthetic sounds are also included. There is a distinctive GM logo which can be found on compliant instruments.

There are 128 sounds, plus a drum kit mapping, in the basic GM sound-set. The sounds are divided into 16 categories or 'groupings', each containing eight sounds:

- keyboards
- chromatic percussion
- organ
- guitar
- bass
- strings
- ensemble
- brass
- reeds
- pipes
- synth lead
- synth pad
- synth effects
- ethnic
- percussion
- sound effects.

The numbering and names of the groupings and sounds are given in the GM specification. The drum kit (percussion) receives MIDI messages on channel 10, whilst the remaining channels can be used for any purpose: the allocation of channels to parts is left to the user. The GM specification requires that the polyphony of the GM playback device should be a minimum of either '24 fully dynamically allocated' notes, or 16 'dynamically allocated' notes for instrumental sounds and 8 notes for percussion. GM modules cannot always meet this requirement. GM playback devices should also be 16-part multi-timbral.

The many GM sound-sets which are produced by manufacturers are all intended to sound as similar as possible, so that a MIDI file created using one GM device will sound much the same when played back using another GM expander. This extends to how they combine together, how they respond to MIDI performance controllers like pitch bend, velocity and volume. The result is a largely uniform range of instruments which can reproduce the same sounds in a predictable way.

The actual sounds that are included in the GM sound-set are not intended as anything more than a means to an end: making a form of 'paint by numbers' music, where the MIDI file contains the information for both the musical events and the musical sounds. With minor differences, a given MIDI file will produce very similar results on any GM playback device. For use as a synthesis medium, GM offers very little. The sounds are chosen for their ubiquity of application, and many GM playback devices do not offer any editing of the sounds – sample replay is frequently used for producing the GM sounds. GM playback devices can be used in combination with synthesizers to produce composite sounds.

GM playback devices can range from professional synthesizers, through small expander modules, to personal computer (PC) sound cards and even software-only versions. Two manufacturers have introduced their own additional extensions to the basic GM specification, and these allow extra functionality to be provided by specifying additional sounds and controls: Yamaha's XG and Roland's GS systems are similar in many respects, but are not completely cross-compatible.

In 1999, an enhanced superset of the original GM specification was released. GM2 was designed to enable better exploitation of the powerful facilities of the synthesizers of the time, and the GM specification was renamed GM1. In contrast, 2001 saw the publishing of GM Lite, a reduced specification intended for use in devices like mobile phones. These new specifications do not change the basic concept behind GM: define a standard set of synthesis capabilities across sound generating devices in order to maximise the chances of getting acceptable and consistent playback of music files.

6.6.1   Music files

Standard MIDI files (often suffixed with the file extension .SMF) need not use GM, but in practice, most .SMF or .MID files will use the GM1 sound and drum mappings. GM Lite is a pragmatic solution to the limited synthesis features of some devices: like mobile phones, and so is usually used only as part of a process of converting conventional GM files to suit the specific devices. GM2 is backwards compatible with GM1, but has necessarily limited compatibility in the other direction.

The idea of producing music files which will play on a wide variety of different end devices is not new. These fall into three main types:

* waveform
* waveform and representational
* representational.

MIDI files are representational. They describe the notes which are played, the sounds to be used, and how the notes should be played. MIDI files can thus be seen as a sophisticated form of abstracted music notation, since the performance information is explicit, and can be edited. Changing the sound of a bass guitar requires only a change of the mapping of the sound: either a different sound or a change to the settings of the synthesizer producing the sound. The flexibility of representational files can result in the substitution of a very different sound: a whistle to replace the bass guitar or drum sounds. These changes are playback-only modifications: the file itself normally remains unaltered.

In contrast, *MP3, WAV, AIFF* and *other recorded files* are actually just audio files, where the actual *waveform* of the sound is stored, albeit sometimes in highly compressed formats. Audio files do not store anything about the notes which are played, the sounds used or how the notes should be played, other than by capturing the actual sound.

Audio files are just records of the sound, and performance information is implicit and cannot easily be edited. If you do not like the sound of the bass guitar in a recorded audio file, then it is very hard to change it. If a technique is available for replacing the bass guitar with another sound, then this would require the creation of an altered audio file.

MOD files and MIDI DLS files are mixtures of the waveform and representational types. Both mix notation systems with samples of sounds. The notation is used to control the playback of sounds which are just samples of the required sounds. These files can produce very good reproductions of recorded music, but with very small file sizes. Mixed representations of music are good when synthesis techniques are limited. Vocal or spoken performances are one example where MIDI files used to be severely limited by the lack of vocal synthesis capability, but Yamaha's Vocaloid vocal synthesizer software, released in 2003, showed that even singing could be synthesised, albeit by a technique which mixes waveform and representational approaches.

Some music programming languages like CSOUND can replay audio, or control sample replay or synthesis from notation, and so can utilise waveform, mixed or representational types of music file.

The future of music files seems to be the mixed type. Audio file compression is moving increasingly towards an approach where the coding technique analyses the content to derive notation-like control signals that are used to control sample-like fragments of audio. MIDI files are increasingly using samples rather than synthesis for part of their sound generation.

6.7   On-board effects

In the 1970s built-in effects processors were the exception rather than the rule. Reverb, chorus and echo were added to the output of the synthesizer in the studio by using 'out-board' effects units. Built-in effects processors first began to appear in polyphonic hybrid synthesizers in the 1980s as a low-cost method of disguising the use of single digitally controlled oscillators (DCOs) instead of dual VCOs. String and piano sounds were significantly improved by the use of small amounts of chorus effect and in fact, this spelled the end of the 'string machine' as a separate instrument. Whilst simple effects processors were commonly added to hybrid and S&S polyphonic synthesizers throughout the 1980s, it was not until the end of that decade that digital FM synthesizers were fitted with built-in effects processors. The 1990s have seen the effects unit change from being a simple 'afterthought' to an integral part of the synthesizer by the start of the 21st century.

6.7.1   *Effects history*

Although effects processors have been used to process most electronic (and amplified) musical instruments for many years, the addition of effects processors to synthesizers has been a gradual evolution.

The effects types which follow are organised into a rough chronological order of introduction.

- reverb
- chorus
- echo
- automatic double tracking (ADT)
- phasers and flangers
- ring modulation
- exciters, compressors and auto-wah
- pitch-shifters
- distortion.

Reverb

Reverberation is the effect produced by almost any acoustic environment: a short delay (the 'pre-delay') is followed by a series of echoes from the boundaries (the 'early reflections') and then echoes of those echoes (reverberation) which decay in volume (the 'reverb time') as energy is transferred from the audio to the environment. Small containers have short delays whilst large rooms have long delays. The timbral quality of the reflections and reverberation are determined by the boundaries of the environment: smooth, shiny concrete walls will give long reverberation times and only slight high frequency attenuation, whilst sound absorbent material like curtains will give short reverberation times and significant high frequency attenuation.

Sounds without reverberation sound empty and synthetic – the human ear expects to hear sounds played in real spaces, and removing the effect of the space can sound very 'wrong'. Synthesizers without reverberation emphasise this perception, so adding reverberation is almost essential unless the synthesizer will be heard in a naturally very reverberant acoustic environment. Many commercial synthesizers suffer from having too much reverberation on their preset sounds – based on the assumptions that 'more is better' and that the synthesizer will be auditioned using headphones. (It is easy to show that large amounts of reverberation can detract rather than enhance a sound: listening to a piano being played in a ballet or dance rehearsal room with lots of mirrors to reflect the sound is an excellent tutorial exercise.)

In the 1970s, reverb usually meant either a spring-line reverberation unit or a tape-based fedback flutter echo unit. The spring-line was prone to mechanical interference (they were microphonic and knocking the casing was audible in the audio output), whilst the tape machines were prone to mechanical failure (tape loop splice failures or wearing out of components because of long periods of continuous use). In the 1980s, analogue bucket-brigade delay lines were replaced by digital effects, and the quality of reverberation improved markedly. The 1990s saw sophisticated digital signal processing (DSP)-based reverberation algorithms, with features like early reflections, pre-delays, and room size simulations. Reverberation has now become so low in cost to implement that it has become largely ubiquitous: it is provided even on low-cost minimalistic GM modules.

On a synthesizer, reverberation can also be approximated by using an envelope release curve which starts out decaying rapidly, but then lengthens the release time as the level or volume of the audio signal decreases.

Chorus

The chorus effect is a cyclic detuning of the sound, mixed with the original sound. This cause the same type of phase cancellations that are associated with several performers playing the same notes on acoustic instruments: the violin section in an orchestra is a good example.

Chorus is normally achieved by delaying the audio signal slightly in time by a few tens of milliseconds, and then changing the time delay dynamically. This has the effect of changing the pitch as the delay time is altered, and so produces the detuned chorusing effect. Chorus can also be produced by deliberately detuning two VCOs, oscillators or sounds.

Echo

Echo is a repetition of the original audio signal after a specific time delay, usually hundreds of milliseconds or more, and at a lower volume. It simulates the effect of having a remote large object from which audio sounds will be reflected. Rapid echoes are often called 'flutter echoes', whilst the timbral quality of echoes is largely determined by the object from which the audio sounds are reflected.

Echo is produced by time delaying the audio signal, and feeding back the output to the input via an attenuator to produce additional repeat echoes. Echo can also be simulated by retriggering envelopes using a low frequency oscillator (LFO).

ADT

ADT is just echoes with a very short delay time: tens of milliseconds. It can be used to 'thicken' some sounds, although because it produces a fixed set of cancellation notches in the output spectrum, it also produces timbral changes that sound 'metallic' in timbre. Turning off the LFO modulation in a chorus effect produces ADT. If the delay is exactly one bar long, and the music being played is very repetitive, then ADT can produce a smearing or blurring effect when the music changes.

Phasers and flangers

Phasers and flangers are variations on the chorus effect, with a mixing of a undelayed with a delayed audio signal, but with feedback from the output to the input. Phasers use a *phase shift circuit*, whilst flangers use a *time delay circuit*. In both cases, cancellations occur when the delayed and undelayed audio signals are out of phase, and so a series of narrow cancellation 'notches' is formed in the audio spectrum. The spectrum looks like a comb, and these filters are sometimes known as 'comb' filters. As the phase shift or time delay is changed by an LFO, the notches move up and down in frequency. Phasers produce notches which are harmonically related because they are related to the phase of the audio signal, whilst flangers produce notches which have a constant frequency difference because they are related to the time delay.

Ring modulation

Ring modulation takes two sounds and produces sum and difference frequencies based on the frequency content of the input sounds (see Section 2.4.3). In effects processors, ring modulation normally uses an LFO or audio oscillator for one source sound, and the signal to be processed as the other. Ring modulation can make major changes to the timbre of a sound, normally making it metallic or robotic in sound. It is particularly useful for processing drum and percussion sounds, whilst for special effects it can provide sounds which are often suitable only for science fiction genres.

Exciters, compressors and auto-wah

By the 1990s, the effects processors which were being incorporated into leading edge commercial synthesizers were similar to the top models of stand-alone effects processors from the same manufacturers. The range of effects programs expanded to include additional effects which were not based around time delays, such as exciters (small amounts of frequency-dependent harmonic distortion), compressors (dynamic range reduction) and auto-wah (envelope following/triggered filters). By the middle of the 1990s the first effects processors designed specifically for use with 'multi-timbral' synthesizers began to appear.

Pitch-shifters

Pitch-shifting is a variation on the detuning that takes place in a chorus unit, and is sometimes called *harmonisation*. Instead of cyclically changing a time delay, the audio signal is stored at one rate and read out at another which produces a fixed pitch-shift. Slight pitch-shifts produce subtle chorus effects, whilst larger pitch-shifts produce transpositions (often at the loss of audio quality). Feedback from the output to the input can produce sounds which transpose up or down repeatedly from the initial input.

Distortion

Guitar-type 'fuzz' ranging from subtle harmonic distortion through to gross distortion can improve electric guitar and solo synthesizer sounds. Small amounts of distortion can also help to enhance any subsequent effects processing of sounds with limited harmonic content: filters, phasers and flangers can sound very poor if they are processing sounds which are little more than simple sine waveforms.

Distortion effects have very different results when used with monophonic and polyphonic sounds. Monophonic sounds are affected timbrally, with additional harmonics being added to produce a brighter sound. Polyphonic sounds are affected melodically, since the distortion produces a large number of new frequencies which are often not

Figure 6.7.1 Effects summary. (i) Reverberation and echo can be produced using a tapped delay line. (ii) Chorus, flanging, phasing and pitch-shifting can be produced using a delay line whose delay time is modulated using an LFO. (iii) Exciters and distortion use non-linear amplifiers to produce harmonic distortion. (iv) Compressors and auto-wahs use an envelope follower to filter or attenuate the signal.

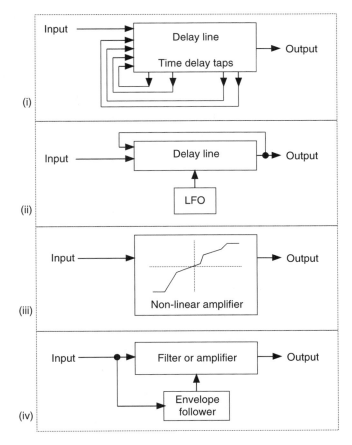

harmonically related to the input frequencies. These 'intermodulation' products are similar to the FM and ring modulation effects described in Section 2.4.3. In some circumstances, the results are useful: guitar power-chords, for example. But the result is often discordant: sounding rather too much like noise.

6.7.2 Effects synthesis

Until the early 1990s, the majority of effects processors which were built into synthesizers were still designed as an additional feature rather than as an integral part of the synthesizer – and they were usually little more than simple reverb or chorus units. This is most apparent when the topology of the effects processors is examined: before the 1990s, most had a single monophonic input but produced stereo outputs. The output of the synthesizer section thus needed to be combined into a single channel before being processed by the effects, despite most synthesizers producing stereo outputs directly. True stereo effects, where the input and output are stereo, and the effects are part of the synthesis, only began to appear in the 1990s (Figure 6.7.2).

Using effects only for post-processing of sounds produced by a synthesizer is probably a consequence of the use of simple reverb or

Figure 6.7.2 (i) Mono input effects processors are added into the output of some synthesizers, but they cannot use the stereo output, and so are mixed in with the output of the Pan processor.
(ii) Stereo input effects processors can be integrated into the synthesizer, including using LFO and envelope generators (EGs).

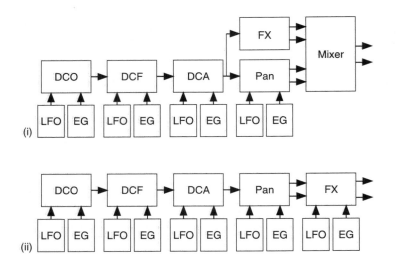

chorus units in early polyphonic hybrid synthesizers. As the effects have increased in complexity and become an essential part of the timbre-forming modifier section of the synthesizer, the need for linking the effects to the synthesizer controllers has correspondingly increased. By having some of the important effects parameters controllable in real time from the synthesizer, the effects processing can be made an integral part of the synthesizer. Chorusing or flanging whose depth changes with filter cut-off or panning which also changes the reverb time are just two of the many possibilities which are often not possible with stand-alone 'out-board' effects processors which are not part of a synthesizer. This idea is not new: many modular analogue synthesizers have dynamically controlled effects, but it has only recently become available on commercial digital synthesizers.

When the effects become part of the 'sound', then the 'traditional' isolation between the effects and the rest of the synthesizer does not work. Instead of choosing a sound, and then choosing an appropriate effects setting, the sound and the effect need to be linked together. This is feasible whilst the synthesizer is only producing one sound, but poses a problem when it is being used multi-timbrally. Whereas most synthesizers can be split into 'voices' which each produce one note of a specific timbre by multiplexing the same circuitry, an effects processor requires a separate complete dedicated audio processing section for each timbre that it is processing. Most stand-alone effects processors are designed to process just one stereo signal, rather than anything up to 16 stereo signals simultaneously.

This means that an effects processor intended for use as part of a multi-timbral synthesizer is not just a minor revision of an existing stand-alone – it requires additional real-time control inputs and the duplication of a large proportion of the audio processing. A partial solution developed during the late 1980s and early 1990s was to provide several simple effects processors and to provide independent inputs and outputs via effects send and return controls in much

Figure 6.7.3 Some multi-timbral synthesizers use global FX busses to enable any of the parts to have an effect applied to them. This means that only one effect is available globally. More sophisticated synthesizers may provide two or more separate effects processors, but it is rare for there to be the same number of effects as parts.

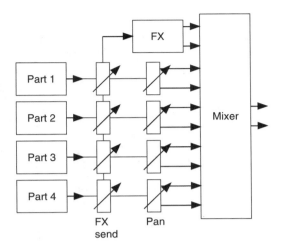

Korg's 1995 Trinity workstation featured multi-timbral effects, and so a sound could have the same effects applied when it was played on its own, or as part of a complex multi-part stack of sounds.

the same way as on an audio mixing console. Since many effects processors were already being designed to produce several simultaneous effects, this avoided any need for a major reworking of existing stand-alone designs. For multi-timbral synthesizers, this meant that two, three or perhaps four separate effects would be available, but that these were shared amongst all the parts. The topology of the effects processors was often predetermined by their original conception as a series of effects, with the result that a chorus or echo effect would always be connected to a reverb, and only limited combinations of chorus and reverb would be available, and then only globally (Figure 6.7.3).

This limitation was first addressed in the mid-1990s by providing global reverb and chorus effects processors, and additional separate processors which could be applied to individual parts. This 'global and individual' approach changed the nature of the effects processing inside a synthesizer away from the stand-alone 'out-board' studio use and started the development of a new type of effects processor. This process was completed when the first 'multi-timbral' effects processors were introduced into high-performance workstations in the second half of the 1990s. Although still incorporating some global reverb and chorus effects, separate effects processing for each part meant that the effect could be treated as part of the modifier section of the synthesizer (Figures 6.7.4 and 6.7.5).

6.7.3 Software effects

In the 1970s and early 80s, the majority of effects processing used analogue circuitry. The late 1980s and mid-1990s saw this being replaced by digital signal processors as part of the general trend to replace analogue circuits with digital processing. By the late 1990s, the processing power of ordinary microprocessors was such that they could carry out the same audio processing as the specialist DSP chips of 5 or more years before. This meant that a personal computer could be used as an audio processor, although a number of key elements

Figure 6.7.4 Some synthesizers allow access to the individual sections of multi-effects, so that a chorus followed by a reverb can then be used as a chorus and reverb as well as just a reverb. Individual effects processors that can be inserted into just one part are also found in some synthesizers.

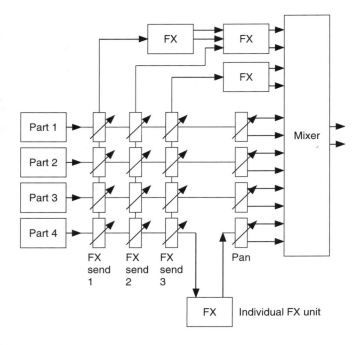

Figure 6.7.5 Fully integrated multi-timbral effects allow each part of a synthesizer arrangement to have the same effects when played individually or as part of the whole.

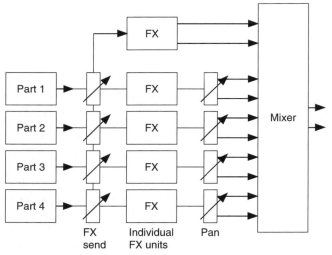

needed to be in place for this to change from a specialist application to the norm. These triggers included:

- audio input/output capability
- MIDI and audio sequencers, rather than just MIDI
- sufficient processing power in the PC
- fast hard drives and cheap random-access memory (RAM).

It was not until personal computers acquired these capabilities that the scene was really set for audio effects to move from hardware to software. From the vantage point of the 21st century, it is hard to appreciate just how limited the basic audio capabilities of many personal computers were in the early 1990s, or how slow the hard drives were, or how expensive RAM memory was ... but by the middle of the 1990s, the available processing power was sufficient to allow audio processing, and one final piece of technology changed things forever: the 'plug-in'.

Taking a cue from the 'plug-in' method of allowing extra functionality to be added to software which had proved so popular in graphics applications like Adobe Photoshop, the audio effects 'plug-in' was created to allow third-party developers to create and market their own audio processing software for use inside audio sequencers. A 'plug-in' is just software, but instead of being independent and stand-alone, it requires a larger piece of 'host' software that provides it with the inputs, outputs and other essential services so that it can work effectively. It is really just a software equivalent of a piece of out-board studio effects equipment.

In the case of a 21st century audio sequencer, the generic audio mixing and routing support is part of the audio sequencer, but the effects are typically provided via a 'plug-in' architecture. The creators and publishers of the audio sequencers include support for 'plug-ins' because it means that they can allow users to customise and enhance the software themselves. It also allows a wider range of functionality to be available, and is therefore especially suited to specialist or unusual requirements. In about 10 years, the third-party marketplace for audio 'plug-ins' has grown from reverb effects to complete 'soft' recording systems.

The 'plug-in'

The software environment which is required in a 'host' to provide the support for an audio 'plug-in' is in six main parts:

- initialisation
- audio input
- audio output
- control signals
- user interface
- termination.

The initialisation tells the 'plug-in' software to start up: this involves operations like setting up its initial state, requesting RAM memory, registering with the host software, and starting operation. The audio input and output can be streams of audio data or blocks of memory containing audio data. The control signals can be a wide variety of types: MIDI commands, general-purpose digital signals, trigger signals or digitised voltages. The user interface allows the 'plug-in' writer to provide a graphical interface for the end user to control the settings of the 'plug-in'. The termination causes the 'plug-in' software to unregister from the host software, free up memory, and finally, to cease operation.

Simple digital effects like reverb are programmed from a chain of time delays. Audio signals are passed through this chain and the outputs from several parts of the chain are mixed back into the chain, normally earlier in the chain. These feedback loops have gains of less than one, and so the audio signals fade away. Other effects are merely conversions of analogue effects into their digital equivalents and expressed in software rather than dedicated signal processing hardware. Synthesizer and sample-replay 'plug-ins' produce just an output audio signal.

Most users will only audition 'plug-ins' in order to choose their favourites, but user-programmed 'plug-ins' can be produced from a variety of resources ranging from high-level graphical tools to low-level code compilation.

'Plug-in' history

One early ancestor of software synthesizer 'plug-ins' was Digidesign's Turbosynth, a software synthesizer released for the Apple Macintosh computer in 1988 which used a graphical user interface (GUI) to assemble audio oscillator and modifier modules together to produce sounds. Turbosynth produced audio sample files as its main output.

Digidesign's Pro Tools audio editing and sequencing software, released in 1992, gradually acquired a number of hardware add-ons to provide additional audio processing and sample-replay capabilities, and these were consolidated in 1995 with the launch of TDM, a digital bus designed to allow more flexible use of hardware expansions. In 1996, Steinberg released the first version of VST, their audio 'plug-in' standard, and other sequencer manufacturers followed with their own different and incompatible 'plug-in' formats. Many 'plug-ins' are platform-specific, meaning that they will only run on a PC or a Macintosh, and some formats are also platform-specific. The early 21st century has seen some attempts to provide single 'universal' audio 'plug-in' formats that will work across platforms.

6.7.4 Using effects

The user interface metaphor which is almost always used for the output of multi-timbral synthesizers, especially in the context of effects processing, is the mixing console or mixer. This is an example of the way that synthesizers, sequencers and computers have gradually incorporated external equipment. First the effects were built in, and then the mixer that would have mixed together several synthesizers becomes part of a multi-timbral synthesizer, and finally the mixer becomes the output mixer for the sequencer and synthesizers.

The mixer takes the outputs of the voices and mixes these parts down into a stereo signal. Individual pan and level controls, and sometimes equalisation are included, but except on synthesizers with 'multi-timbral' effects processors, the effects provision is often implemented as several effects send and effects return controls. These are summed across the parts and sent to the specific effect, and then returned via the return level control; in some cases only the send control is implemented since the stereo outputs of the effects are mixed into the main stereo output and the effects output level control can be used to replace the effects return control. Individual send and return controls are used when an effects processor has one or more independent effects processors rather than a full 'multi-timbral' capability.

Making effective use of effects involves careful choice and frugality: using a deep flange on everything may sound impressive at first, but this initial impression soon fades:

- **Reverberation** is very good for providing a sense of space and size, as well as placing instruments closer to (less reverb, with short pre-delay and short reverb time) or further away (more reverb, with a longer pre-delay and longer reverb time) from the listener.
- **Chorus** can add extra movement to a static pad sound.
- **Echo** is useful for enhancing rhythmic elements, and can be used to provide syncopation effects by arranging the echo time so that it either falls on, or in between beats.
- **Flanging** is a very distinctive effect and needs to be used sparingly, although it can also be used as a chorus-type effect if the depth and feedback are kept deliberately low.
- **Pitch-shifting** can be useful as a subtle chorus replacement, but with feedback it can be used as a means of producing glissando or portamento effects.
- **Distortion** can improve the realism of electric guitar-type sounds, although this has become a cliché.

6.8 Editing

There are two contexts in which editing of a sound can take place: live changes to parameters whilst the instrument is being played in a performance; and programming sounds in preparation for a performance. Although both can use the same controls, it is more usual for the live changes to be made using the performance controls provided by the synthesizer's keyboard, wheels, pedals, etc. Programming sounds uses more detailed controls which can often be accessed from the front panel or via MIDI messages from a computer.

There are two principles of editing which the author uses:

- never use a preset sound
- never the same note twice.

The avoidance of factory preset sounds is the result of striving to sound different and to use new sounds. Some synthesists go to the extreme of clearing all the user memories in newly acquired instruments to force them to create new sounds.

One example of this variation technique is sometimes called 'micro-tempo changes'. Instead of using a drum machine to provide a precise metronomic tempo, a sequencer control or conductor track is used to make small changes to the

Avoiding playing the same note with exactly the same timbre or intonation is an attempt to try and emulate the unpredictability and imprecise control of the real world. Professional orchestral musicians spend years learning to repeatedly produce performances with very tight control over the tone and expression, and this requires considerable skill and practice. The results are impressive, but still have minor imperfections that are a key part of the 'realism' of the sound.

tempo, often less than one beat per minute of variation within a bar or phrase. In longer song structures, the chorus might be played several beats per minute faster or slower, or there might be a gradual acceleration or deceleration of tempo throughout a song.

By emulating these slight variations in a synthesizer using editing, a mathematically precise performance can be turned into something much more human and hopefully appealing.

6.8.1 Live parameter control

Making changes to a sound during live performance requires that the edits be easy to make and constrained to a specific parameter. This is usually set up by assigning a performance controller like a modulation wheel or foot pedal to a parameter which can be controlled using a MIDI controller message. The performance controller can then be used to change that single parameter with no concerns about changing any other parameters inadvertently. This is frequently used to provide live control over filter cut-off or a global envelope release time control when this is not available as a front panel control. Some synthesizers allow front panel controls to be set up in a similar way, although this is often as part of the sound editing facilities and is more prone to accidental editing of other parameters.

6.8.2 Programming sounds

Programming sounds is normally achieved either from the front panel or via MIDI messages from a computer running specialised editor software. The front panel method has the advantage of being rapid and requires no additional equipment, but the display is normally small and not well suited to the display of large numbers of inter-related parameters or detailed graphical information. Computer-based editing has the advantages of a larger screen and a more sophisticated user interface, but requires a computer and extra cabling, and the speed is limited by the screen update speed and the transfer of editing commands to the synthesizer. Both methods have their strengths and weaknesses; in general, the smaller the screen, the greater the advantage of the computer-based method for the beginner, whilst for the advanced user who requires quick edits, the front panel may well be preferred (Figure 6.8.1).

6.8.3 Editing techniques

There are a number of techniques which are used by sound programmers to produce useful sounds. Most of these involve an iterative process where a parameter is changed, and the result noted, and then another parameter is altered, and its effect on the sound is listened to. One essential requirement is that the programmer has a mental model of how the method of synthesis works, so that appropriate parameters can be changed. Without any clear understanding of how the synthesis technique produces sounds, then the programmer would effectively be changing parameters at random. A critical audition of the results of any of the computer-based editing programs which can produce sounds merely by randomising the parameter values will quickly reveal that the majority of the sounds produced by this method are not usable.

Figure 6.8.1 'Front panel' editing involves a very direct interface between the user and the synthesizer – the front panel controls alter the parameter value which changes the sound. Computer-based editing uses the computer to display the parameter values which has been obtained from the synthesizer by using MIDI messages. Any changes to the parameters are made on the computer screen using a mouse or keyboard commands, and the changed parameters are transmitted to the synthesizer using MIDI messages.

The other mental model that is often required is the mapping of how the synthesis technique works to the way that the user interface is organised. This is closely related to the mapping difficulties of user controls to extracted parameters in resynthesizers. Many mental models of synthesis techniques are based on the flow of the sound from the source through the modifiers and to the audio output, whereas the user interface of many synthesizers is based on a hierarchical set of pages which are not always related to the sound flow, and are sometimes very unrelated. Trying to maintain two sets of different mental models at once can be tiring and stressful, and this can be a strong contributing factor to avoid carrying out any editing.

Because S&S synthesizers normally contain a large number of preset sounds, both as programmed sounds, and as the underlying samples, it is very easy to use them as replay-only machines, rather than synthesizers. For this reason, they are often regarded as being harder to program, since avoiding the clichés which are provided by the samples and modifiers is often not very easy. Most users of S&S instruments will be more concerned with making minor changes to the sounds, so that they can be used in a specific idiom. The techniques for achieving this are somewhat different to other types of synthesizer, since they involve just changing the coarse modifier. Using an S&S instrument as a synthesizer rather than a sample-replay device involves much more detailed use of the facilities which are available. The examples which follow are thus designed to illustrate how even an instrument which is designed to replay samples can be used as a true synthesizer, with specific emphasis on making quick edits to a sound.

Sample changes

Choosing a different sample probably the simplest technique for making changes to an S&S sound. The 'sample selection' parameter is located and then changed. This changes the raw sound source of the S&S sound, and so should radically change the overall sound. The source samples are normally arranged in groups with similar timbres, so that several piano samples will be followed by a range of string samples, which will, in turn, be followed by another group of sounds. Choosing a different sample from the same grouping will only make subtle changes to the final timbre, whereas a sample from a different group may have a very big effect on the final sound.

Most S&S synthesizers have two sample sources which are processed separately and then combined either at the filter or at the output. This means that it is simple to leave one of the two samples alone, and just change the other. For example, many piano sounds are chorused by detuning two piano samples: by replacing one of the piano samples with another sample, the result is two different instrument sounds with the same envelopes, but detuned away from each other. The user is not restricted to choosing another piano sound, and in fact, there is no need to use a percussive sound at all; often string or brass samples, or even single-cycle synthesizer waveforms can make a very effective contrast to a piano sample. Just leaving the detuned piano preset and changing both piano samples for other samples can produce some unusual timbres.

Often, the mixing of two contrasting timbres can give a new 'composite' instrument which has some elements of the two component timbres, but has a new character all of its own. The relative levels of the two components may need some adjusting to maximise this effect – the volume of each is changed until the sound 'gells' into one new timbre. It is a very striking effect: two separate sounds suddenly become one. This technique is widely used in orchestral music.

If a 'familiar' timbre is mixed with an unusual or 'alien' timbre, then the extra information in the 'alien' sample can give a sort of 'halo' around the existing sound: it expands and enhances, often in an unexpected way. Adding bell-like timbres to electric piano sounds can emphasise the metallic nature of the sound and give it a new 'edge'. Often, a slight change to the envelope of the 'alien' sample section can enhance the result; reducing the sustain to zero and shortening the decay time will give a short 'blip' of the 'alien' sample at the start of the sound. Increasing the velocity sensitivity of the 'alien' part, so that it only sounds when a key is played hard, can produce a more realistic feel to the sound, since it mimics that way that harmonic structures tend to become more complex as the velocity of playing increases. Preset sounds which use velocity fading or switching to make changes to a sound depending on how hard they are played are very good for exploring a combination of 'familiar' and 'alien' samples; 'slap bass' sounds are a good place to start an exploration of this type of editing.

A common technique is to make up a sound from the attack part of one sample, and the sustain part of another. This first found commercial success in instruments like Roland's D50 synthesizer, where the technique was somewhat confusingly called linear arithmetic (LA) synthesis. Changing the samples for the attack or sustain can produce large changes to the sound. Even swapping the two around, so that the attack comes from the sustain sample, and the sustain from the attack sample, can give some unexpected timbres.

It should be apparent from the above examples that editing S&S sounds is not the same process as with most other methods of sound synthesis. The limited set of available samples means that more sophisticated editing techniques are required if the resulting sound is to avoid being merely a replay of one of the existing samples. The use of envelopes to control separate parts of the sounds is a particularly useful method.

Pitch changes

Pitch can easily be overlooked when editing a timbre. Shifting the pitch of a sound up or down an octave is merely a transposition, but some very useful and interesting timbres can be found if you explore the boundaries, which means going as low in pitch, or as high in pitch as possible. Shifting the pitch of samples down low usually produces dark and mysterious timbres, whilst shifting the pitch upwards can produce aliasing effects and noise-like sounds.

How the samples are constructed can also be important. Short loops can give a pitched sound, and drum sounds can provide huge resources of suitably short sounds, provided that they can be looped. Longer looped sounds produce pitches when they are pitch-shifted upwards, but the complex rhythmic patterns that they produce when shifted down can form a useful background to a pad sound. If possible, fixing the pitch of this type of sample when it is transposed by a large amount, so that it does not track the keyboard, is the best idea, since then the rhythmic part stays at a constant tempo, and does not speed up as notes are played towards the right-hand end of the keyboard.

Pitch envelopes are often overlooked by inexperienced sound programmers. Some brass sounds will have a quick rise to the note, but the only other commonly encountered use is a much more exaggerated pitch chirp on a lead sound, when it is often labelled as a 'funny' sound. But there are quite a few more creative opportunities with pitch envelopes. Subtle slow envelopes on vocal sounds can give a very realistic sound, especially if this is mixed with a more conventional 'choir' sample. Putting a pitch envelope onto only one of the two parts of a sound can give a very useful 'detune'-type chorus effect at the start of a note, and which is then followed by the two parts of the note coming into tune for the sustain portion of the envelope. By tuning the envelope range, so that it covers one or more octaves, it is possible to create sounds which have octave doubling for only part of the envelope.

LFO modulation of pitch is often restricted to providing vibrato. But a slow LFO applied to one part of a sound will then produce a cyclic 'detuned in-tune detuned in-tune' effect, which can be very effective on string sounds, particularly if this is layered with a more 'realistic' string timbre. Using a sawtooth waveshape on the LFO is one of the elements of many string-like synthetic timbres.

Envelopes

Envelopes tend to have the reputation of being suitable only for detailed editing. Quick edits are restricted to merely changing the attack or release times. But many S&S instruments provide more than one envelope, and it is often possible to change the envelope which is being used by a part very easily. The quick edit advice is thus to use the wrong envelopes. Shortening or lengthening attacks, decays and releases can turn sustained sounds into percussive sounds, and vice versa. String timbres sound very different if they have piano-type envelopes.

For an S&S instrument with two separate parts to the sound, contrasting envelopes can be used for the two parts. This opens up possibilities like using the sustain part of a string sound to provide the attack, and the attack sample (looped) to provide the sustain sound. Since these sounds can be layered with presets, long, slow envelopes can be used to provide evolving sounds which are much more interesting to listen to than static pads. Using envelopes to cross-fade from one sound to another is usually used just to paste together an attack and sustain samples, but this can also be used with contrasting sounds, and altering the cross-fading can provide more dynamic changes. Sounds which vanish in the middle and then reappear in a different form may appear quirky, but they are very useful for adding a bit of interest into washes and introductions. They can also provide some interesting song or melody ideas when they are used with a sequencer, since the timing will probably not coincide with the sequencer playback, and this can create complex poly-rhythms.

Filters

The filter controls probably take the second place after the sample choice, as the main focus of quick edits. The filter frequency is probably the most powerful of all the modifier control for determining the timbre of the sound. Filtering should be used to try and remove or enhance the sample 'fingerprint'. Removing parts of the spectrum of the sample may well disguise it, whilst emphasising other parts of the spectrum may also help to lose some of the distinctiveness. The more complex the filter, the more control that it will give over the harmonic content of the sound, so simple low-pass filters are not ideal. Low-pass, high-pass and band-pass filters are a good starting point, whilst notch and comb filters can be very useful, and any other shapes are a bonus.

As a general rule, the resonance control acts as a selectivity control – the higher the resonance, the more changes the filter will make to the sound. This is especially noticeable with a low-pass filter which has a resonance control which gradually makes it more and more like a band-pass filter: the sound becomes more interesting and synthetic sounding as the resonance approaches the point where the filter is about to self-oscillate.

The overuse of low-pass filters can tend to give a very thick and mushy sounding bottom end to sounds, whilst high-pass filters tend to 'thin out' the sound. Filters should always be used in combination with the other elements of the sound, rather than hoping that they can provide a stand-alone sound on their own. The one exception to this is the 'synth brass' filter sweep sound, where an attack decay sustain release (ADSR) envelope is used to sweep the filter cut-off frequency and process a sawtooth cycle or brass sample. The resulting sound is one of the strongest synthesizer clichés of the mid-to-late 1970s!

Summary

Although these modifications to sounds have been described in the context of S&S synthesizers, almost all of the techniques also apply to samplers. Many of the techniques also apply to other methods of synthesis, particularly subtractive, hybrid and FM synthesis (Figure 6.8.2).

6.8.4 Using factory sounds

Commercial synthesizers are typically supplied with a set of 'factory presets'. These are sounds whose purpose is to demonstrate the capabilities of the instrument to a potential purchaser. They are often well suited to this purpose, but rarely suited to subsequent use when

Figure 6.8.2 Some quick edits on an S&S synthesizer.

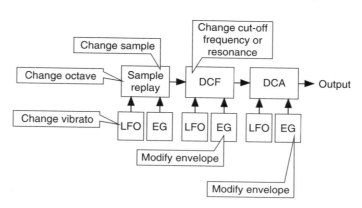

combined with other instrument sounds in a mix. Editing factory sounds is a good way of learning how to edit a synthesizer, and can also be used to produce more widely applicable sounds.

The basic techniques for making factory sounds useful are relatively straightforward: often a combination of reducing the amount of effects which have been applied, and changing the envelope and modulation. Here are a few typical symptoms and the required correction:

- **Buried in reverb**: Reduce the 'wet'ness of the reverb mix or the depth of reverb effect.
- **Drowned in chorus and detune**: Reduce the detuning of the oscillators, and reduce the depth of LFO modulation in the effects.
- **Excessively wide stereo image**: Adjust the pan of the components of the sound so that they are not panned hard left or right.
- **Auto-pan**: Sounds which move around in the stereo sound image excessively are interesting once. Reduce the depth of LFO or envelope modulation of the panning or stereo position control.
- **Slow rise and fall**: Very languorous strings and lots of chorus can sound very impressive on pad sounds, but the attack and release times are often too long. An expression or volume pedal is a better and more controllable way of getting the same sort of crescendo effect.
- **Polyphony stealing**: Long release times can make block chords sound impressive on their own, but can swamp other instruments in a mix. Reducing the release time and increasing the velocity sensitivity of some of the components will give a much more expressive sound which does not use up excessive amounts of polyphony, and which will work better when part of an accompaniment.
- **Filter resonance**: Filter oscillation can be overused, and the very distinctive accentuation of harmonics at high resonance settings can also become a cliché. Lower settings of resonance, particularly when the resonance is controlled by keyboard note position, can be a much more powerful, controllable (and imitative) effect when used in performance.
- **Filter modulation**: Cyclic sample and hold modulation of filter cut-off frequency at about 4 Hz has become an overused cliché. Much more useful and interesting is a much slower modulation, or one which is related to keyboard velocity.
- **Echo and timed samples**: Repeated sample loops with widely panned stereo echoes can produce some very impressive demonstration sounds. But every user of the instrument will also be thinking exactly the same thing, and using such an immediately recognisable sound is not very imaginative (at best). Any use of these highly distinctive sample loops is probably best avoided entirely, or else they need to be altered so drastically that they cannot be easily recognised: large pitch changes are one technique.
- **Hyper-reality**: String buzzes, harpsichord jack-clicks and other ephemera of real-world instruments tend to be over-emphasised

in factory sounds. But they quickly pale when used with other instruments or for solo work of more than a few bars. Increasing the velocity sensitivity, or using keyboard note position to reduce the ubiquity can make these sounds much more acceptable.

- **Stacks and one-note chords**: Fifths, octaves and rich, thick sounds with lots of harmonics can be impressive in a demo. But parallel chords can quickly lose their appeal with repetition, and they restrict the possibilities for harmonisation, cadences, suspensions, etc. Simpler, purer sounds, which can be used to make up chords when required, are much more useful in wider contexts.

6.8.5 *Managing edited sounds*

Although producing, controlling, manipulating and arranging sounds is the 'foreground' activity of the synthesist, a number of associated background management functions need to take place in order for these performance-related activities to happen. Since sounds and music are the major resources for a synthesist, they need to be catalogued, stored and made available for subsequent retrieval.

Although a large number of synthesizers now offer digital storage of their operating parameters, this is not always possible for analogue instruments or for some effects processing equipment, nor for custom-built apparatus. Notes on these items may well be stored on paper in notebooks or binders, but the principles of resource management remain the same.

Sorting

Part of the process of learning how to use a synthesizer is the production of sounds. Some synthesists prefer to start their investigations from the preset sounds which are provided, whilst others ignore or remove any presets and work from the basic 'initialised' sound: which is normally a simple sine wave.

The programming of sounds can be approached in many ways. A methodical exploration of the available parameters may suit one synthesist, whilst an iterative process based on making changes to the parameters more or less at random and observing their effects may work for another person. In this learning phase, a large number of sounds are typically produced, and these can either be used as finished sounds, or they can provide the raw material for further programming. After some programming, the synthesist begins to develop a mental 'model' of the operation of the synthesizer, and this will be confirmed by further programming: followed by checking of the expected results against the reality. Once this stage is complete, the synthesist should feel confident of their ability to produce almost any required sound.

Regardless of their origin, the sounds which are produced by the programmer need to be filed in a way that allows them to be reused at a later date. Few synthesists have the ability to rapidly produce a specific sound when required: instead, they use these pre-prepared sounds or edit them in context. In order to file the sounds they need to be sorted and categorised, and a number of systems can be used to provide this function. Perhaps the most useful technique in cataloguing the sounds and music is to split the pool into categories. There are many different methods of naming sounds: ranging from ones which describe them in generalised terms like slow, smooth, bright or dark, whilst others use references to conventional instruments: brassy, piano-like, flute-like or bell-like. Some composers have used references to the moods or emotions which the timbres suggest. Most of the commercial software packages use a mixture of generalised terms and references to specific instruments.

Whilst categories are useful for grouping sounds and retrieving them at a later date, they can be unwieldy for everyday use. Simple names for sounds are a much better way of ensuring that they will be remembered easily in day-to-day use. Most synthesizers use some sort of 'bank and number' system to numbering their sounds: often based around banks containing 8- or 16-numbered sounds. The banks can either be numbered or assigned letters, so 4-12 and B5 are both typical examples. Numbers of these forms are not readily associated with sounds, and so the naming of sounds is very important. There are a number of methods used for naming sounds. The simplest is to attempt to describe each sound as accurately as possible given the limitations of the number of character which can be used. Since this is normally between 8 and 16 characters, names like 'chorused electric piano with bell-like attack sound' will need to be abbreviated. One useful technique is to divide the names into two parts. The first two or three characters are used to indicate the rough grouping of sound: piano, string, brass, etc. The remaining characters can then be used to provide more detailed description, albeit in an abbreviated form. This method produces names of the form 'EP:ChrBelAtt' where the descriptive part is split into meaningful parts by using capital letters instead of spaces.

More egotistical methods of naming sounds exist. These sometimes use any additional available characters to provide a graphical name: '- = [V] = -' being a typical example. Other programmers use in-jokes or clever plays on words: one rather obvious example being 'sdrawkcab' for a sound which gives the impression of being played backwards. Foreign language words are often popular because they have a mystique due to their unfamiliarity, or they may further plays on words: 'bête noire' (literally French for 'black beast', but English for a weakness) and 'koko wa?' (Japanese for 'What is this?') are two examples. For this type of name, the more memorable a name is, then the more effective it is.

Finding

Once filed away with a suitable set of category words, name and perhaps grouping, the sound is ready to be used. Finding the right sound involves using the reverse process of naming: deciding on the type of sound to be used, and then choosing category words and perhaps groupings so that a 'short-list' of potential candidate sounds can be produced. Auditioning of these sounds can then begin, with the user testing each sound in context with other sounds which will be used. Once the sound-set has been chosen, then the sounds will probably be stored as a single entity temporarily whilst the project is in progress.

Librarians

Software which carries out the naming, categorisation, grouping, sorting and finding functions is called *librarian software*. A number of programs exist for this purpose, and they are often associated with programs which can be used to edit sounds, called *editors*. Two types of these programs exist. Generic or 'universal' versions can be used with almost any synthesizer or related equipment since their behaviour can be programmed and new equipment can be added merely by loading in new programs. Specific versions are designed to be used with only one synthesizer, although more limited they are often lower in cost than the generic versions.

Storage

Storing sounds can be achieved in several ways. Most synthesizers have internal RAM memory storage, often with additional card-based or floppy disk-based storage. System exclusive MIDI messages can be used to send and receive the same sound data, and this method is used by librarian software. Librarians can then store the sound data on floppy, hard or optical disk storage. In general, the storage of important sound data should be in several different locations and on different media, so that the chances for losing sounds are minimised whilst the possibilities for recovering sounds are maximised.

6.9   Questions

1. What types of control over a performance does synthesis provide?
2. Outline three approaches to stacking.
3. Describe three methods of layering sounds.
4. What is hocketing?
5. What is the difference between multi-timbrality and polyphony?
6. Outline a simple note assignment strategy. How could this be improved?
7. Describe the major types of audio effects and their uses.
8. Describe the differences between an effects unit intended for use in a monophonic synthesizer and one designed for use in a multi-timbral polyphonic synthesizer.
9. Outline some of the ways of controlling a synthesizer during performance.
10. Which editing techniques would be common to an S&S synthesizer and a subtractive analogue synthesizer?

Time line

Date	Name	Event	Notes
1934	John Compton	UK patent for rotating loudspeaker	
1939	Hammond	Hammond Solovox, monophonic 'synthesizer'	British Patent 541911, US Patent 209920
1958	Charlie Watkins	Charlie Watkins produces the CopyCat tape echo device	
1969	Robert Moog	Mini Moog was launched. Simple, compact monophonic synthesizer intended for live performance use	Hugely successful, although the learning curve is very steep for many musicians
1969	Peter Zinovief?	Electronic music studios (EMS) produced the VCS3, the UK's first affordable synthesizer	The unmodified VCS3 was notable for its tuning instability
1975	New England Digital	Synclavier was launched, the first 'transportable' all-digital synthesizer	Expensive and bulky
1976	Lol Creme and Kevin Godley	The Gizmotron, a mechanical 'infinite sustain' device for guitars	A variation on the 'bowing with steel rods' technique
1978	Electronic dream plant	Wasp synthesizer launched. Monophonic, all-plastic casing, very low-cost, touch keyboard, but it sounded much more expensive	Designed by Chris Hugget and Adrian Wagner
1979	Fairlight	Fairlight Computer Musical Instrument (CMI) announced. Sophisticated sampler and synthesizer	The start of the dominance of computers in popular music
1979	Roland	Roland space-echo launched, used long tape loop and has built-in spring line reverb and chorus	A classic device, used as the basis of several specialist guitar performance techniques (for example, Robert Fripp)
1981	Moog	Robert Moog was presented with the last Mini Moog at NAMM in Chicago	The end of an era
1983	Yamaha	MSX Music Computer: CX-5 launched	The MSX standard failed to make any real impression in a market already full of 8-bit microprocessors
1984	Kurzweil	Kurzweil 250 provides 2 Mbytes of read-only memory (ROM) sample playback	Memory size was very date dependent …
1984	Sequential	Sequential launch the max, an early attempt at mixing home computers and synthesizers	A complete failure, too early for the market
1984	Sequential circuits	SixTrak. A multi-timbral synthesizer with a simple sequencer	The first 'workstation'?
1984	SynthAxe	MIDI-based guitar controller with separate strings, trigger switches and fretboard switches	Nice design, but very expensive
1985	Digidesign	Sound designer software was released for the Macintosh, early computer-based recording and editing software	Digidesign were eventually acquired by avid, a video editing company …

parameters like brightness or release time, and which may well need adjusting in the course of a performance. With all these possible controls, the synthesizer player can begin to look rather reminiscent of a one-man band.

7.1.1 Performance

Regardless of how the control is achieved, the important consideration is that the player of a synthesizer should be able to interact with the instrument in a way that allows expression to be conveyed to the listener. The more natural and intuitive that the controls are, the more that they need to avoid imposing limitations on the player. Although the keyboard imposes many limitations on the player in performance, it is still the most common interface, and careful use of the additional performance controllers can minimise the problems.

Familiarity with the instrument's controllers can help considerably. The use of a single 'master' keyboard is useful in this context because the player will be comfortable with the keyboard and its controls, and will not need to try and locate wheels or pedals in poor lighting conditions on stage. Careful preparation of the music can help to identify points where additional performance controllers like pedals, or even alternative controllers like a string or wind controller may be required. For a player of an orchestral instrument, the instrument and its controllers are one and the same, and so no consideration is required for how to accomplish a specific musical result, whereas a synthesizer and its controls can be separate and can be changed to suit the player.

Some of the possible controllers are described in the remainder of this chapter.

7.2 MIDI control

7.2.1 MIDI

The wide adoption of the musical instrument digital interface (MIDI) has meant that it has become the dominant method of digital control for hardware. Even in software standards like VST for audio 'plug-ins', MIDI-style controllers are still assumed.

Although MIDI (Rumsey, 1994) provides a large number of controllers, the basic underlying assumption is that a keyboard will be used to provide the main source of pitch and performance information; since the pitch and velocity parameters are tied together in the note on and off messages, whilst volume information, pitch bend, modulation and other parameters all require separate continuous controllers.

MIDI is capable of detailed and precise control in some circumstances, but it has some limitations which reflect its keyboard-based origins. MIDI is poor at specifying the transitions between notes, which makes controlling guitar sounds difficult because the reuse of a string cannot be controlled properly, and attempting to control polyphonic vibrato requires the use of several channels of MIDI's mono mode, which is not widely supported by commercial synthesizers. Glissando and portamento effects are only controllable for parameters such as time via MIDI, and MIDI expects that each change in pitch associated with a note will initiate a new envelope.

MIDI provides two ways to control synthesizer (or sampler) parameters: controller messages and system exclusive messages.

7.2.2 MIDI controllers

MIDI controllers are a sort of general case of the pitch-bend or pressure MIDI messages. They are performance controls: modulation, expression, vibrato and many more. There is provision for a large number of possible MIDI controllers: more than 64 000, but only a few are typically used. The most common is the modulation wheel (controller number 1), often found next to the pitch-bend wheel on the left side of a synthesizer keyboard.

Controller messages have 3 bytes. The first byte identifies the message as a controller message, and also indicates which of the 16 MIDI channels the controller is on. The controller number is the second byte. The 128 possible controller numbers are organised into three basic types:

- **continuous controllers**, for example modulation wheels, pedals, etc.
- **switches** (theoretically just two values: on and off)
- **mode messages** to control note reception conditions.

Continuous controllers are divided into 14 bit (from 0 to 63) and 7 bit (from 64 to 95). Switches are a special case of a 7-bit continuous controller, and true on/off switches occupy a few numbers (from 64 to 69). Controllers 96–119 are split into two sections: the first covering 96–101 is the *registered* and *non-registered* parameters section, where huge numbers of parameters can be assigned to MIDI controller messages without using up valuable 14- or 7-bit controller numbers. In fact, it is possible to use just controllers 6 and 38 to control the most and the least significant bytes (MSB and LSB, respectively) of all these extra parameters.

In summary, the controller numbering looks like this:

0–31	14-bit controllers (MSByte)
32–63	14-bit controllers (LSByte)
64–69	Switches (pedals and foot switches)
70–95	7-bit controllers
96–101	Registered and non-registered parameters
102–119	Undefined
120–127	Mode messages

Not all controllers numbers are assigned to specific devices like pedals, wheels etc.; in fact, most are not. Those that are seem to have been driven by historical precedent.

14- and 7-bit controllers

MIDI controllers come in several varieties. Some act as switches for just on and off controls: sustain pedals, for example. Some act as 7-bit controllers with 128 different values: volume pedals, and even some slightly esoteric sustain pedals. But there are also 14-bit controllers where detailed control is needed. Using 14 bits allows up to 16 384 different values, which is enough for most purposes, and in fact, very few manufacturers take advantage of 14-bit controllers. For volume pedals, 128 values is often too much precision, and only 8 or 16 different volume settings are actually used in some cases. Modulation wheels often use all 128 possible values to give smooth transitions for vibrato, etc. The most common use of 14-bit precision is the pitch-bend message, which is technically not a controller at all! The ultimate limit on the number of values is the switches, where there are effectively only two values that are used.

The 7- and 14-bit controllers can work together. The range (maximum to minimum values) of a MIDI controller is normally thought of as being 0–27, which makes sense for a 7-bit controller, but what about 14-bit controllers? Where are those 16 000 values? In fact, the 14-bit controllers 'fill in' the gaps between the 7-bit values, which provide much finer resolution. So the 7-bit controllers give coarse control, whilst the 14-bit controllers allow much more precise adjustment.

The 7- and 14-bit controllers share messages. The first 32 MIDI controllers appear to be 7-bit controllers, but they are actually just the most significant part of a 14-bit controller which uses another controller number as the fine resolution part. Using computer-speak, the 'most significant byte' and 'least significant byte' are referred to by acronyms: MSByte and LSByte. The MSByte controllers can be used on their own as 7-bit controllers, whilst the LSByte controllers just add 32 to the MSByte controller number, thus volume (MSByte controller number 7) can also be finely tweaked with LSByte controller number 39, if your equipment supports this feature.

Note that an MSByte controller message on its own resets the LSByte value to zero.

Controller numbers

- **Controller 1**: Modulation wheel, MSByte
  Wheels are not the only source of control for this message. Levers, joysticks, sliders and pressure-sensitive plates can all provide the physical control.
- **Controller 2**: Breath controller, MSByte
  Breath controllers convert breath pressure into a control signal. They are especially useful for controlling monophonic synthesizers which are playing melodies.
- **Controller 4**: Foot controller, MSByte
  Usually a pedal, although some foot controller inputs will accept a control voltage instead.

- **Controller 5**: Portamento time, MSByte
  This control can be used as a switch, or as a continuous control over portamento time.
- **Controller 6**: Data entry, MSByte
  The data entry front panel control may be a slider, a rotary dial or a keypad.
- **Controller 7**: Main volume, MSByte
  Often only the MSByte is used to control volume, which can cause 'zipper' noise because of the large jumps in volume caused by just 127 steps in volume.
- **Controller 8**: Balance, MSByte
  This controls the volume balance between two sounds, and is taken from 'organ' terminology.
- **Controller 10**: Pan, MSByte
  This controls the position of the audio in the stereo sound field.
- **Controller 11**: Expression controller, MSByte
  Typically a pedal, the expression controller provides a volume boost in addition to the main volume, and is intended to be used for accenting.
- **Controller 12**: Effect control 1, MSByte
- **Controller 13**: Effect control 2, MSByte
- **Controllers 16–19**: General-purpose controllers 1–4, MSBytes
  These are intended to be general-purpose controllers, which means that they will be defined by individual manufacturers to suit specific parameters in their instruments.
- **Controllers 32–63**: LSBytes
  Controllers 32–63 are the LSByte equivalents to the above controllers.
- **Controllers 64–69**: Switches (7-bit controllers)
  64   Damper pedal (sustain)
  65   Portamento on/off
  66   Sostenuto
  67   Soft pedal
  68   Legato
  69   Hold 2
  These are actually 7-bit controllers, although the most MIDI equipment will expect them to be only two values: on and off.
- **Controllers 70–74**: Sound controllers 1–5
  Controllers 70–74 are the defined sound controllers. These have a default descriptive name, which is shown below, and are designed to provide simple 'global' quick editing facilities. They are intended to allow slight changes to a sound, probably during a performance, rather than editing actual parameter values permanently.
  The names are guides to suggested usage, but they can be redefined to suit a particular application. They are in two groups:
  Timbre controls
  70   Sound variation
  71   Timbre/harmonic content
  74   Brightness

Envelope controls
72   Release time
73   Attack time

- **Controllers 75–79**: Sound controllers 6–10
  Controllers 75–79 are the undefined sound controllers. These do not have any predefined name, and can be assigned freely by a manufacturer.
  75   Sound controller 6
  76   Sound controller 7
  77   Sound controller 8
  78   Sound controller 9
  79   Sound controller 10
- **Controllers 80–83**: General-purpose controllers 5–8
- **Controller 84**: Portamento control
  This is an unusual controller in that it indicates the note number from which the currently 'portamento'ed note started.
  Controller numbers from 102 to 119 are undefined.
- **Controller 120**: All sounds off
  Controller 120 is the all sounds off message. This is an improved version of the all notes off message, since it forces any sounds which are currently playing to stop as quickly as possible.
- **Controller 121**: Reset all controllers
  Controller 121 is used for the reset all controllers message, a sort of all notes off but for controllers instead of notes. Although this message can be interpreted as meaning just the MIDI controllers, it actually indicates that the current state of all controllers, pitch bend, modulation wheel and pressure should be returned to the reset state (typically zero for most controllers, with the exception of the pitch-bend wheel, which resets to its centre position (no pitch bend), and the volume control (controller number 7), which returns to full volume (127)).
- **Controller 123**: All notes off
  The all notes off message is designed for use by sequencers when a sequence playback is stopped whilst some notes are still playing, although it often seems to be used to indicate when all the keys on a keyboard have been released.

Finally, remember that the MIDI controllers are channelised, that is, the data is sent using a specific MIDI channel.

7.2.3   *System exclusive*

The system exclusive message is unusual because it is the only MIDI message which has special bytes which indicate the beginning and the end. All other MIDI messages start with a special byte which indicates the type of message which follows, and also includes the MIDI channel number. The length of a message can normally be determined by knowing what type of message it is, but system exclusive (normally abbreviated to sysex) messages can be of any length, and so the start and stop bytes are used to indicate the limits of the message.

The sysex provides a carefully designed 'loophole' to the MIDI specification which allows manufacturers to have extra information in their

Table 7.2.1 Some MIDI controller numbers

14-bit MSBytes	
0	Bank select
1	Modulation wheel
2	Breath control
4	Foot controller
5	Portamento time
7	Volume
8	Balance
10	Pan
11	Expression
12	Effect control 1
13	Effect control 2
7-bit contollers	
64	Hold, damper, sustain
65	Portamento
66	Sostenuto
67	Soft pedal
91	Effects 1 (ex external effects depth)
92	Effects 2 (ex tremolo depth)
93	Effects 3 (ex chorus depth)
94	Effects 4 (ex celeste depth)
95	Effects 5 (ex phaser depth)
Non-registered and registered	
6	Data entry MSBytes
38	Data entry LSBytes
96	Increment data
97	Decrement data
98	Non-registered parameter number lsb
99	Non-registered parameter number msb
100	Registered parameter number lsb
101	Registered parameter number msb
Mode	
121	Reset all controllers
122	Local control on/off
123	All notes off
124	Omni off
125	Omni on
126	Mono on
127	Poly on

own specific formats embedded in MIDI messages whilst still retaining full MIDI compatibility.

The byte which immediately follows the sysex start byte is used for the identification of the owner of the message, and it is commonly called the manufacturer's ID number. This byte is rather like the address on an envelope: the envelope is ignored by everything except the correct addressee, who can open it and see what is inside.

The actual contents of the sysex message are entirely up to the individual manufacturer, but typically it consists of sound data, editing or control functions. The format of these messages is left entirely to the manufacturer and although most manufacturers maintain a reasonably consistent standard within their own products, there is almost no commonality between manufacturers. Often, the format includes a few bytes which indicate the intended destination device; this ensures that only editing or control messages which are applicable to a specific device are acted upon. Most sysex formats have 1 or 2 bytes which indicate which parameter value is being changed, and then 1 or 2 bytes for the value of the parameter itself. To ensure that the message has been correctly received, some formats also include a checksum byte which can be used by the receiving instrument to determine if an error has occurred during transmission of the MIDI data.

Sysex editing and control messages are normally short: less than 10 bytes in length. Longer messages are normally complete sets of data bytes for a sound, or a set of sounds. Information on the contents of the sysex messages for a specific synthesizer or sampler is often included with the owner's manual.

7.2.4 ZIPI

ZIPI (*Computer Music Journal*, Vol. 18, No. 4) is a proposal for an alternative interface for connecting a performance controller to a synthesizer. It provided much more detail and flexibility in the way that pitch and performance information are conveyed. In particular, it avoided having any assumptions about the type of controller that would be used; so there were no keyboard-specific limitations. ZIPI was not intended as a replacement for MIDI, more a means of transporting performance control from a player's instrumental controller to a remote synthesizer. The exact format of the player's controller is not defined: it could be a keyboard, stringed instrument or wind instrument.

ZIPI's high speed protocol for transferring performance controller information was superceded by Firewire and USB2.0, and no commercial examples using ZIPI were produced.

7.3 Keyboards

The music keyboard has a distinctive appearance that is widely used as a visual metaphor for music or pianos. The interlaced black and white keys are widely used for note selection, as well as starting and ending envelopes.

Keyboards provide two major outputs: *discrete pitch information* about the notes which are being played; and *event information* about the start and end of note events. Both of these are produced by pressing down on the keys and releasing them. The convention is that pressing down on the keys produces both a note pitch indication and a note start event. Releasing the key allows it to return to its rest position and produces a note end event.

'Velocity' measures the speed of the key movement from the 'off' to the 'on' state, or the 'on' to the 'off' state.

Some keyboards also provide additional information on the velocity with which the key is pressed or released; this is known as either 'attack velocity' or 'release velocity', although sometimes the word dynamics is used as a synonym for velocity. When a key is held down and extra pressure is applied, some keyboards will produce information about

Figure 7.3.1 An analogue synthesizer keyboard uses a chain of resistors, and pressing a key closes a switch which connects a bus to the chain of resistors. Only six switches are shown, but a real keyboard would have 30 or more keys. The position in the chain determines the output voltage. A separate switch (not shown here) produces the gate signal and a trigger pulse.

this 'after-touch' or 'key pressure'. These are both covered in the next section.

7.3.1 Polyphony

The simplest keyboards are monophonic. A chain of resistors connected in series with a voltage applied across them can be used to output a voltage whose value is proportional to the position on the chain. By using the keys of a keyboard to switch a contact wire onto the chain of resistors, a keyboard can be used to provide a monophonic note output. By using a second switch contact to produce a note status indication, note event information can also be produced. This is the type of keyboard which was used in early analogue synthesizers (Figure 7.3.1).

By extending the basic design of the monophonic 'analogue' keyboard, it is possible to produce a keyboard which will produce two voltages: one for the highest note which is being pressed down, and the other representing the next lowest note which is held down. Such 'duophonic' keyboards are often found on analogue performance synthesizers.

Keyboards based around chains of resistors are normally designed so that they have so-called 'top-note' priority. This means that if several keys are held down, then only the highest voltage will be produced. This allows a two-handed performance technique to be used where notes are held with the left hand whilst additional notes are played staccato with the right hand. Each time the right hand is released the pitch jumps down to the note being played by the left hand. Duophonic keyboards provide separate storage for these two voltages and allow two-note chords to be played instead.

Polyphonic keyboards can be produced by extending the 'resistor' chain method, but in practice, digital scanning techniques are used. A counter is used to send a pulse to each of the key contacts in turn, and any keys which are held down will allow the pulse to be switched onto a common bus-bar contact, and the keys which are held down can be determined by examining their relationship to the counter. The keyboard scanning is thus based on time-division, and the rate at which the keyboard is scanned determines the minimum timing resolution with which a key press can be measured. To minimise the required wiring, and to align with byte-wide scanning chips, most scanned keyboards arrange the keys in the form of an '$N \times 8$' matrix rather than a linear form (Figure 7.3.2).

Figure 7.3.2 A scanned keyboard uses a matrix of switches which connect one of the outputs of a scanner (it cycles round each of its outputs, putting a logic one signal on each in turn) to a detector and decoder circuit. One switch can be used to produce note and gate information. Velocity information would require an additional switch or a changeover switch.

Scanned keyboards produce note pitch and event information directly from a single contact switch. The polyphony and note priority (if a monophonic keyboard is being emulated) are controlled by software. For velocity information to be produced, a second switch, or a changeover switch are required so that the time between the start of key movement and the end of the key movement can be determined.

7.3.2 Types of keyboard

Keyboards come in two major forms: *organ type* and *piano type*. Although both are fitted to synthesizers, the organ type tends to be used in low-cost and mid-range products, whilst the piano type is more often found in 'master' controller keyboards and 'top-of-the-range' or 'flagship' products.

Organ-type keyboards have light, hollow, plastic keys and the black keys normally slope downwards away from the front of the keyboard. Organ keys tend to be about 140 mm from the front of the keyboard to the back of the playable key surface. The key action is fast and very light, with the key being returned to the rest position by a spring. Organ-type keyboards sometimes have small weights attached to the underside of the front of each key to provide additional weight when they are used as master keyboards; this is intended to emulate some of the feel of a piano type of keyboard. Organ-type keyboards as used on synthesizers are normally either five octave (61 keys) or just over six octaves (76 keys): frequently referred to as an 'extended' keyboard.

For use in small portable 'fun' keyboards and synthesizers, a variant of the organ-type keyboard is often found. This has narrower keys (an octave fits into the space of an 11th on a normal keyboard) and the playable surface of the keys is only about 80 mm from front to back.

Piano-type keyboards have heavier keys which are normally wood covered by a plastic moulding. The black keys normally have flat tops. Piano keys are about 150 mm from the front of the keyboard to the back of the playable key surface. The piano 'action', a complex mechanical arrangement of levers and weights, was originally designed to transfer the downwards movement of a key into the hammer striking the string, but in this application only the mechanical 'feel' is required since there are no strings. In a piano action, the return of the key to the rest position is not a simple spring; it is related to the bounce of the hammer on the string. Piano-type keyboards are often referred to as 'weighted' because of the heavy keys and the feel of the 'action' relative to an organ-type keyboard. Piano-type keyboards normally have at least 76 keys, and frequently more.

7.4 Keyboard control

Although the basic keyboard produces only pitch and event information, the action of pressing the key and then applying pressure to the key can be used to produce additional information.

'Velocity' is the name given to the output from the keyboard which represents the rate at which the key is pressed or released. A pair of contacts is used to measure the time that the key takes to travel from just after when it is first pressed to when it is almost fully down. Changes in velocity whilst the key is being pressed do not normally affect the velocity, since the measurement is based on just the time difference between the switching of the two contacts. Attack velocity is measured when the key is first pressed, and is used to control, the dynamics of the sound: often the level and/or the timbre, whilst release velocity is normally used to control the length of the release segment of the envelope. Release velocity is only rarely implemented in keyboards, although many synthesizer sound generators will respond to release velocity.

'After-touch' or 'key pressure' is the name given to the additional pressure which is applied to a key when it is being held down. The key requires a certain amount of pressure to overcome the spring which pulls it back to the rest position when the key is released, and the after-touch measures any additional pressure which is applied in excess of this. After-touch can be a global parameter where the highest pressure applied over the entire keyboard is used as the output value, or it can be measured individually for each key. These are called 'monophonic' and 'polyphonic', respectively.

7.5 Wheels and other hand-operated controls

The early analogue modular synthesizers used rotary control knobs to control the majority of the non-switched functions. Knobs were thus used for the first controls for pitch bend, and modulation controls were merely the output level controls for LFOs. But knobs were not very satisfactory for live performance use, and a number of alternative approaches were tried for providing the same sort of pitch-bend and modulation control, but in a more intuitive way. The four major contenders were:

- wheels
- levers
- joysticks
- pressure pads.

7.5.1 Wheels

Wheels are a development of the rotary edge controls which turn a rotary control onto its side and use a disc instead of the knob. Pushing the disk forward or backward turns the rotary control. The disks used range from 40 to 80 mm in diameter, and only about a third or a half of the disk protrudes above the panel surface. Some disks have a textured or grooved edge, whilst others are smooth; some manufacturers make them out of clear plastic with polished edges. A small semicircular cut-out or detent provides a reference point for the position of the disk. The disk movement normally uses about 90 degrees of rotation, which is a quarter of a complete circle.

For pitch-bend purposes, the detent is in the centre of travel of the rotary control, and so the pitch can be changed up or down: the rotation is about ±60 degrees away from the centre point of the pitch bend. A spring arrangement is normally used to return the pitch-bend wheel back to the centre position, and this is reinforced by the use of a second detent and a sprung cam follower which slots into the detent to provide a tactile centre position into which the wheel clicks. Pressure on the wheel is required to overcome the detent and the springs before it will move. For modulation use, the detent is with the disk full towards the player, and modulation is added as the disk is pushed forward. No spring return is used with most modulation wheels.

Although most wheels are located so that they move away from or towards the player, some wheels have been designed which move from side to side, so that to bend the pitch upwards the wheel is moved to the right, whilst for pitch bend downwards the wheel is moved to the left. Modulation is added by pulling the wheel towards the player, or pushing the wheel away from the player.

7.5.2　Levers

The lever replaces the disk with a short lever with a length of about 50 or 60 mm. The rotational movement is normally less than a wheel: a maximum of 90 degrees and a minimum of about 45 degrees. As with wheels, the pitch-bend lever is sprung so that it returns to the central detent position.

7.5.3　Joysticks

Joysticks combine the pitch-bend and modulation functions into a single control. Both side-to-side pitch and forwards/backwards modes have been used, with modulation using the opposite axis. Additional control has been provided on some joysticks by allowing the joystick to be rotated or moved up and down.

7.5.4　Pressure pads

Soft rubber pressure pads and metal plates fitted with strain gauges have been used to provide pitch-bend and modulation controls on some synthesizers. Separate pads are required to bend the pitch up and down, whilst the modulation pad can only be used to add modulation interactively – it cannot be set to a value and left, unlike the modulation controls provided by wheels and levers (Figure 7.5.1).

The two main controllers that wheels and similar devices are used to control are pitch bend and modulation.

Figure 7.5.1 Wheels, levers and joysticks are all alternative ways of presenting a user interface to a rotary control. This diagram shows a cross section through a wheel and a lever.

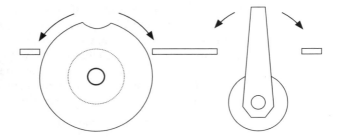

7.5.5   Pitch bend

Continuous control over the pitch is achieved by using a 'pitch-bend' controller. These are normally rotating wheels or levers, and usually change the pitch of the entire instrument over a specified range (often a semitone or a fifth). They produce a control voltage whose value is proportional to the angle of the control. Pitch-bend controls normally have a spring arrangement which always returns the control to the centre 'zero' position (no pitch change) when it is released. This central position is often also mechanically detented, so that it can be felt by the operator, since it will require force to move it away from the centre position.

7.5.6   Modulation

Modulation is controlled using rotary wheels or levers, where the control voltage is proportional to the angle of the control. Modulation controllers are not normally sprung so that they return to the centre position. Some instruments allow pressure on the keyboard to be used as a modulation controller.

There have been several attempts to combine the functions of pitch bend and modulation into a single 'joystick' controller, but the most popular arrangement remains the two wheels: pitch bend and modulation.

7.5.7   2D and 3D controllers

Two-dimensional (2D) controllers, in the form of pads, rather than joysticks, have become popular in early years of the 21st century, and are built into some keyboards and performance effects units. The Korg Kaoss pad is one example that started out as an audio effects unit, but has been extended for use with twin record decks by DJs, and even into the video effects or video synthesizer area too.

Three-dimensional (3D) controllers like the Theremin and more recent ultrasonic Doppler-shift controllers all provide control by moving the hands around.

7.6   Foot controls

Foot controllers or foot pedals are rotary controls which are operated by the player's feet. A flat rectangular plate is covered with a ribbed

Figure 7.6.1 Foot controllers have a restricted angle of rotation, that is, about 30 degrees. The rotation is sensed by any of a number of methods: mechanical linkage to a potentiometer, optical sensors or magnetic rotation sensors. The spring provides a 'soft' end point which can be used to provide additional volume for expression purposes by using additional pressure on the foot control.

or textured rubber surface and allowed to rotate and thus controls a rotary control. Pedals provide a control voltage which is proportional to the angle of the pedal: the pedal is hinged at about one-third of the way along the upper plate. Although usually associated with volume control, they can also be used as modulation controls. When used as a volume control, foot pedals normally have a smooth travel over about two-thirds of the rotary range (about 30 degrees total travel) and then a spring is used to provide additional resistance to further movement. This allows extra volume to be added for specific expression purposes (Figure 7.6.1).

Not all foot pedals use rotary potentiometers to sense the angle of rotation. Opto-electronic and magnetic rotation sensors can also be used.

Some pedals can also be used as pitch-bend controls, with springing to return the pedal to a central default position where no pitch bend is produced.

7.6.1 Foot switches

Foot switches are foot-operated switches. They can be produced in several forms. Some operate like the sustain pedals on a piano, where a short metal lever is pressed downwards to activate the pedal. Others are like small variants on foot pedals, whilst a third type is merely push button switches. Whereas foot pedals are continuous controllers, foot switches normally have two values only, although there are some rare multi-valued variants used to control sustain on pianos. Foot switches are used to control parameters such as sustain, portamento, etc., but they can also be used to select sounds on synthesizers and to start or stop drum machines.

7.6.2 Foot-operated keyboards

Organs sometimes feature keyboards which can be operated by the feet of the performer. The use of foot-operated keyboards for synthesizers has been less successful. One notable example is the Taurus bass pedal, a single octave keyboard produced by Moog in the 1970s, and designed to be played by the feet.

7.7 Ribbon controllers

The ribbon controller is a variation on the pitch-bend wheel. Instead of a rotating wheel, a flexible conductive material is held above a metal plate, and when pressed with a finger, the material touches the plate and a voltage is produced. (Although the ribbon hides what is actually happening from sight, it is rather like holding a guitar string down with a finger.) Moving the finger on the material changes the voltage, and so the controller can be used as a pitch-bend control. Unlike a pitch-bend wheel, it is possible to jump in pitch by pressing a finger onto the ribbon, and then removing the finger causes a jump back to the default pitch. Ribbon controllers have the advantage of mechanical simplicity and small size, but the material does tend to wear out.

There are two main types of ribbon controllers:

- **Short ribbon controllers** which have a silver-coloured cloth-like appearance, are about 100 mm long and 15 mm wide, and have a central raised indication. These have been used in Moog, Yamaha and Korg synthesizers.
- **Long ribbon controllers** which have a black flocked plastic appearance, are about 300 mm long and 10 mm wide. The pitch changes relative to the starting point and the amount of movement along the ribbon by the finger. They are thus more suited to producing portamento and glide effects rather than pure pitch bend. The Yamaha CS-50/60/80 series of polyphonic synthesizers used this type of ribbon controller (Figure 7.7.1).

Figure 7.7.1 Ribbon controllers are actually a variation on a potentiometer or slider volume control on a mixing desk. Instead of having a slider which moves up and down a resistive track, the ribbon is moved onto a metal plate and so produces a voltage which is proportional to its position.

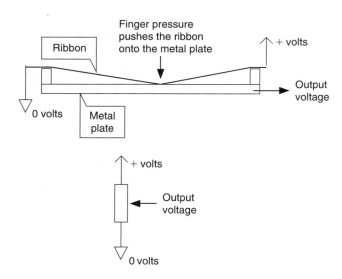

Korg's Prophecy performance mono-synth extended its use of a ribbon controller: it provided additional rotary movement control by mounting the ribbon on a 'log'-like wide modulation wheel: as with the joystick, this combined two-axis control remains the exception rather than the rule.

7.8 Wind controllers

Wind controllers come in two forms: *breath controllers* and *wind instrument controllers*.

Breath controllers are simple pressure measuring devices which are blown into by the player. The output voltage is used to control either modulation or the envelope of a sound by replacing the volume control.

Wind instrument controllers use a controller which is based on a wind instrument like a flute, clarinet or saxophone and convert the finger presses on the keys into pitch information, and the blowing into additional modulation and volume information. One or more selectable modified forms of Boehm fingering are normally used, with extensions to cope with the larger pitch range provided by a synthesizer sound generator. Additional control keys allow octave switching, portamento control and sustain effects to be controlled from the wind controller.

Wind controllers are particularly effective when they are used to play lead line and solo melody sounds.

Although wind controllers are normally monophonic, some provide the ability to hold notes and then play additional notes on top, thus allowing the building up of sustained chords (Figure 7.8.1).

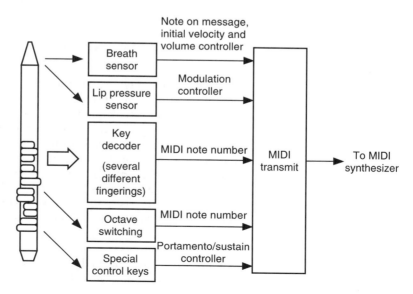

Figure 7.8.1 Wind controllers provide sensing for breath pressure and lip pressure (which are converted into MIDI controller messages for volume or modulation), and convert the keying into MIDI note numbers. Special octave switching keys and control keys (portamento/sustain) are also normally provided.

347

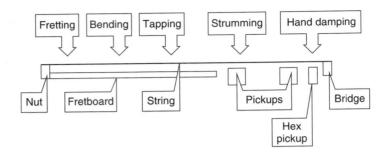

Figure 7.9.1 A guitar player can pluck, strum, tap or bend any of the six strings, which may be open or fretted. This cross section through a guitar shows some of the major performance features and techniques.

7.9 Guitar controllers

Given the enormous interest in using electronics to change the sound of the electric guitar, with a huge range of effects pedals being available, using a guitar as the controller for a synthesizer seems like an obvious development. Unfortunately, using a guitar for this purpose is not easy. There are a number of technical problems whilst using a guitar as the interface between a player and a synthesizer (Figure 7.9.1).

The sound that is heard from an electric guitar is produced by the steel strings vibrating over small magnetic coils in the pickup. This induces a current in the pickup coils and this is then amplified to produce the guitar sound. Normal pickups are designed to 'pickup' the vibrations from all the strings, but in order to process each string's pitch separately, the pickup needs to produce individual signals for each string. This is normally achieved by using a 'hexaphonic' or hex pickup, which has separate coils for each string.

Even with a separate signal for each string, extracting the pitch that a string is playing is far from straightforward: the pitch is dependent on the position at which the string was fretted, as well as any string bending or use of the tremolo arm, and it may also depend on how the string was plucked or strummed; for example, the player may have deliberately produced a harmonic rather than the pitch which might be expected.

In addition, the player can control the volume of each string by the amount of effort which is used in plucking or strumming the strings. This needs to be determined by the guitar controller so that appropriate control signals can be transmitted to the synthesizer. But the player also has additional control like string damping, which is achieved by placing the hand lightly against the strings near the bridge. The timbre that the strings produce can be varied by altering the position at which the string is plucked or strummed, that is, the closer the position is to the bridge, the brighter the sound. Strings may also be tapped onto the fretboard to produce a sound, in which case the position is nowhere near the bridge at all. All of these performance techniques, which are available to the guitarist, make the task of converting a normal electric guitar into a suitable controller for a guitar synthesizer very difficult (Figure 7.9.2).

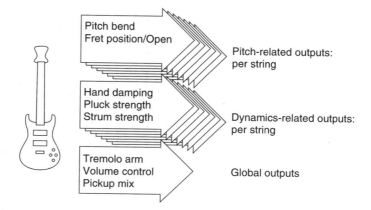

Figure 7.9.2 A guitar controller produces a large quantity of performance information.

Approaches which have been tried include the hexaphonic pickup with sophisticated signal processing to try and extract the pitch; wiring up the frets so that they can be used as electronic switches so that the fretting can be monitored; and even using acoustic radar signals along the strings so that the position of the plucking and fretting can be determined. None of these methods offers a complete solution, but development work is ongoing, and the limitations of the guitar as a controller seem to be gradually diminishing as the complexity of the solutions rises.

Rather than use a normal guitar as the controller, some manufacturers and experimenters have chosen to use a controller which has some elements of a guitar, but deliberately modified so that they suit the conversion process. As well as the wiring of the frets so that the fret position can be determined, some of these designs separate the strings which are fretted from those which are plucked or strummed, and others provide keys to trigger the string events instead. Most of these solutions are also technically complex, and none has been commercially successful.

One of the most useful results of guitar synthesizer research is arguably the hex pickup. By taking the six separate audio signals and applying distortion effects to them individually, a range of effects can be produced which do not have the large amounts of inter-modulation distortion present when a normal guitar pickup signal is distorted. Instead the result is a bright, synthetic sound which has many of the expressive capabilities of the guitar, but without the problems of chords producing too much distortion. Some guitar synthesizers have used this hex signal to drive the strings via heads similar to the record heads on a tape recorder, which allows the vibration of the string to be captured and amplified, opening up possibilities for increased sustain.

7.10 Advantages and disadvantages

Each type of controller has advantages and disadvantages.

On a large scale, the three major controllers are the keyboard, the wind controller and the guitar controller. The keyboard is good for producing complex polyphonic performances based on notes and dynamics, but not so good when expression is required. The wind controller is excellent for detailed monophonic melodies, but not suitable for polyphonic use beyond simple sustained chords. The guitar controller is good for simple polyphonic performances using notes, dynamics and expression, but it has a polyphony limit of six notes, and has limitations in the complexity of the notes used because of the span of the single hand which is used to fret the notes.

Drum controllers are normally used to provide information on dynamics, triggering and pitch. Each drum pad is normally assigned to a fixed pitch, although this can sometimes be altered by using the velocity to change the pitch. Drum controllers are good for controlling percussion sounds, but the limitations of two hands restrict the polyphony to two or four notes at once, and so the complexity of the note patterns is limited by the handling of the drumsticks.

On the small scale, the individual controllers are the wheels, levers, pressure pads and pedals. Wheels and levers are largely similar, although wheels are much more common than levers on commercial synthesizers. Joysticks seem to be less popular than wheels, at least, manufacturers seem to prefer fitting wheels: joysticks tend to be used for mixing between sounds using a vector mixing approach. Pressure pads are rare and require an artificial separation of the pitch bend into two pads. Pedals suffer from problems of response: the pedal is much heavier than the wheel or lever and so this tends to limit the speed at which the foot can move it. The control of a pedal with a foot does not seem to have the same degree of precise control as with a hand on a wheel or lever.

After-touch or key pressure has severe limitations as a pitch or modulation controller because it is limited to use only when the key is pressed down. This limits its use and the speed with which notes can be played whilst attempting to use after-touch. One technique, which can be applied when monophonic after-touch is used, is to use the right hand to play the melody notes, and use a finger on the left hand to play a related bass root note or a pedal bass note and use this finger to control the after-touch. Since monophonic after-touch applies to the entire keyboard, this allows expression to be added to the right hand, albeit the right hand is apparently moving too quickly to be able to apply after-touch. Breath controllers or foot pedals can be used instead of the left hand to produce similar effects.

7.11   Front panel controls

Controllers have had a marked influence on synthesizer design. As well as the performance controllers described in the majority of this chapter, a synthesizer also has front panel controls. The knobs, switches and sliders which are used to control the synthesizer in non-realtime performance are just as important to the function of the instrument, and they have changed significantly in the lifetime of the synthesizer.

Early analogue modular synthesizers used knobs and switches to set the operating parameters of the modules. Interconnections were made using patch-cords which were plugged into front panel sockets. This imposes a number of constraints on the use of such a synthesizer: the patch-cords can obscure the front panel, making it difficult to see the settings of the knobs and switches; and it also makes it awkward to change the patching rapidly.

As synthesizers developed, both monophonic and polyphonic versions streamlined the patching, reducing it down to the voltage controlled oscillator/filter/amplifier (VCO/VCF/VCA) format. Switches could now be used to control the routing of controllers like LFOs and envelopes, and the front panel could be designed so that it represented the signal flow through the synthesizer. The result was the front panels of the late 1970s and early 1980s, where the front panel used a large number of knobs and switches, and occasionally some sliders. In most cases, each knob or switch controlled a single parameter, which meant that learning the front panel layout told you how the synthesizer produced the sounds, and what to adjust to change a sound.

A much more minimalistic arrangement was introduced by the first digital synthesizers. The Yamaha DX7 provided one knob and a lot of buttons. The buttons were assigned so that there was usually one button per parameter, although this was complicated slightly by having several different modes, where the buttons had different meanings. Even so, for most editing, you pressed a button to select the parameter with one hand (the right), and adjusted the value with the other hand (the left). This two-handed approach to editing can be very fast, but it requires the co-ordination of both hands and the eyes, and a considerable amount of concentration.

Computer-based editors can be used to replace this minimalistic interface, but screen redraws with large numbers of displayed parameters tend to be slow. The advantage of using this type of interface is that it is easy to scale to a rackmounting expander module with a limited front panel area since few controls are required and the multiple selection buttons can be replaced by a single parameter selection knob or slider. As this two-handed editing interface has been developed, the slider has been augmented by additional controls like a rotary dial and increment/decrement buttons.

When larger liquid crystal displays (LCDs) became available, the emphasis changed from the button selection of parameters to using the display itself. By arranging a row of assignable buttons or softkeys underneath the display, the selection of parameters could be achieved by reusing the softkeys. LCDs gradually developed from character-based to full graphics capability, which allowed the screens to become more and more the focus of the editing, with an increasing number of softkeys clustered around the increasingly large display.

With large displays, the front panel controls became less specific to the synthesizer, and sound selection keys, numeric keypads and play/edit mode buttons replaced the named parameter buttons. This placed an increasing load on the display, but allowed the front panel to stay static whilst the software could be changed and updated.

Figure 7.11.1 Front panels have developed with technology.
(i) In the 1970s, the layout reflected the structure of the synthesis method.
(ii) In the 1980s, the use of a single slider control with parameter selection buttons introduced a minimalistic interface.
(iii) The 1990s saw a move towards softkeys and displays, with the displays increasing in size and more softkeys being added as the decade progressed.
The late 1990s and early 2000s saw three trends: a return to the minimalism of the 1980s for low-cost instruments; a mixture of display-based softkeys and softknobs for mid-range instruments; and touch screens for high-end instruments.

1970s 'form = function' front panel

1980s 'minimalistic' front panel

Early 1990s 'softkey-driven' front panel

Mid-1990s 'the display is everything' front panel

2000s 'softknobs' and touch screen with dedicated function areas

As less and less controls were required when the display became dominant, front panel space was then available for additional performance controllers: track-balls and joysticks are two examples of methods used to permit rapid real-time control of parameters from the front panel of a synthesizer (Figure 7.11.1).

The mid-1990s saw the first introduction of touch screens with sophisticated graphical interfaces. Although novel, and stylistically impressive, the sparsely populated front panels tend to force the use of the touch screen with only limited alternative methods of controlling the synthesizer. Early touch screens were slow and awkward in their response to touch.

The late 1990s and early 2000s saw three trends:

- A return to the minimalism of the 1980s for low-cost instruments.
- A mixture of display-based softkeys and softknobs for mid-range instruments, often organised into functional groupings.
- Touch screens for high-end instruments. Combined with small number of dedicated controls for specific functions.

The first colour touch screens appeared in the early 2000s, and one expectation is that these will gradually move down to mid-priced instruments.

Front panel design continues to evolve, and has moved away from hardware to solutions which increasingly depend almost entirely on software: thus mirroring the developments in synthesis techniques.

7.12   Questions

1. Describe some alternative synthesizer controllers to the music keyboard.
2. Give examples of performance parameters being produced by controllers?
3. Compare and contrast a guitar controller, a wind controller and a keyboard.
4. What are the two main types of keyboard controllers?
5. What is the difference between attack velocity and release velocity?
6. What limits the use of after-touch by a performer?
7. What are the differences between a wheel designed for pitch bend and one designed for modulation?
8. Why is a guitar a difficult instrument to use as a synthesizer controller?
9. What are the limitations of drum controllers when they are used to control a synthesizer?
10. Outline the history of synthesizer front panel controls.

Time line

Date	Name	Event	Notes
1920	Lev Theremin	The Theremin, patented in 1928 in the US. Originally called the 'aetherophone'	Based on interfering radio waves
1969	Robert Moog	Mini Moog was launched. Simple, compact monophonic synthesizer intended for live performance use	Hugely successful, although the learning curve was very steep for many musicians
1974	Lyricon	The Lyricon was the first commercially available wind controlled electronic instrument	
1976	Lol Creme and Kevin Godley	The Gizmotron, a mechanical 'infinite sustain' device for guitars	A variation on the 'bowing with steel rods' technique
1977	ARP	ARP Avatar Guitar Synthesizer	Monophonic synthesizer with hex pickup (for 'hex fuzz')
1977	Roland	GR-500 Guitar Synthesizer	Hex pickup and string drivers gave 'infinite' sustain
1979	Realton	Variophon launched the simple electronic wind instrument	
1980	Roland	GR-300 Guitar Synthesizer	

Date	Name	Event	Notes
1982	Roland	Roland SH-101, a monophonic synthesizer with an add-on hand-grip performance-oriented pitch-bend and modulation controller.	Notable for its range of colour finishes: red, blue and grey
1983	Yamaha	BC1 breath controller launched as add-on for CS01 mono-synth	A small silver box that you grasped with your teeth!
1984	SynthAxe	MIDI-based guitar controller with separate strings, trigger switches and fretboard switches	Nice design, but very expensive
1985	Yamaha	FC7 foot pedal launched. Used optical rotation sensor	
1987	Stepp	DG1 digital guitar, an ambitious guitar synthesizer/controller	
1988	Yamaha	WX11 wind controller launched. Designed to be used with WT11 synthesizer expander	Low-cost replacement for the WX7 predecessor
1988	Casio	DH100 Digital horn launched. Very low-cost wind controller	
1991	Yamaha	BC2 breath controller launched. Improved version of BC1 built into a headset	Significantly improved appearance over the BC1
1995	Roland	GI-10 Guitar-oriented hex pickup pitch-to-MIDI converter	

8 Performance

8.1 Synthesis live

The live reworking of sound has evolved from the playback of simple pre-recorded sequences, through 'scratching' LPs, to sophisticated performances. In the process, sound synthesis and sampling have moved from a 'back-room' activity that happened slowly and meticulously, to an interactive improvised performance. Both extremes still exist: the 'studio album' may still take years of careful, painstaking, detailed assembly; but when used in performance, the technology has increasingly been used for much more transitory and immediate material.

8.1.1 The rise and fall of keyboard stacks

When synthesizers were room sized, then the synthesist went to the synthesizer. This is analogous to the conductor going to the orchestra: the orchestra requires many resources, takes time to set up and is expensive to run. The orchestra and the synthesizer both require considerable effort to be expended by the conductor and the synthesist in order to realise the music.

Although they were both released in 1969, the design approaches of the Mini Moog and the EMS VCS-3 were very different. The Mini Moog was more performance oriented and so was used for melodies and bass lines, whilst the VCS-3's patch-panel flexibility was more suited to sound effects – especially with the patch memory plug-ins.

This situation only changed when more portable instruments like the Mini Moog and Electronic Music Studios (EMS) VCS-3 were produced. These were more suited to live performance and were affordable and accessible to ordinary musicians, in contrast to the small number of pioneers who had been performing live with modular synthesizers up until this point.

Perceptions and expectations change with time and context. In the late 1960s, audiences were familiar with seeing keyboard players playing a single instrument that made just one sound: either a piano or organ. Keyboards were either solo instruments (accompaniment to a soloist) or in a band context, they were typically used as backing instruments, providing chordal and rhythmic backing. The 1970s saw the introduction of additional keyboards and a gradual change to 'stacks' of keyboards, where one would be arranged above another. Unfortunately, the physical design of most synthesizers or keyboards is not well suited to being placed on top of another, and over the next 10 years, rapid development of keyboard 'stands' provided ample scope for the raising and angling of keyboards to suit the demands of performers.

Some mono-synths may have been intended to be used in conjunction with other keyboards, but the physical design did not match this.

The first 'stacks' were more modest: often just a Mini Moog on top of a Fender Rhodes or Wurlitzer piano, and in fact, the rounded top of the Fender Rhodes was eventually replaced with a flatter, more stack-friendly version. By the end of the 1970s, the well-equipped performer

Even with a flat top keyboard underneath, the front-to-back depth of a Mini Moog or ARP Odyssey made it almost impossible to place in a good and stable playing position.

Performers such as Rick Wakeman and Herbie Hancock were probably almost as famous in the 1970s for their use of multiple stacked keyboards on stage, as they were for their music.

might typically have a 'keyboard stack' containing a dedicated string section emulation called a *string machine, string synthesizer* or a *string synth*, an organ, an electric piano and a monophonic synthesizer (frequently a Mini Moog). Serious professionals would also have a portable electric replacement for a real piano: Yamaha's CP70 'electric' piano was a ruggedised piano where the contact-microphone-equipped harp section could be removed from the keys, action and hammers, and so was easier to transport.

The early 1980s saw a gradual replacement of the string machine and the organ with polyphonic synthesizers, but the biggest two changes came with the Yamaha DX7, which also became the first instrument that most people encountered which had musical instrument digital interface (MIDI) as standard. The Sequential Prophet 600 may have had MIDI, but it was just another polyphonic synthesizer, whilst the DX7 changed the keyboard stack forever.

The DX7 had one preset sound which triggered a sudden change in the established stack, as well as the electric piano industry: preset 11, called 'E. Piano 1'. Although not a perfect emulation of an electric piano, this single preset changed the keyboard stack, and introduced a whole new category of musicians to electronic musical instruments. Musicians who had been using big and heavy electric pianos replaced them with the light and slim DX7, and hence hastened the rapid development of a variety of keyboard stands. But the many players of home organs finally had access to a single keyboard which could sit on top of an electronic organ and supply polyphonic versions of a wide variety of additional instruments with a much higher fidelity than many of the organs at the time. In particular, the DX7 excelled at sounds which were both reasonably realistic, and difficult to achieve with analogue instruments: vibes, glockenspiels, harpsichords and bells. It was also very affordable and offered 16-note polyphony, which was more than generous at the time in comparison to the more usual four-, six- and eight-note polyphony of the analogue poly-synth alternatives. As a consequence, the DX7 sold in large numbers to both markets, and became a runaway best seller for Yamaha. But the DX7 also created a solid user base for MIDI, and changed the synthesizer from an expensive, specialist rarity into an affordable, general-purpose, mass-market workhorse.

As MIDI became established, then there were other casualties in the late 1980s. String machines and organs were gradually replaced by polyphonic synthesizers, and even keyboards themselves disappeared to produce sound modules designed to be driven exclusively via MIDI from a 'master' controller keyboard. The 'live performance' keyboard stack quickly became just one or two master keyboards connected to a 19-inch rack of sound generating modules connected to a mixer. In the studio the playability of physical keyboards meant that the ratio of keyboards to modules was higher, but the 'one keyboard: one sound' era was gone forever. The word 'stack' does live on: it has become common usage to refer to a number of sounds all triggered from one keyboard as a 'stack'.

Figure 8.1.1 Stack evolution.

In the 1990s, the keyboard stack and the keyboard player moved away from the limelight on stage as guitars, vocalists and the ability to dance became fashion essentials. Keyboards and synthesizers moved backwards into the darker part of the stage, as vocalists and dancers became the visible 'stage presence'. Increasingly powerful synthesizer modules meant that the 19-inch rack required less and less space for individual modules. By the end of the 1990s, the laptop computer had become a viable alternative for some types of music, and it joined the DJ's pair of decks as a familiar live performance setup.

In the first years of the 21st century, the trend away from keyboards and hardware increased, with phrase sequencers consolidating their position as a reworking of the hardware sequencers of old, but this time with drums, bass, accompaniment and even vocal backing via samplers all on board; and with ever more powerful laptop computers as an alternative.

8.1.2 Playing multiple keyboards

Looking back almost 50 years, it is now difficult to appreciate just how fundamental and far-reaching the effect of placing one keyboard alongside, or on top of, another keyboard was to keyboard players. There are four main types of performance criteria that changed:

- sounds
- polyphony

- playing technique
- controllers.

Sounds

Having more than one keyboard available means that the keyboard player can produce more than one sound, and can even play two sounds simultaneously. For a keyboard player who played piano, this was a fundamental change in mental attitude: the sound palette was no longer just the piano. More crucially, it allowed the keyboard player additional control over how they sounded, and how they played the notes. For example, if a piano and organ were available, then a chord could be played on the piano or on the organ, or could be doubled on both, or any split of notes could be played on the two instruments, or any note from an inversion of a chord could be played on the other instrument. This opens up a number of new ways of emphasising harmonies, holding suspensions and hocketing arpeggios; and it allows the keyboard player much greater control over how the music is arranged.

In the case of a mono-synth being added to a piano or organ sound, then the contrast of timbre is very strong because of the familiarity of the piano and organ sound contrasted to the more unfamiliar timbre of the mono-synth. The analogue synthesizers in the late 1970s and early 1980s predominantly used low-pass filters, and this means that they only pass high frequencies when the filter is wide open. The result can be a sound spectrum which emphasises the lower frequencies. In contrast, the digital synthesizers from the mid-1980s onwards, like the DX7, often produced outputs with more formant-style or band-pass spectrums, and so could be used in combination with analogue synthesizers and still be heard in the mix.

It is interesting to note that the most successful keyboard of the late 1980s, the Korg M1, and arguably the most sophisticated keyboard of the early 1990s, the Korg Wavestation, both had low-pass filters without resonance.

The other characteristic that was widely exploited with the early analogue synthesizers was the resonance of the low-pass filters. When a resonant filter is just at the point of breaking into self-oscillation, then a very distinctive tone is produced which is not normally found in natural instruments. Excessive use of the unusual quickly renders it boring and familiar, at least for one cycle of musical fashion.

Polyphony

Pianos and organs are naturally polyphonic. In fact, since you can play all of the keys simultaneously, and every key will produce a note, then they could be termed omniphonic. In contrast, most synthesizers will only play a limited number of notes simultaneously. The very first synthesizer designed to be used as a live performance instrument was the Mini Moog, and this was intended to be a monophonic solo instrument. The keyboard player would thus have to learn a rather different way of playing a keyboard in order to use an instrument which could only produce one note at once. The task was complicated because the design of the keyboard circuitry of early monophonic synthesizers was such that they always played the highest note being held down. This meant that if a chord was played, then the highest note would be the only one that would sound, assuming that the fingers all pressed the keys at the same time. If not, then the synthesizer would play one or more grace notes ascending up to the final highest note. A similar set of short notes would also appear if the fingers were not removed from the keys simultaneously. For the same reasons, legato playing of runs of notes would have different timing when ascending or descending the keyboard, because the higher note would always sound, regardless of any other notes which were being played.

The initial reaction to all these unwanted notes and timing changes was for the keyboard player to pick at notes with the fingers of the right hand with a slight staccato to avoid any overlapping notes. With practice, this initially unfamiliar technique could be mastered, although this was only the first of several specialised performance techniques that would be required to make the most of a mono-synth. The second technique is the deliberate use of the left hand to hold down notes whilst the right hand plays. When the right hand is not playing a note, then the mono-synth will play the left-hand-held note. This is rather like the open string on a guitar that sounds when the string is not held against a fret, except that in this case, the left hand can select any note to act as the 'open' note. In many cases, the left hand would play the root note of the chord, and the combined effect of both hands playing would be almost like two separate instruments: the right hand playing a relatively static bass note; whilst the left hand played the melody.

A more demanding mono-synth technique uses both hands playing staccato, but with only one note being held down at once, with control of the note that is sounding passing continually from one hand to the other. Skilful use of both hands in this way can produce startlingly complex runs of notes from a mono-synth, although the modern alternative of using a polyphonic synthesizer and a sequencer to store the notes is far easier, if less impressive. Perfectionist synthesists might consider adding an exercise based on this dual staccato monophonic playing to their warm-up routines ...

Using both hands to play a mono-synth which can only play a single note may seem extravagant, but two-handed playing is almost essential in order to exploit the full expression capabilities offered by a mono-synth. This will be explored further in the sections below.

Playing technique

The keyboards found on analogue mono-synths were based on organ keyboards, and so were light, springy and responsive. They were thus familiar to organ players in one way, although achieving the right balance between the fixed volume mono-synth and the 'swell pedal' controlled organ was not always easy.

For piano players, the analogue mono-synth represented a keyboard with no action and almost no weight, and most importantly, no velocity sensitivity. This again meant that the balance between the fixed volume mono-synth and the velocity-sensitive piano was critical.

In both cases, the simple solution was to adjust the output level of the mono-synth, and this is another reason why both hands are frequently required to play a mono-synth. Making quick adjustments to volume also looks good on stage, since it is easily misinterpreted by an audience as a far more demanding technical adjustment.

Additional complexity arose when polyphonic synthesizers became available in the late 1970s. Poly-synths normally have organ-style keyboards, but are velocity sensitive. So piano and organ players are both faced with an instrument which has an unfamiliar user interface.

In an attempt to find a solution that would please both types of player, some manufacturers started to add small metal weights to the underside of the keys on polyphonic synthesizers and MIDI master controller keyboards. The Poly Moog and Kawai K5 are just two examples of this type of 'weighted' organ-style keyboard. The 'weighted' keyboards were not popular, since they did very little to emulate a real piano action. Many manufacturers now put piano action keyboards onto pianos, poly-synths and MIDI master controller keyboards, especially where keyboards with more than a five-octave span are fitted, or where the piano action will be more familiar to the player.

It is worth noting the differences between velocity (how quickly you depress a key) and after-touch (how hard you press the key once it is being held down). After-touch is normally monophonic, and so the whole of the keyboard is affected by how you press a key down. Velocity is normally polyphonic, and so each key press can control the timbre of that note. If a keyboard is not velocity sensitive, then since each key press would produce the same velocity value, it could be considered to be 'monophonic' in terms of velocity sensitivity, but this is never normally used in specifications.

After-touch is a keyboard controller, which was unfamiliar to both piano and organ players, and it comes in two variants: *monophonic* and *polyphonic*. Polyphonic after-touch, where each key has independent sensing of the pressure applied to whilst the note is held down, is very rare. Paradoxically, the Yamaha CS80, one of the first commercial poly-synths, had polyphonic after-touch. Monophonic after-touch uses a single pressure-sensitive bar under the whole of the keyboard, and pressing down on any key produces a global after-touch sense.

Tying to use after-touch to control the timbre of a note during live performance is not straightforward, especially in fast fluid runs, and there are a number of techniques that can be used to overcome this problem. Some players developed a two-handed technique for use with monophonic after-touch-equipped keyboards. The right hand plays the melody notes without any attempt to press the keys to produce after-touch effects, whilst the left hand plays the root note or another harmonically related note as a drone or accompaniment. If the after-touch is not very sensitive, then a variation is to hold the key and the underside of the keyboard casing between the thumb and other fingers of the left hand and use this grip to activate the after-touch. A simpler solution is to use a modulation wheel or foot pedal to achieve control over the timbre.

In general, a synthesizer with a five-octave keyboard is likely to have organ-style keys, whilst a keyboard that is wider than five octaves is more likely to have a piano action (Table 8.1.1).

Controllers

Portamento

In a variation of the two-handed 'open' low-note technique mentioned above, portamento can be added to a performance on a mono-synth by

Table 8.1.1 Keyboard features

	Velocity sensitive	After-touch	Piano action
Organ	No	No	No
Piano	Yes	No	Yes
Mono-synth	Yes	Yes	No
Poly-synth	Yes	Yes	For wide keyboard version

deliberately playing a low note, followed by a higher note to emphasise the glide effect. The portamento effect is less noticeable when subsequent notes are played close together. In performance, this two-handed 'leap' is sometimes replaced by a variation where one hand spans an octave width with the low note initially held down, and then the hand pivots to play the note one octave up and emphasise the portamento. In this way, the audible effect is similar to the portamento being controlled by a foot switch, but requires less co-ordination of hands and feet. Later instruments sometimes provided 'fingered' portamento where staccato notes were unchanged, but legato playing would add in portamento – an acknowledgement of the early performance technique.

Sustain pedal or foot switch

Organs lack a sustain pedal, and the effect of a sustain pedal on a real piano is different from the sustain pedal or foot switch on a synthesizer. Sustain on a synthesizer is equivalent to holding notes down, and so can be used as part of playing technique for producing drone notes or held chords, whilst the hands move to another keyboard to provide accompanying notes to the held note or chord. On a darkened stage, sustain pedals and foot switches can be difficult to find, and so many keyboard players will adjust some sounds so that they have long release times, thus removing the need to use a sustain pedal to lengthen notes. This programming technique is particularly effective on pad sounds with long attack times, but spectacularly ineffective for percussive sounds where the sustain is used to add legato to specific transitions between notes for effect.

Pitch bend

The pitch-bend control as a performance control first appeared on analogue mono-synths like the Mini Moog in 1969. Although it is possible to bend the pitch of some hammered mechanical instruments where the key remains in contact with the string (the clavinet is a notable example), pitch control via a wheel was not part of pre-existing piano or organ playing technique back in the late 1960s. As with most mono-synth performance techniques, the approach that evolved used both hands, and consists of the right hand playing the keyboard, whilst the index and middle finger of the left hand are used to control the modulation and pitch-bend wheels. The amount of pitch bend applied to notes, and the direction of pitch change were initially derived from listening and observing guitar players. The general rules are:

- Normal setting of pitch bend is a semitone.
- Pull the pitch down by a semitone, then play the note as you restore pitch back to normal.

- When you hold a note in the middle of a phrase, bend the pitch up and down again by a semitone or less.
- When you hold a note at the end of a phrase, bend the pitch down and up again by a semitone or less.

Pitch bend is often applied in place of a grace note, especially in percussive sounds where retriggering the note would cause an undesired repeat of the start of the note.

Although the pitch-bend wheel has become almost a standard, there are still some manufacturers who have replaced it or augmented it in various ways. The Multi Moog used a ribbon controller, whilst two-axis joystick controllers have been used by Korg.

Modulation

The modulation wheel is often used to apply vibrato or tremolo to a sound, and the normal point of application is when a note is held in the middle of a phrase, at the same time as an upwards pitch bend is being applied. Although the modulation wheel is almost always assigned so that it produces vibrato or tremolo, it can also be used to control parameters like filter cut-off or other timbral changes, or even effects mixing, pan position or low frequency oscillator (LFO) speed. Keyboard players use very specific frequencies for vibrato and tremolo, and tend to set them to fade in automatically at about the same time. Assigning the modulation wheel to LFO speed with an auto-fade modulation setting allows the modulation wheel to adjust the speed of the vibrato or tremolo instead. A small change of the rate of LFO modulation can be very effective, and is also used in Leslie speaker emulations, where it simulates the non-instantaneous change in speed of the motor as it changes between the slow and fast settings.

Front panel controls are there for two reasons. One is for programming sounds. The other is for adjusting sounds whilst playing. Using front panel controls as controllers can be a very effective way of adding extra expression or variation into a performance. Changing the detune of oscillators can change the mood of a bass sound, and following the mood of the music by adjusting or 'riding' the filter cut-off can produce very flowing lead-lines. Unlike the regular repetition of an LFO modulation, human generated changes to front panel controls can be much more irregular, or restricted to bar or phrase divisions.

8.2 The role of electronics

The introduction, in 1969, of performance-oriented 'extra', electronic keyboards which were intended to be used in conjunction with another 'main', more traditional, mechanical keyboard, is very significant. From this point onwards, musical performance involving electronics has changed and evolved more or less continuously through to the present day. The keyboard gradually moved from being an unseen accompaniment instrument to somewhere much closer to centre stage in the 1970s and 1980s, and has since then moved back into the shadows as the guitar has returned to popularity, and as the two

Keyboards are not the only instruments which can appear static and disconnected when used live. Drum machines played by hitting drum pads are very visual, but programming one live on stage is less visually interesting. In fact, the connection between pressing buttons on a control panel and the drumming sound which is then produced will not be immediately apparent to many people in an audience ...

decks plus DJ has become a synonym for the use of sampling and sequencing. Even guitar-like 'keyboard controllers' with shoulder straps did not succeed in reversing this trend. Sometimes deliberate misdirection can be successful, as in the use of a guitar synthesizer to play drum sounds which was used by Bela Fleck and the Flecktones in the 1990s.

The guitar has also been changed by electronics. The electric guitar is much like the electric piano; a passive electromagnetic pickup (essentially a microphone) connected to an acoustic musical instrument. Just as the sound of electric pianos could be altered by phases and flanger 'effects' boxes, so could the guitar, although the fuzz box and wah-wah pedal also extended the tonal range and performance possibilities of the guitar considerably. But the tactile user interface of the guitar was less suited to replacement by electronics than the keyboard, and so the evolution of the guitar synthesizer has been slower and less far-reaching than the keyboard synthesizer. The electric guitar of the 21st century is fundamentally very similar to a Les Paul original from the 1950s, whilst the same is not true for most keyboard instruments.

Drums are another example where electronics has taken the physical instrument and changed it beyond all recognition, but in this case, the original physical drum has still survived, albeit augmented and often replaced by its electronic offspring. In many ways, the 21st century has seen the descendants of the drum take over almost all of the roles that the accompaniment section of drums, bass and rhythm backing used to occupy, leaving just lead vocals and solo instruments. The electrification of the drum is therefore very significant.

8.3 Drum machines

Traditional drum kits are large, heavy and loud. They also require considerable time to assemble, and recording drums and percussion in a studio can be a time-consuming, exacting process. But drum and percussion sounds are the perfect accompaniment and contrast to the strongly pitched sounds produced by keyboard-based synthesizers, and so the application of electronics and synthesis to creating drum sounds has a long history.

8.3.1 History

The earliest electromechanical devices to produce drum sounds as a rhythmic accompaniment were tape based, and were probably derived from the practice of splicing tape into a loop so that it would play repeatedly. Harry Chamberlin produced a few of the first purpose-built stand-alone tape-loop rhythm units in 1949: the Rhythmate 40. This type of tape playback unit is the basis for the many later tape replay devices like the Chamberlin and Mellotron.

Ten years later, in 1959, the Wurlitzer organ company released the 'Sideman', a rhythm unit which had a rotating disc to actuate electrical contacts that timed the 12 rhythms using 10 drum sounds produced with valve filtering and shaping circuitry. This was a reworking of a musical box: a disc instead of a drum with pins as the timing mechanism combined with sophisticated sound generating circuits to replace the metal tines. It is interesting to note that the technology of the time was very much based around combinations of electrical motors to provide rotary motion, mechanical linkages, magnetic induction for signal generation and valves electronics for signal processing. The same technology was used in organs, rotary speakers and drum machines.

After another 10 years of incremental development, including the Donca-Matic DA-20 produced by a Japanese company called the Keio Organ Company, or Korg, the transistor replaced both the electro-mechanical discs and the sound generation circuitry. One of the first products made by Roland in 1972 was one of these early transistor rhythm units: the TR-33. In the 1970s, organs quickly acquired rhythm units, and development was rapid, with rhythm units gradually moving from the home organ into other areas of music. Roland's TR-77, from 1972, was one of these crossover products that was used by non-organists, featuring on several hit records. In 1975, PAiA, a built-it-yourself electronics kit company, produced one of the first programmable drum sets with eight drum sounds. Hobbyist electronic magazines at the time were full of kits for drum machines, analogue synthesizers and audio processing equipment of all kinds. It probably comes as no surprise to the reader to discover that the author of this book was a active builder of many of these devices from the early 1970s onwards, and went on from this to repairing synthesizers commercially in the late 1970s.

In 1978, Roland launched their first user-programmable drum machine, the CR-78 (CompuRhythm). This was large, being housed in a wooden box that was almost a cube in shape, and it echoed the styling of the early tape and disc-based rhythm units. A year later, the TR-808 was released, and this was very differently styled, being intended for use by synthesizer users rather than home organ players. Although not a huge success initially, it provided limited control over the drum sounds themselves, and complete user programmability with a very clear interactive display of when drum sounds would play: a row of switches and light-emitting diodes (LEDs), with the switches selecting when a drum would sound, and this being indicated by the LED being lit. When the pattern plays, the LEDs light up in sequence as time scans across the switches. This type of intuitive interface has been widely adopted for subsequent drum machines and other live performance devices.

In 1981, the TR-808 ceased production, and a scaled-down version, the TR-606, with a smaller case, chrome styling and a simplified user interface, was released by Roland as the drum part of a pair of devices. The other device was the TB-303, a dedicated sequencer driving a bass synthesizer. This linking of a drum pattern with a bass sequence was intended to replace most of a rhythm section for guitarists and keyboard players, but it was actually the starting point for the later phrase sequencers or 'groove boxes' (see Section 8.7).

8.3.3 Drum machine operation

Figure 8.3.3 shows a typical low-cost drum machine. The velocity-sensitive drum pads are at the front, arranged in a keyboard pattern of 'black' and 'white' notes. These pads are also frequently used to control the operation via menus and enter values for parameters by acting as a numeric keypad, and so care needs to be taken when using them to determine which mode they are in. Most drum machines divide the operation of the drum machine into modes like:

- song creation (chaining patterns)
- pattern creation (recording patterns)
- instrument settings (drum sounds and mappings)
- MIDI settings (inputs and outputs, clock sync …)
- play mode (live pads),

and the use of the pads may well be different in each of these. When the drum machine is actually playing, then this 'play' mode often forces the pads to become live and able to manually trigger the drum sounds. This is very useful for removing some of the repetition from long sequences of the same pattern by adding in some additional hi-hat or snare hits. Of course, if the pattern and song memory allow, it is a better approach to produce several slightly different patterns and chain them together.

The patterns which are found on a drum machine are very dependent on the current musical fashion. Early rhythm units were intended as accompaniment to organs, and so had dance names: like Waltz; Bossa Nova, Rock'n'Roll; Mambo; Cha-cha; Beguine; March; Tango; Fox Trot and Rhumba. From the 1970s to the end of the century, drum machines reflected the fall of progressive rock, the rise and fall of disco, and the

Figure 8.3.3 Typical low-cost drum machine.

rise of dance music. The music market has become divided into a number of separate areas, with little crossover between them, and drum machines have become locked into specific parts of these areas. So an early 21st century drum machine aimed at the high-tech market might have no reference at all in its patterns to traditional ballroom dances or guitar-based music, instead providing patterns based on musical genres like: Techno, House, Breaks, Trance, Hip-hop, Trip-Hop, Drum'n'Bass, Ambient. Home organs, though, now have built-in drum and rhythm facilities which reflect a wide range of musical influences.

For the synthesist, the factory preset rhythms are merely an illustration of the basic use of the drum machine, and replacing them with variations or new patterns is as much a part of the creative process as creating new sounds on a synthesizer. As Section 8.7 shows, the drum machine is rarely the stand-alone device, which it once was, and composition now encompasses the whole of percussion, rhythm accompaniment and melody.

Recording drum patterns can use any of three metaphors:

- In **real-time recording**, the pattern loops continuously round its bar length, and any drum pads which are played will be played on that beat and bar in subsequent repeats. The drum machine thus behaves like a simple tape recorder, although a time-saving convention is that a recorded beat can be erased by holding the same pad down for the repeat when the pattern loops around.
- In **step recording**, the pattern can be advanced manually by one beat at a time, and for each beat, the pads can be pressed to control which drum sounds happen on that beat. This is useful for complex drum patterns or for transcribing a pattern from a score.
- In **grid recording**, the pattern loops continuously round its bar length as in real-time mode, but this time the pads are all assigned to the same drum sound, with each pad determining when in the pattern the sound will occur. The pads thus become on–off toggles for the drum sounds on specific beats. This method is useful for musicians with a strong visual feel for drum patterns.

All of these methods are reinforced by the liquid crystal display (LCD), which shows one or more sets of drum sounds as either a line with blobs to represent drum hits, or a grid to represent several drum sounds simultaneously (Figure 8.3.4). This 'blob' display is the same as the grid mode pad layout. When working with a drum machine, the sound, the display and any feedback from LEDs on the front panel or pads should all be gathered by the performer as inputs that provide different aspects of information on the drum pattern. Keyboard synthesizers and sequencers tend not to provide as much information, and so the drum machine can be a valuable resource when performing.

Once one or more patterns have been recorded, then songs are created by chaining patterns together. The default is often set so that playing a song with just one pattern set will repeat that pattern, but songs are fully described by setting the pattern for a bar, then the number of repeats of that bar, or alternatively, the bar at which

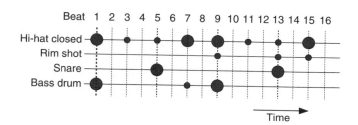

Figure 8.3.4 Drum grid (the blob size indicates accented beats).

the pattern changes. The representation of songs in drum machines is not always as good as the pattern grid, and many drum machines merely provide a list of patterns against bar numbers. Again, more recent performance devices have improved on this, and software sequencers are generally much stronger in their graphical representations of song structure.

8.4 Sequencers

Early humans had two main ways of making sounds: the mouth and the hands. The mouth could sing, whistle and pop, whilst the hands could click, bang, slap and hit just about everything with anything that they could grasp. Given the right location, then environmental effects like echo could provide hours of slap-back entertainment for mouth and hands. With the right stimulus, a dog could be persuaded to stop its plesio-rhythmical barking, and to howl in an approximation of accompaniment instead. And when several mouths and hands were gathered together, then the resulting human orchestra had huge possibilities ... as well as enormous organisational barriers. Perhaps speech is evolution's solution to trying to get a capella singers to do four-part harmony or attempting to persuade the percussion section to play three beats against four beats ...

There seems to be an in-built dissatisfaction with solo music. Whilst many people can appreciate a neat melody, there is something about adding a warm rolling bass line, a splashy bright percussion track and a silky smooth pad that makes it so much more complete. But people are also lazy, and so becoming a composer, writing down lots of parts *and* conducting an orchestra is probably too much like hard work. What is needed is something that produces music automatically and semi-independently. Ideally this would be a compliant, intelligent, skilled fellow musician with infinite reserves of patience, but this specification would need to be open to compromise. And that is where the sequencer comes into play, or is that 'into play'?

8.4.1 Beginnings

The wind-chime may have started out as a bird scarer, or vice versa, but it certainly offers a very primitive method of making music automatically. Once set on this mechanical path, then human ingenuity quickly explores the possibilities: friction bowing provides the hurdy-gurdy, and water or steam power armed with cams and punched control tablets opens up almost every conventional instrument to the on-demand replay of stored performances. Clock-making skills can be re-purposed into musical boxes, reprising the chime sounds of the wind-chime. But mechanical ingenuity has limits, and although it is possible to construct player pianos, steam organs and musical boxes with user-definable control mechanisms, most musical instrument retailers are not full of customers looking for them. Replaying fixed patterns is okay as far as it goes, but …

Electronics changes everything. As with rhythm units, once you realise that there is no longer any need for the physical movement of a mechanical device, then there are less constraints, and you can easily achieve some very sophisticated control possibilities. Relaxation oscillators are an example. They are simple two-transistor circuits where the frequency of the sawtooth waveform that it produces is related to the current flowing through the circuit. Change the current that flows through a relaxation oscillator and you have a simple way of controlling the pitch. But best of all, if you put a low frequency relaxation oscillator in the wire that supplies the current to another relaxation oscillator, then the pitch changes at the rate set by the slow oscillator. Stack three or more of these oscillators together and you have the sort of device which makes most people produce comments like: 'stop that noise!' But to the ears of a synthesist that cacophony can be something much more significant: with just a few knobs it is possible to produce a vast range of complex rhythmic warbles, whistles and whizzes. Minimal compositional effort, no score-writing and no performers to conduct!

Relaxation oscillators have severe limitations. But if you transform that current control technique into one based on voltages, add a keyboard stolen from an electronic organ modified to produce fixed discrete voltages, and mix in lots of other electronic processing goodies, then you eventually get an analogue sound synthesizer. Extending that idea of using one oscillator to control another: a low frequency square wave gives possibly the simplest electronic automation that has a musical purpose rather than merely sounding like an alien siren: the trill. In the limiting case, this is just two notes repeated one after the other which matches the two levels in the square wave perfectly. Trills sound best when the interval is an integer number of semitones, and although this sounds simple to achieve, skill and expense are needed to make it happen, and to keep it happening. Much more challenging is how to produce longer and more complex sets of control voltages or currents.

The multiplexer in the drum machine schematic (Figure 8.3.2) behaves in much the same way, but with drum triggers instead of control voltages.

The answer is to produce a sequence of control voltages using a counter or multiplexer: essentially a switch which selects different control voltages, and which moves cyclically round those voltages, repeating the 'sequence' of voltages. Connect those voltages to an oscillator where pitch is controlled by the voltage, and you have a sequencer which will repeat that sequence of notes. The easiest way to make a multiplexer in the 1970s was to mis-use chips used in the computer industry, and the simplest of these had eight outputs. The result was a sequencer which produced sequences of eight notes, and by connecting two together, sequences of 16 notes. In one of those curious serendipitous coincidences, it turns out that a 1- or 2-bar, 8- or 16-note sequence is probably the shortest sequence that is quite interesting to listen to for more than a few bars, as well as being economic to produce as a circuit. It is these 16-note sequences that form the basis of much of the electronic music of the 1960s and 1970s, plus more recent retro revivals.

Analogue step sequencers are prone to problems with setting the pitch of the notes. The obvious approach is to put a control knob for each step of the sequence, but then each knob needs to control pitch over several octaves, which makes them very sensitive. It also means that changes in temperature or accidental knocks can all too easily detune a sequence. One answer to this problem might be to try and produce a derivative of the keyboard divider chain or resistors, and replace the continuous pitch control knobs with switches. But the solution that became adopted was much neater, and it has much the same effect.

What is required is to change the pitch knob from one which produces continuous changes in pitch as it is moved, to one which works chromatically: jumping from one semitone to the one above or below. Producing a circuit that does it this is rather like a specialised type of analogue-to-digital converter (ADC) where the pitch knob is the analogue input, and by setting the digital output so that each semitone produces a voltage change of 1.00057779 volts. Notice that even though we require an analogue step sequencer, the beat counter and multiplexer are digital, and so is the pitch quantiser or chromatic converter. This illustrates the difficulty of making a pure analogue performance device and shows that many are a hybrid of analogue and digital circuitry.

Step sequencers are very effective on stage, especially in darkness when the scanning of the LEDs across the 16 beats is visible. But 16 steps is also limiting, and in 1977 Roland introduced the first computer-based hardware sequencer with significantly larger storage: the Roland MC-8. Roland called it a *computer music composer*. This was an expensive, professional device consisting of two boxes: one containing the digital eight-track sequencer and the other containing the digital-to-analogue converters (DACs) to produce control voltages. The MC-8 was very straightforward to use, but rather tedious. The programmer entered step times, gate times and control voltages individually into the tracks by entering numbers into a numeric keypad. Copying a pasting was an innovative addition to the feature set, and this improved the entry of numbers. It was also possible to enter notes using an analogue synthesizer keyboard to generate control voltages, but this often required detailed editing with numbers to sort out the timing. Storage of completed sequences was via a compact audio cassette in much the same way as the home computers of the time, and was restricted to just 5300-note events in the expanded 16-Kbyte RAM models (1100 in the original 4-Kbyte RAM version). From the viewpoint of the 21st century, it is hard to believe that anyone could ever have used audio cassettes to store digital information, and patience and perseverance were valuable allies.

Although there were other computer-based sequencers from other manufacturers available before the MC-8, they were based on keyboard entry or did not have the depth of control or synchronisation facilities of the MC-8. Roland improved on the MC-8 with subsequent release like the MC-4 and the MC-202, a novel device from 1983 consisting of a two-track sequencer and a single voltage controlled oscillator (VCO) synthesizer not unlike the SH-101 'sling it over your shoulder' performance synthesizer. Perhaps the ultimate expression of this line of hardware 'enter by numbers' sequencers was the MC-500 from 1986 and the revised MC-500 II from 1988, with standard 3.5-inch floppy disk storage.

Yamaha's QX1 computer-based sequencer in 1984 added dedicated keys for note lengths, but had a 5.25-inch floppy disk for storage, which was in a proprietary format. Ten years later the effect of two standardising forces meant that a sequencer would have a 3.5-inch floppy disk for storage, and that it would read and write standard MIDI files. If you have a sequencer which can be used to create MIDI files, as well as read them, then adding sound generation turns it into a much more complete device, since in one box you can listen to the sequences without any need to connect the sequencer to an external sound source. Drum sounds were also standardised by GM, and so when Yamaha stopped making stand-alone hardware sequencer when they released the QY700 in 1996, it was a sequencer with floppy drive, instrumental sound source and drum sound source. In fact, it actually added a third element: song chaining via preset phrases, rather like a drum machine (Figure 8.4.1).

It should be noted that early hardware sequencers have a tendency to go out of date primarily because of the storage media used, and secondly for the on-board memory size and features. Longevity of storage devices is not a feature of the computer industry: one of the few exceptions being the 3.5-inch floppy disk.

In the studio, the gradual transition from hardware sequencers to software began in the mid-1980s with Apple Macintosh computers. For live performance then the computer has probably been seen more for set dressing than serious use. Computers are not made for the stage. Unreliable electricity supply, interference from lighting controllers and a lack of anywhere flat to put a mouse are all factors that weigh against the computer, but probably the most significant one is the time it takes for a computer to restart after a power failure. A hardware sequencer can be back in operation after a few seconds, whilst for a computer it could be several minutes. What changed in the 1990s was the availability of laptop computers which were powerful, small, light and portable, and which could run from internal batteries. Power failure was no longer a problem, and musicians had a sequencer which could move seamlessly from the studio to the stage.

Section 8.9 covers software sequencers.

A variation on the hardware sequencers is the MIDI data recorder or MDR. Early hardware sequencers were not well suited to receiving

Figure 8.4.1 Hardware evolution (Smart media = flash memory).

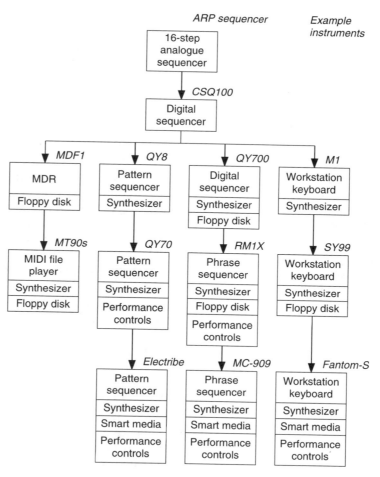

large MIDI system exclusive (sysex) dumps, and the MDR was a purpose-built device which was designed solely to record MIDI data and then play it back. With minimalistic controls derived from tape recorders (play, record and stop) mixed with computer filing systems (next file and previous file) and storing to 3.5-inch floppy disks, most MDRs are simple and functional.

MDRs which can interpret and play out MIDI files are called MIDI file players, and these can be used to provide backing tracks from the many floppy disks of pre-recorded MIDI file songs by using external MIDI sound sources. Some MIDI file players incorporate a GM-compatible sound source, and these provide a single piece of equipment that can be used as a general-purpose accompaniment unit.

8.5 Workstations

The late 1980s saw a change from the 'pure' synthesizer to the workstation: a combination of sound source and sequencer intended to form a single compositional device.

Although only conceptually combining two functions: sound generation with sequencing, a music workstation actually provides a larger number of distinct capabilities. These capabilities are normally implied by the word synthesizer or sequencer, but it is worth enumerating them in order to illustrate the changes required to merge them into a single coherent unit.

The sound source needed to be multi-timbral, provide piano, orchestral and band instruments, as well as drums, percussion, special effects and synthesizer sounds, with a velocity and after-touch pressure-sensitive keyboard, and an effects unit.

Yamaha allegedly returned to earlier research on adding pulse code modulation (PCM) sample playback to the DX7 FM synthesizer, in order to produce the SY77 FM and advanced wave modulation (AWM) (sample replay) workstation in 1989.

The sound-set provided by many synthesizers of the time needed to be widened to meet this specification, and two specific instruments usually needed to be added: drums and piano. Analogue synthesizers are not well suited to these instrument requirements (particularly the polyphony and multi-timbrality required by drums), and frequency modulation (FM) synthesizers were not well suited to providing the drum sounds, but sample replay was, and so S&S became the default sound source.

The sequencer needs to be a multi-track digital event recorder that can record, store, recall and replay the musical information for the composition: note events (pitch, timing, duration and volume), controller events (including note timbre, pitch bend and modulation), drum events, timing events, effects settings, drum patterns, songs, sounds and setup data, as well as complete workstation status data. Ideally, storage should be in a removable form: floppy disk (1990s) or flash memory cards (2000s).

Hardware sequencers in the 1980s were capable of recording the MIDI sysex information for all of these events, but some integration was needed to simplify this storage, particularly for storing the complete workstation setup.

8.5.1 History

One of the first sampler-based workstations was the E-mu Emulator II (EII) in 1984. This used companded 27.7-kHz sample rate, 12-bit samples which were stored as 8-bit samples, and had a 16-track (eight internal plus eight external MIDI) sequencer. The 1987 Emax extended the sample rate to 42 kHz. Both of these workstations lacked an internal effects unit, and were limited in polyphony and multi-timbrality. Perhaps a better description is a 'sampler with sequencer'.

By choosing a very different sound generation technology, Korg were able to release the M1 in 1988. This was arguably the first commercially successful workstation, using S&S with (at the time) a huge 4 Mbyte read-only memory (ROM) as its sound source, with a sequencer and effects unit. The M1 was a huge success for Korg, and it paved the way for the S&S domination of the synthesizer market for almost the whole of the next 10 years. One of the key elements in its success was the high quality of the factory preset sounds and the samples in the ROM. Hidden away in the sequencer is a very interesting feature: phrase-based sequencing built up from short patterns. This is very much a forerunner of phrase sequencers (see Section 8.7).

Roland's 1988 D20 was an S&S synthesizer with sequencer, and a floppy drive, but still called itself a synthesizer. A year later, Roland released the W-30 music workstation, a sample-based keyboard that also had on-board removable storage (a 3.5-inch floppy disk drive), whereas the Korg M1 only had a RAM card slot for storage. Having a floppy disk for storage has become a standard feature of workstations, whilst synthesizers generally lack them. As the 21st century progresses, it seems likely that flash memory cards will replace these floppy drives, although unlike the 3.5-inch floppy, there are so many different and incompatible formats that we are almost back in the 1980s, when the 3.5-inch Sony floppy faced competition from sizes of 2-, 2.5-, 2.8-, 3-, 3.25- and 4-inch alternatives.

The Korg M1's success led to the inclusion of a sequencer and effects in many subsequent instruments, and by the end of the 20th century the stand-alone synthesizer had become something of a rarity. Instruments which are described as synthesizers often include sequencers, and in many ways, the work 'workstation' has started to mean a professional expandable controller synthesizer/sampler/drum machine with more than a five-octave keyboard combined with a sophisticated sequencer and a storage device. Synthesizers tend to have five-octave keyboards or smaller, and are less broad in their sound-sets, but more specialised in their sound generation.

At the opposite extreme to this definition of 'workstation' the start of the 21st century has seen the development of a number of specialised desktop devices which have some of the elements of a workstation, but not all. Typically they have the appearance of a drum machine, and this is reinforced by a drum machine-style set of pads arranged in a single-octave music keyboard layout. Inside they contain a synthesizer, drum- or sample-based sound generator, and a simple 16-step, pattern-based sequencer. In many ways they are drum machines that happen to play pitched sounds instead of drum sounds, and in fact, many of them also provide simple sample-replay facilities for drums and other accompaniment sounds to augment the main sound generator. Yamaha's 'Loop Factory' and Korg's 'Electribe' are two examples of these low-cost alternatives to 1U rackmount modules.

Many manufacturers have taken hardware sequencers intended for the professional market, and added high quality GM-compatible sound sources to produce keyboard-less workstations. For example, Yamaha's last stand-alone sequencer was the 3.5-inch floppy disk-equipped QX5FD which was released in 1988, but was followed in 1990 by the compact QY10, the first of a series of combined sequencers and sound sources. The top of the range desktop QY700 also provides a good example of another trend towards miniaturised and portable versions of existing products. The QY70 and, the enhanced feature version, the QY100, released in 2001, are almost the same specification as the QY700, but a fraction of the size and can be battery powered. The QY series of 'workstations' are single pieces of equipment which combine sequencing with drum and instrument sounds, and are very powerful compositional aids.

E-mu have approached the sequencer plus sound source combination from the opposite direction – they have built on their studio sound source experience and added sequencers and features intended for live performance use. The latest Proteus-series S&S module, the Proteus 2500 includes a sequencer, but the MP7 and LX7 (and the drum/percussion variant, the PX7) Command stations are rugged desktop devices that incorporate a number of live performance controllers to make them more immediate and interactive outside of the studio.

Another genre of music workstation is based around the low-cost fun/home keyboards that are the opposite of the professional workstation. These have taken elements of home organs like auto-accompaniment and made them accessible to people with limited musical ability and funds. By taking one of these keyboards and giving it more specialised contemporary dance-oriented sounds and drum patterns, Yamaha created the DJX keyboard in 1998, followed by a keyboard-less version which used a compact disk (CD)-style rotary controller to emulate a DJ operating record decks. Although low cost and limited in their sounds and facilities, these can be used in performance, and the techniques used are very transferable to more professional devices. The mis-use of low-cost musical devices has always happened, but sophisticated electronics means that some of them are now very viable for use in real performances.

One notable instrument which sits on the boundary between the synthesizer and the workstation is the Korg Karma. This has the sounds, drums, keyboard and sequencer to make it a workstation, but it also has some very sophisticated patented automatic note generation facilities that make this a performer's instrument with a very unique way of augmenting playing. The 'generated effects' are like an 'intelligent' arpeggiator that can take the notes you play in a held chord plus special trigger buttons as the starting point for transpositions and other harmonic expansions, or that can use the held notes as they are and retrigger them in rhythmic patterns, or that can use a rhythmic pattern as the basis for drums of other note sequences, or all of these in combination. The breadth and complexity of Karma mean that any description is going to be incomplete. The best description might be that Karma does for performance what a synthesizer does with sound.

Digital audio workstations are the latest adoption of the workstation name. These are combinations of sequencer, motorised-fader mixer, effects, hard drive and CD-writer, and are sometimes called *hard disk recorders* or *multi-track studios*.

8.5.2 Using workstations

Workstations can be used in many ways. They can be used as synthesizers by ignoring the sequencer, or as sequencers by ignoring the sound generator, or even as master controller keyboards or drum machines. But they excel at rapid composition because of their integration – there is no need to wire anything up with MIDI cables or connect audio into a mixer. Familiarity with the operation of the workstation is also a key enabler to working at speed, and the importance of learning it thoroughly is just as important as with any musical instrument.

The starting point for composing music using a workstation can be a specific sound, a drum pattern, or a short sequence of notes. Many people use workstations as musical notepads by capturing ideas and storing them away for later development, or as phrases to be used in live performance as the basis for extemporisation. One of the key techniques here is to use the storage facilities of the workstation to support and facilitate performance – use tracks and memories so that variations and builds are instantly available, rather than trying to retain a wide variety of favourites. Learning to throw away unused sounds, patterns and sequences to make room for new material can actually be a compositional aid as well, but make sure to store the unused material for the future too.

Workstations are very good at providing accompaniment for live playing. An arrangement with a drum pattern, block or arpeggiated chords and a walking bass line can be used as the backing for singing, guitar playing or even playing a solo melody on the workstation. Muting one of the elements of a performance can enable that part to be worked on or if a human performer is available. The same workstation setup can thus be used with no additional musicians, or with any number, just by muting the appropriate parts as the relevant performer becomes available.

Transferring a composition from a workstation to a computer sequencer in order to make detailed edits, or to increase the available polyphony, or to provide for more diverse instrumentation, is not always straightforward. Exporting the song as a MIDI file and importing it into the computer sequencer often requires post-processing of the information in order to adjust it for the different instrumentation. Differences in timbres and velocity response can change the feel of an arrangement. Pitch-bend, modulation and other real-time performance control may behave differently with alternative instrumentation. Since many workstations provide dedicated additional sequencer tracks for external instruments via MIDI, copying internal tracks across to these external MIDI tracks, can ease the inclusion of additional instrumentation because A–B comparisons can be made using track mutes.

8.6 Accompaniment

An accompanist can be the piano that supports the singer or solo instrumentalist; or an orchestral backing for a piano soloist in a concerto. Solo piano was the accompaniment for silent black and white films at the start of the 20th century, and the start of the 21st century still sees singer–songwriters accomplishing the demanding and exacting skill of accompanying themselves on their piano as they sing and play simultaneously.

A duet on a piano is one form of accompaniment, and has some of the function of a sequencer, except that sequencers are happy playing boring repetitive parts that might tax some performers of a duet. Drum machines might be seen as a replacement for a drummer, except that programming good patterns still requires drumming skills, and there are a number of electronic pads and percussion sensors that allow drum machines to be played or programmed by real drummers. Perhaps the true role of an accompanist is to play not what they are told to play, but what the performer requires. This is much harder, and maybe a descendant of the Korg Karma will feature in this role in the future.

In the past, the drum machine played drum patterns, repeatedly, until the player stopped them. Some organs had a feature which only started the built-in drum machine part when you started playing the keyboard, but the reverse did not seem to emerge: the player stopped the drums at the end of the performance. Changing drum pattern whilst playing was possible, but required single-handed playing on one manual whilst the other quickly pressed the button at the end of a bar. Programming a song into a drum machine turns it from an accompanist to a conductor: if the drum machine starts the chorus and the player has lost a few bars because they did an extra repeat, then they had better play the chorus, because the drum machine is going to continue to play the programmed song sequence.

Organs also feature another accompaniment device: automation that produces walking bass patterns and chordal accompaniment based on the root note played by the left hand and the dance genre selected on the drum machine. Unlike the conducting drum machine, this is under the control of the player, and so an extra repeat does not affect when the chorus is played. This type of automatic accompaniment can be very sophisticated, and is found in home organs and home/fun keyboards, but rarely on synthesizers. Synthesizers may share many common bits of functionality with other musical instruments, but user-programmed accompaniment is their preferred differentiator.

Taking automatic accompaniment, mixing it with drum patterns, and releasing it as software is what happened in the late 1980s with PG Music's 'Band-in-a-Box'. Once a minimalistic song representation of chords and melody as in a busker's fake sheet has been entered, then choosing a song style creates a complete multi-part arrangement with drums, bass, chorded backing and even extemporised melodies. By muting some of the parts, you get either as much, or as little, accompaniment as you require. Interestingly, this type of automatic generation of accompaniment also appears in some workstations, especially the small portable devices like the Yamaha QY100 or the Roland Boss JS5 JamStation, and the idea of a fake sheet is very strongly related to pattern-based drum machine song creation.

One part of the extraction of human accompaniment that is possible happens with drum patterns. The difference between an on-the-beat, equal-volume, simple drum machine pattern and a real drummer's performance is called a *groove*. It is all of those slight timing variations and inconsistencies in volume that help to humanise performances. Capturing grooves allows them to be added to otherwise machine-perfect patterns.

Sampling has become one of the major ways of working with sounds, and software has provided some useful facilities which can aid accompaniment. Pitch extraction can be used to provide control over further processing, like correcting vocal pitching. Even singers who do not need their pitching ability improving can benefit from the creative mis-use of pitch-shifters and harmonisers, as several records in the late 1990s showed.

8.7 Groove boxes

Roland started out calling the MC-303 a 'groove box' but with success has come a price, because the phrase 'groove box' has increasingly become a generic term for any composite device which incorporates a pattern or phrase-based sequencer, drum machine, sound source, effects and live performance controls. Putting all of these components together reflects the increasing integration that happened during the 1990s, and the result is a powerful stand-alone performance tool.

The idea is very simple. The performer creates a number of phrases, and then puts those phrases together to produce the song in performance. Because the phrases will loop repeatedly unless you select a new one, then the structure and length of the song is not fixed. Repeating a line or two, or missing put part of a verse, is no longer a problem.

This type of functionality is not restricted to just hardware unit. In the 1980s and early 1990s Opcode's Vision sequencer software had a feature which allowed you to label phrases with letters of the alphabet, and then chain them together by typing the letters on the computer keyboard. Typing 'abacab' could actually be used to create a verse/chorus/verse/break type of song structure very quickly.

Groove boxes vary in their design and detailed implementation. Roland have a broad range from small and simple to large and complex, and a very different D-Beam ultrasonic controller. Yamaha took a gradual approach, starting with the RM1X's S&S voices, a phrase-based sequencer and a large display. The SU700 took similar sequencer functionality but married it with a sampler. The RM1X and the SU700 were then combined and the specification adjusted to produce the RS7000. E-mu took solid sounds and lots of polyphony, and added a rugged box and sequencer, with the added bonus of plug-in extra sound ROMs. Korg has taken a different approach again with its 'Electribe' series, which are a collection of desktop units that provide a step sequencer with S&S sounds plus either drums, samples, modelled virtual analogue synthesis and more. Boss groove boxes are more oriented towards guitar and bass players, although they do incorporate some 2D controller pads which are useful in other genres.

Phrase sequencing requires pre-planning if it is to be used successfully. Most groove boxes provide a number of controls over the pitch and the selection of phrases, and using these to the full is key. Depending on the type of song or style of music, and how the user wants to work with it, the immediately available phrases need to cover categories like an intro, a verse, a chorus, a break or middle eight, an outro and some fills. Another useful phrase to have in some circumstances is a bar of silence. There are several techniques for building up song phrases, but the simple one is to build up from a drum pattern or build down from a melody, adding accompaniment, a bass line, harmonies and rhythm parts until you have a very dense, full arrangement. The individual phrases are then just this core phrase with different parts muted out. The intro might be just the bass line, or perhaps the drum pattern. The core phrase might never actually be played with all the parts unmuted!

Phrases need not be of the same length. One useful approach is to record two complete sets of verse and chorus or a single repeat of an entire section and to use these when you need both hands to play a synthesizer part on a real keyboard. Making a complete run-through of the song available as a song can be used as a stand-by should something go wrong and you do not have the time to drive the groove box directly.

Most groove boxes allow you to choose the next phrase before the current one has completed, although the Opcode Vision type-ahead can be overly prescriptive if too many phrases are entered at once since this removes any possibility for user control during the performance. Having selected the intro, and then moved to the verse, the pitch control (usually a set of pads or buttons laid out as per a music keyboard) can be used to transpose the verse to whatever 'cycle of fifths' or 'last repeat key change' variant the user chooses. Drum parts are almost always set so that they are not transposed, although this can sometimes be a useful special effect.

Other real-time controls that can be used in performance include ribbon controllers or pitch wheel which can be used to change the playback pitch or speed. Restarting the playback of the phrase before it has finished, or playing it at half or double speed is also found. Some groove boxes allow a pair of adjacent or a run of steps within a phrase to be cycled until the control button is released. Arpeggiators can be used to provide variations to parts, or even to generate bass lines. Muting individual parts of phrases can change their character, and there may be several mute 'memories' available. Using these controls requires that the phrases, their location, duration, key, tempo and purpose are all familiar, and that the performer has mastered the required timing for the controls. Often buttons need to be pressed just slightly ahead of the beat if they are to work correctly, and this requires practice and experimentation. In particular, most groove boxes use several different 'modes' of operation in order to allow all of the controls and selections to be made, and the performer needs to be aware of the current mode before pressing any buttons.

Live performance using a groove box can also be augmented by the use of an effects box or a ribbon type of controller. The Korg KAOSS

Pad 2 is one example that combines an effects unit which is designed to exploit tempo and allows real-time changes to the effects or the groove box sound generation using a 2D touch-sensitive control pad.

8.8 Dance, clubs and DJs

DJs have changed their role over the last few decades. In the 1970s, they were anonymous people who played vinyl records, and the sequencing of the records, plus a little linking patter over the transition from one record to the next, was all that was required for most performances. Most of the vinyl records were singles lasting only about 3 minutes. Despite the short length, with sufficient patter to pad out the gaps between the music, only one deck was required. In the 1980s, more interaction was introduced as scratching turned the turntable from a playback device to a performance instrument. Pairs of Technics SL-1200 Mk 2 turntables connected by a special mixer with a cross-fade slider to mix from one turntable to the other, became the accepted standard equipment.

Transitions between records became more important, and by the end of the 1990s, a DJ was a music maker rather than a mere player of records. The tempo of adjacent tracks would be expected to be the same, and synchronised to each other so that when the cross-fade slider was moved from one turntable, the beats did not syncopate (unless this was the required effect). Scratching techniques would be used to extemporise around the material on the record, and samplers might be used to augment the available sounds from the two turntables. DJs increasingly became creators of music rather than replayers (Figure 8.8.1).

DJs of the 2000s are now skilled, named musical artists, and are capable of performing for several hours with perfect synchronisation between the ever-changing vinyl records on two turntables, hitting exactly the part of the record and being on the correct beat every time. The tools they use are becoming increasingly sophisticated and tailored to the genre: like the specially designed sampler units which can store and replay music or effects on demand. One of the distinguishing features of many pieces of DJ equipment is the lack of any MIDI sockets, something which has become almost a standard part of electronic musical equipment. But some devices can work in both environments: the Korg KAOSS Mixer takes a 2D touch-sensitive effects controller and embeds it in a two channel cross-fade DJ mixer to produce a powerful live performance device.

8.9 Studios on computers

In the 1970s computers were big, heavy and expensive. In the early 1980s the home computer made low-cost computing available to all, but with limited power and only a few pieces of music software. In the second half of the 1980s the Apple Macintosh was the professional musician's computer of choice, although the Atari ST was a far cheaper alternative that saw huge success from hobbyists to

Figure 8.8.1 The workflow of a DJ playing vinyl disks in sequence requires a complex set of activities to be carried out in sequence. The lower section between the dotted lines is repeated with new disks for each repetition.

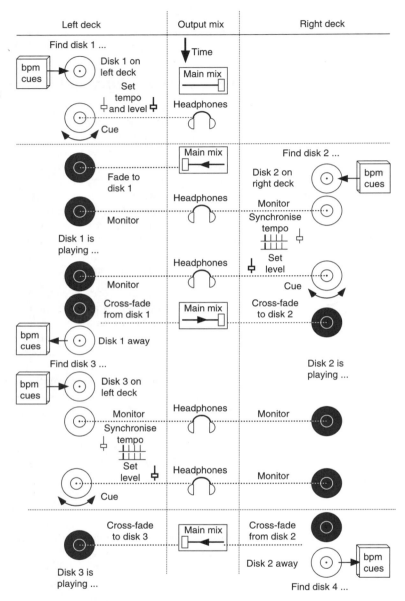

professional use. The first musical applications were simple, but they rapidly developed into sophisticated sequencers that made the most of the large screens and graphical operating systems that differentiated them so strongly from the tiny LED and LCD displays of the hardware sequencers of the time.

As time went by, software sequencers began to converge on a common feature set, with the only differentiators being the details of the user interface, and particularly the metaphors used to represent the music. Sequencers were MIDI only for all but the most powerful computer setups until the early 1990s, when storage and processing began to make audio recording and playback viable. By the end of the 1990s sequencers were able to work with MIDI and audio with almost equal ease, and some people were moving to working only with audio.

The incremental addition of features to match the developing processing power of the computers meant that the major sequencer packages were large and complicated, and new specialist manufacturers appeared with simpler, dedicated sequencers devoted to sample replay (for example, ACID) or a complete replacement for a rack of equipment, a mixer, effects and a sequencer (for example, Reason).

Once audio is integrated into a sequencer, then it can also be used as a virtual sampler. Software samplers based on computer storage and high performance audio input/output cards have removed some of the dependence on hardware samplers, but computers are not well suited to road usage. Laptop computers are more suited in some ways to live performance (size and battery power) and are increasingly used live.

Beyond audio, sequencers next acquired mixing functions, so that the samples could be mixed together before being converted into audio for the final mix alongside any MIDI instrumentation and live instruments. Plug-in architectures allow effects to be added into these mixers, and this gradually developed into plug-in modelling instruments to augment the sample replay. Eventually, the whole of the sequencing, instrumentation, effects and mixing could be carried out in software on the computer, with only the final output being converted to audio. Even vocals can be sampled and incorporated as part of the virtual studio on the computer.

The addition of plug-in capability to sequencers allowed the functionality to be extended only where the user required it, but the extra processing power required can be extensive. Magazine reviews have increasingly been quoting figures based on the number of channels of specific plug-ins as a measure of a computer's capability. Sequencers now typically have a meter which displays the processing power being used, and by implication, how much power is left. This is something which hardware rarely, if ever, revealed (Figure 8.9.1).

The brief history of the software sequencer is thus:

- MIDI
- audio + MIDI
- audio + MIDI + mixing
- audio + MIDI + mixing + effects
- audio + MIDI + synthesis + mixing + effects.

One final thought about the way that MIDI has become an accepted part of the way that computers work is to consider using a computer to play a MIDI file. On a Windows personal computer (PC), the Media Player will use a built-in software GM synthesizer to replay the file, just as if it was an audio file. On an Apple Macintosh, the QuickTime Player does the same replay function through a GM soft synthesizer.

Figure 8.9.1 Software evolution.

For many computer users, standard MIDI files appear to be audio files, because when you 'play' them, they make music. The abstraction that a MIDI file represents is not apparent.

8.10 Performance unravelled

Synthesis in performance is a wide topic, and one where the ongoing developments cut across any attempt to categorise and neaten them. Accordingly, this chapter has noted many times the reuse of devices, of critical success after commercial failure, and of an evolution towards integrated devices and away from traditional hardware. The new hardware seems to be computers, but these can have a relatively high initial cost, issues with stability of operating systems, and frequent software update costs. This makes the true cost of ownership much greater than the initial purchase price, and so there still seems to be a role for hardware. It remains to be seen if the traditional hardware manufacturers can survive in an increasingly software-led world.

The integration of many devices into single integrated performance stations is also significant. In the 1980s, apart from the amplifiers and speakers, there would have been little musical equipment in common between a performance by an experimental electro-acoustic band, a pop band on tour, and a DJ in a nightclub. The first years of the 21st century see much more reuse of equipment, and so the unsigned experimenter, turntablist and band might all use a groove box for part of the set, plus two turntables, plus a laptop running a sequencer or something more exotic …

8.11   Questions

1. When would you use a stack?
2. How has the role of the keyboard player changed since the 1950s?
3. Compare and contrast three examples of electromechanical instruments, with three electronic equivalents.
4. What drum patterns would you expect to find in a typical drum machine from the 1960s, 1970s, 1980s, 1990s and the 2000s?
5. How would you go about composing the drum patterns for a medley of songs over the last half-century?
6. How would you use a 21st century sequencer to emulate a 1970s' 16-step analogue step sequencer?
7. Who would find a 21st century workstation the most familiar: a 1950s organist or a 1950s pianist?
8. How would you use a workstation to produce the live accompaniment for a solo singer?
9. How would you use a groove box in combination with two turntables in a DJ set?
10. Compare and contrast the live performances of two performers: one using mostly hardware and the other software?

Time line

Date	Name	Event	Notes
1949	Harry Chamberlin	Rhythmate 40	A tape loop-based ancestor of tape replay units like the Chamberlin and Mellotron, but this one played rhythms. Housed in a plain wooden box, with controls on the top
1959	Wurlitzer	The 'Sideman' Rhythm Unit	A wooden box by the side of the organ that produced drum sounds. Electromechanical design used rotating disc and contacts to time the 12 rhythms, and valve-based circuits to filter and shape the 10 drum sounds
1963	Korg	Donca-Matic DA-20 Rhythm Unit	The Keio Organ (Korg) company's first major product: designed as an improvement on the Sideman

Date	Name	Event	Notes
1972	Technics	SL-1200 hi-fi turntable. Direct drive	An hi-fi turntable for the serious enthusiast
1972	Roland	TR-33 Rhythm Unit: early transistor drum machine	Drum pattern selection was via dance style: Bossa Nova, Beguine, Samba, etc.
1977	Roland	Roland launched the MC-8, the first 'computer music composer'. A digital eight-part (track) sequencer, with an accompanying converter box to produce analogue voltages	Cassette storage of the maximum 5300-note events
1978	Roland	Roland launched the CR-78 CompuRhythm, one of the first commercial drum machines to provide user-programmability	Housed in a large box that was almost a cube, the CR-78 has a unique appearance: not too dissimilar to the very earliest rhythm units!
1979	Roland	TR-808 Rhythm Composer, an analogue drum machine whose limitations (sounds and tempo stability) were its greatest assets. Widely mis-used live in the hip-hop, techno and house music genres	Saw major success only after it had ceased production in 1981. The TR-909 from 1984 is a Latin percussion follow-on
1979	Linn	LM-1 sampled sounds as a contrast to the analogue drum machines of the time	Although only about 500 were made, this was a hugely influential machine at the time
1979	Technics	SL-1200 Mk 2 hi-fi turntable. Now the definitive 'industry standard' DJ deck (current new model is the Mk 5)	The motor, casing and grounding were improved to give the Mark 2 version
1980	Electronic Dream Plant	Spider Sequencer for the wasp synthesizer. One of the first low-cost digital sequencers	252-note memory, and used the Wasp DIN plug interface
1980	Oberheim	DMX drum machine	Pre-MIDI (although could be retro-fitted) sampled drum machine, using drum sounds in electrically programmable ROMs (EPROMs)
1980	Grand Wizard Theodore	Pioneer of scratching and needle drop techniques for vinyl discs	Grand Wizard Theodore was a New York DJ and one of the first hip-hop producers
1980	Sony	3.5-inch floppy disk introduced for portable data storage	The 3.5-inch Sony floppy faced competition from sizes of 2-, 2.5-, 2.8-, 3-, 3.25- and 4-inch alternatives
1982	Roland	TB-303 'bass line', a monophonic sequencer and simple single-VCO bass synthesizer. Intended originally as an accompaniment device for guitarists knobs	Found increased popularity just after production ceased in 1995. Manual adjustment of the filter cut-off and resonance became the basis of 'acid-house' genre
1982	Linn	LinnDrum, the first commercially successful drum machine to feature digitally sampled drums sounds	An upgraded LM-1 (better sampling rate and some new samples). The 'sound' of the early 1980s is almost all LinnDrum
1983	Sequential	DrumTraks. One of the first MIDI-equipped drum machines	Analogue drums with (for the time) very sophisticated per beat programming of level and tuning
1984	Yamaha	QX1 hardware sequencer	Big, and used 5¼-inch floppy disks. But it was accurate with 384-ppq timing resolution
1984	Roland	TR-909 drum machine	More accenting detail than the TR-808, and shuffle to provide swing. The machine for techno and all forms of dance music

Date	Name	Event	Notes
1986	Yamaha	RY30 drum machine. One of the last conventional 'studio' drum machines from the Japanese manufacturer	Incorporates S&S generated drum sounds, plus a miniature modulation wheel-style real-time controller for volume, pitch, pan, etc.
1986	Roland	TR-505 drum machine	Budget 12-bit sample-equipped drum machine with LCD 'blob' view
1986	Yamaha	RX5 drum machine	Top of the range at the time. Lots of pads, programmable drum pitch and drums sounds on plug-in cartridges
1987	Korg	DDD-1 drum machine	Sampled drum machine with ROM card port for additional drum sounds. Good MIDI implementation
1988	Roland	D20 synthesizer	Included a sequencer and floppy disk storage
1989	Roland	W-30 Music workstation	A sample/S&S-based keyboard workstation with floppy disk storage and small computer system interface (SCSI) port for CD-ROM access
1991	Roland	CR-80 drum machine with special randomiser to simulate human playing	CD quality drum samples in Roland's last stand-alone studio drum machine
1996	Novation	DrumStation drum module	1U rack containing modelled drums from the 808 and 909 stable
1996	Roland	MC-303 Groove box	A combination of drum machine, sequencer, synthesizer and lots of preset and user-definable phrases that can be strung together easily into songs
1997	Jomox	X-Base 09 drum machine	German revisiting of the classic 808 or 909 style of drum machine. Analogue sounds with a fully up-to-date feature list
1998	Roland	MC-505 Groove box	The second generation of the phrase sequencer box. Bigger and better
1998	Konami	Dance: Dance Revolution, an arcade-based dance mat-driven arcade game from Japan	Dance as performance, but for the masses
1998	Yamaha	DJ-X, a dance performance keyboard disguised as a 'fun' keyboard	Followed by a keyboard-less DJ version, the DJXIIB
1998	Sony	Memory Stick introduced	One of the first of a new generation of flash RAM-based memory products
1999	Korg	KAOSS Pad performance effects unit	2D pad-based MIDI controller and effects unit with samples thrown in too
2001	Korg	KAOSS Mixer	DJ-oriented version of the KAOSS Pad effects unit tightly integrated with a scratch mix
2003	Korg	KAOSS Entrancer KP2	An audio and video processor intended for live use
2003	Korg	Electribe MX (EMX-1) groove box. The third generation of Electribe desktop groove machine	Main synth is modelling, but also S&S samples for accompaniment. 3FX units plus valve distortion and 'warmth'

Analysis

9 The future of synthesis

In some ways, the development of sound synthesis is nearly complete. The technology has now reached the point where the quality of the sounds that can be produced is close to the physical limits of the human ear. Improvements still need to be made to the details of the synthesis techniques, particularly to the parameters that are used to control synthesis: the same problem mentioned in the context of resynthesis. Finding viable alternatives to the widely accepted source-modifier model could be one area where there is scope for innovation, but producing intuitive and appropriate control parameters for synthesis techniques like frequency modulation (FM) or granular synthesis is not a simple challenge.

But the major area which still requires large amounts of work is the interfacing between the performer and the instrument. When this is refined sufficiently, then electronic instruments will truly be able to join their well-developed conventional predecessors, perhaps the Theremin provides a glimpse of what is possible, as well as a warning of just how complex the problem is. Digital technology enables software models of real and imaginary musical instruments to be combined with purely synthetic mathematically derived instruments, with almost no foreseeable limit. The future lies in combining the control, sequencing, recording, composition, performance and arranging of these sound-making tools into an integrated whole.

9.1 Closing the circle

It is interesting to return to the opening chapter of this book and to consider some of the other types of synthesis and their 'completeness':

- **3D scene synthesis, or rendering**, has now reached a level of sophistication where true photorealistic scenes are almost possible, and where convincing human beings are tantalisingly close. Advances in this area are likely to reach the limits of human perception in the next 10 or 20 years, if it follows the same time line as sound synthesis. But the area which requires work is how to control the rendering and the animation. In particular, animated and rendered human beings are still some way from being mistaken for real actors.
- **Speech synthesis** continues to make incremental advances, but as with sound synthesis, the main limiting factor now is the control mechanisms, not the synthesis technology.

- **Word processors and authors** are still some way from completeness, and this author expects large amounts of interfacing work to be required for the foreseeable future.

The long-term problem of synthesis thus seems to be one of control and interfacing, rather than one of the generation technology.

9.2 Control

Despite the best efforts of some advertising hyperbole, the future of synthesis does not depend on larger read-only memory (ROM) samples and ever more clichéd sample sets, or even networked compact disk (CD)-ROMs or digital versatile disk (DVD)-ROMs of pre-packaged sounds. It lies with controllable flexible synthesis, and mathematically based modelling techniques seem to be in pole position, although the resulting synthesizer is likely to incorporate multiple methods of control and synthesis combined using software rather than the instrument simulation which the early implementations have concentrated on. Just as all-digital synthesis has been used to emulate even analogue synthesis, so advanced modelling techniques will be able to produce sounds which combine the best of real, analogue and digital instruments.

The current split into solo and accompaniment instruments actually makes a great deal of sense after all, many real-world instruments naturally divide into monophonic solo instruments and polyphonic accompaniment ones. This is especially true of the control interfaces which are used by the performer: accompaniment instruments seem to be naturally limited in their expressive capabilities in comparison to solo instruments. But as synthesis becomes incorporated into equipment which is concerned with producing multimedia, then these divisions will become different software routines within a much larger framework, and the individual differences will become less relevant to the end-user.

The powerful digital signal processing (DSP)-based engines which are used for producing sounds can also be used to process sounds, and hence sound synthesis and hard disk recording will continue to converge. Many personal computer (PC)-based music creating systems now provide a completely digital path from the creation of the audio via a software synthesis, modelling or sample replay, through sequencing, arranging, mixing and effects, and final output as multichannel digital audio. As storage prices continue to fall, these hard disk music recording systems will become more and more attractive, and will continue the move from the top end of the market to the lower end. Stand-alone music recording systems will provide increasingly powerful tools for those who do not wish to build a PC-based system.

MIDI seems to have reached a stable point where its capabilities are adequate for the requirements of most of its users, and a digital audio networking technology like mLAN should eventually enable a similar integration for digital audio. MIDI has been a powerful enabling standard whilst the abstraction between notes and sounds was forced by low bandwidth serial digital connections, but now looks increasingly outdated in a world where the separation of synthesizers, mixers, effects and the sequencing of these elements is no longer necessary.

The ideal end result will be a DSP-engined computer, which could well be a multimedia PC, since the ultimate aim of the forthcoming generations of microprocessors chips seems to be to mix DSP-type functionality with more conventional processors so that the resulting composites have the speed and flexibility to suit a variety of tasks with diverse needs. Although the two main processor contenders at the turn of the 20th century had a number of built-in hardware 'speed-ups' for specific tasks, the start of the 21st century is seeing a number of emerging manufacturers who seem to be taking this approach much further. It may be that the future is not a composite machine which does some processing on-chip and the numerically intensive calculations on-DSP, but one where the same processor chip is used several times for different purposes, perhaps with the ultimate being a reconfigurable system where the capabilities are limited by the number of processing chips and associated memory, and not by arbitrary physical limitations like processing speed or specialised command sets.

The bringing together of all the software from synthesizers, samplers, mixers, effects and sequencing is increasingly happening at the basic 'entry' level, with the result that sophisticated music creation capability is becoming a mass-market commodity, rather than something which requires complex customisation. This should allow live performance instruments to become more responsive to the playing skill of the user. If we call the arpeggiators, phrase recorders and split keyboards of the performance-oriented keyboards of the 20th century the first generation, then the Korg Karma is arguably the first of the second generation of instrument which provides this type of enhanced interactional playability. It is unlikely to be the last, and nor is this type of functionality likely to remain at a price where only professional musicians can utilise it.

So, by early in the 21st century, the synthesizer will have changed into something much more like the 'orchestra in a box' of public perception. In fact, it is likely to be much more than that, since the audio will be part of a larger 'multimedia' facility which can work with pictures and moving images with the same ease as audio. This brings together all of the types of synthesis which can be imagined, and some which may not be expected. Software 'expert' systems are likely to be able to do most of the work of composing and arranging in the widest sense, which means that the role of the end-user may well become more akin to that of a mixture of performer and director. The creative input from a human being at the beginning and the end of making music (and moving pictures) will still be essential – the processes in-between may well be devolved to computers.

The 'synthesizer' will thus no longer be a simple musical instrument which requires large amounts of additional equipment to produce music, but will be a smaller and much more capable creative assistant. It may even be so different in appearance that many of the musicians of the 20th century would not recognise it or they might mistake it for a laptop computer.

9.3 Commercial imperatives

In order to understand why a future synthesizer could be so very different, the role of the synthesizer in terms of both making music and making money needs to be considered.

Fundamentally, musicians want to make sounds. Making sounds can be accomplished using a number of technologies, and sound synthesis has typically used electronic means to produce sounds. As science and technology have developed, so the available methods that are available to designers have changed, and this has, in turn, influenced the way that synthesizers have evolved.

Synthesizers in the 1950s used analogue circuitry because it was cheap, widely available and well understood, whereas digital technology was new and expensive. Subtractive synthesis is a cost-effective solution to the problem of producing lots of sounds with a minimum of complexity, at least for monophonic sounds, and so it became widely used. The music keyboard was a cheap, widely available and well-understood piece of organ technology, and thus was used for playing interface of the first synthesizers. To keep the cost down, monophonic instruments were designed, and they were well suited to producing melody lines that suited the progressive rock of the time, whilst polyphony or modular instruments were rare and expensive. As digital technology began to become suitable for tasks like scanning keyboards and providing sets of control voltages, then microprocessors were used to add functionality to analogue subtractive synthesis, but the sound generation was still analogue, and polyphony was produced by using multiple monophonic synthesizers.

FM provided the first cost-effective way to use digital circuitry to produce sounds, since it requires a small ROM containing a numerical representation of a sine wave, and some numerical processing. By using that same circuitry repeatedly for each part of the FM generation process, and for several notes, then generating FM sounds in real-time became possible with low-cost digital chips. FM also provided larger polyphony than analogue synthesizers, and the polyphonic analogue synthesizer faded away into an item for collectors only.

Samples and synthesis (S&S) continues this theme of making the most of available resources by maximising the use of ROM or random-access memory (RAM) with short looped samples, and then processing those samples initially through analogue filters and enveloping, but later through digital filters, enveloping and effects. As the price of memory has fallen, so it has become less and less of an expense, and so the number, sophistication and quality of the on-board sound samples has increased over time.

With very low-cost storage, sample replay can provide large ranges of sound without great expense, and in such a device, the savings produced by synthesis are not really required any longer. Since the majority of users require just the sounds, then boxes with keyboards and sounds from memory chips are what economic pressure will produce. The start of the 21st century thus sees the predominant technology used to produce sounds from keyboard instruments being a strongly sample-replay-oriented S&S technology, one which requires very little synthesis capability to produce the sounds, and one where the majority of the users will never carry out anything more than minor edits to the sounds. It could be argued that this type of instrument is almost not a synthesizer at all, but merely a sophisticated sample-replay device perhaps with simplified and abstracted editing controls to ease the task of making changes to the sounds. The commercial drive to produce instruments which meet the needs of the playing musician is unlikely to do anything other than continue along this path of increasingly sophisticated replay capability for live performance usage.

However, whilst the instrument used on stage to produce sounds might not be a true synthesizer any longer, this is not the case for software synthesis. The software sequencer has expanded to encompass both the sound generation and the sound mixing extremes of its operating environment. Generating sounds in software has limitations set by the budget of the end-user, not the synthesizer designer, and so the resulting synthesizer 'plug-ins' can use esoteric and unusual synthesis techniques whose complexity and processing power requirements would prohibit their use in a mass-market keyboard instrument. The future of synthesis is thus not in the keyboard instruments which continue to be called 'synthesizers' even though they are largely sample replayers, but instead, it lies with software synthesis engines which are themselves components in software-based 'studios'. The laptop computer could well become the 21st century's preferred device for synthesising music.

The second edition of this book has already required considerable rewriting of this section, and future editions of this book will almost certainly be very different in their content as well!

Time line

Date	Name	Event	Notes
2007	Invation	PolyClassic will be launched. All the classic instruments of the last 100 years	Modelling-based polyphonic synthesizer intended for live performance use
2009	Black Disc	The Black Disc, or BD, will be launched	A 5-cm disc with 20 times the storage capacity of a DVD
2010	Vocalit 4 released	The industry-standard synthetic vocal performance software will get a major rework	A Vocalit performance will be in the top 10 of the charts at the time of the release
2012	Generic Synthetic Guitar Systems	The GSGS BTAG synthetic guitar will be launched to critical acclaim	The BTAG stands for 'better than a guitar', and will get rave reviews
2014	Advanced Playing Technology	Band4U (B4U) 37 will be released	B4U has already had 4 number 1 hits in the last year
2015	Pippin Music Systems	Mars Cube will be released	15-cube sound generation workstation with virtual keyboard, modelled orchestra library and IET-772 storage interface
2017	The Supergroup Group	Recreate 42 modelled performers from the 1960s, 1970s and 1980s	Triggers a 50 years ago retro boom

References

The first edition of this book contained references to books in this section, and these are repeated and updated here. But the Internet has changed the way that many people look for reference material, and my advice now would be that readers of this book also consider using the many World Wide Web (WWW) resources available on the Internet as an alternative to printed books. Manufacturer information is easily available on the Internet, and has formed the major reference source for the second edition of this book.

Books

Bogdanov, Vladimir, Chris Woodstra, Stephen Thomas Erlewine, John Bush (editors) (2001) *All Music Guide to Electronica: The Definitive Guide to Electronic Music (AMG All Music Guide Series)*. Backbeat Books, ISBN 0879306289.

Buick, Peter and Lennard, Vic (1995) *Music Technology Reference Book*. PC Publishing (Technology).

Clayton, George B. (1975) *Experiments with Operational Amplifiers*. MacMillan (Hybrid).

Capel, Vivian (1988) *Audio and Hi-Fi Engineers Pocket Book*. Butterworth-Heinemann (Newnes) (Technology).

Cary, Tristram (1992) *Illustrated Compendium of Musical Technology*. Faber and Faber (Technology).

Chamberlin, Hal (1987) *Musical Applications of Microprocessors*. Hayden Books (Digital).

Chowning, John and Bristow, David (1986) *FM Theory and Applications for Musicians*. Yamaha Music Foundation (Digital).

Colbeck, Julian (1985–) *Keyfax 1, 2, 3, 4, 5, ...* (Ongoing). Virgin/Music Maker/Making Music (General).

Cook, Perry, R. (2002) *Real Sound Synthesis for Interactive Applications*. AK Peters Ltd, ISBN: 1568811683 (Physical modelling).

De Furia, Steve (1986) *The Secrets of Analog and Digital Synthesis*. Ferro Productions/Hal Leonard Books (Analogue, digital).

Forrest, Peter (1994) *The A–Z of Analogue Synthesizers, Part 1: A–M*. Susurreal Publishing (Analogue).

Kettlewell Ben (2001) *Electronic Music Pioneers*. ArtistPro.com, ISBN 1931140170.

Lee, Lara, Reynolds, Simon, and Shapiro, Peter (editors) (2000) *Modulations: A History of Electronic Music: Throbbing Words on Sound*, Distributed Art Publishers, ISBN 189102406X.

Mellor, David (1992) *How to Set Up a Home Recording Studio*. PC Publishing (General).

Newcomb, Martin (1994) *Guide to the Museum of Synthesizer Technology.* The Museum of Synthesizer Technology (Analogue).

Pellman, Samuel (1994) *An Introduction to the Creation of Electroacoustic Music.* Wadsworth (General).

Pierce, John R. (1992) *The Science of Musical Sound.* Scientific American/Freeman (General).

Prendergast, Mark (2001) *The Ambient Century: From Mahler to Trance: The Evolution of Sound in the Electronic Age.* Bloomsbury, ISBN 0747542139, ISBN 1582341346 (hardcover editions) ISBN 1582343233 (papercover).

Reynolds, Simon (1998) *Energy Flash: A Journey through Rave Music and Dance Culture* (UK Title, Pan Macmillan, ISBN 0330350560), also released in US as *Generation Ecstasy: Into the World of Techno and Rave Culture* (US Title, Routledge, 1999, ISBN 0415923735 (*If only this book contained technical details of how the music described was produced!*)).

Roland Corporation (1978) *A Foundation for Electronic Music.* Roland Corporation (Analogue).

Roland Corporation (1979) *Practical Synthesis for Electronic Music.* Roland Corporation (Analogue).

Rumsey, Francis (1994) *MIDI Systems and Control.* Focal Press (Digital).

Schaefer, John (1987) *New Sounds: A Listener's Guide to New Music,* HarperCollins. ISBN 0060970812.

Sicko, Dan (1999) *Techno Rebels: The Renegades of Electronic Funk.* Billboard Books, ISBN 0823084280.

Tooley, Michael (1987) *Computer Engineers Pocket Book.* Butterworth-Heinemann (Newnes) (Digital).

Vail, Mark (1993) *Vintage Synthesizers.* GPI/Miller Freeman (General).

Watkinson, John (1988) *The Art of Digital Audio.* Focal Press (Digital).

Some of the more obscure or 'out of print' books above are included because they may either be located in libraries or may become available via a republishing facsimile.

I also thoroughly recommend the Focal Press audio books, of which this book is just one example!

Magazines and journals

Greenwald, Ted (Magazine article) Samplers laid bare. *Keyboard Magazine,* March 1989 (Digital).

Computer Music Journal (Quarterly journal) The MIT Press, Fitzroy House, 11 Chenies Street, London WC1E 7ET. Telephone: 071 306 0603; Fax: 071 306 0604 (History, analogue, digital, sampling).

Sound on Sound (Monthly magazine) SOS Publications Limited, Media House, Burrel Road, St Ives, Cambridgeshire PE17 4LE. Telephone: 01480 461 244 (History, analogue, digital, sampling).

Keefe, Douglas H. (Journal paper) Physical modelling of wind instruments. *Computer Music Journal,* Vol. 16, No. 4, Winter 1992 (Digital).

Karplus, K. and Strong A. (Journal paper) Digital: synthesis of plucked string and drum timbres. *Computer Music Journal*, Vol. 7, No. 2, Summer 1983 (Digital).

Cook, Perry R. (Journal paper) SPASM, a real-time vocal tract physical model controller. *Computer Music Journal*, Vol. 17, No. 1, Spring 1993 (Digital).

Sullivan, Charles R. (Journal paper) Extending the Karplus–Strong algorithm to synthesise electric guitar timbres with distortion and feedback. *Computer Music Journal*, Vol. 14, No. 3, Fall 1990 (Digital).

Julius O. Smith III (Journal paper) Musical applications of digital waveguides. Stanford University Center for Computer Research in Music and Acoustics, *STAN-M-39* (Digital).

Julius O. Smith III (Journal paper) Physical modelling using digital waveguides. *Computer Music Journal*, Vol. 16, No. 4, Winter 1992 (Digital).

White, Paul (Magazine article) General MIDI. *Sound on Sound*, August 1993 (Using synthesis).

ZIPI (Journal papers) *Computer Music Journal*, Vol. 18, No. 4 (Digital).

Other sources

I have deliberately avoided using Internet uniform resource locators (URLs) to indicate links to web pages in this book. URLs are often long and complex, which results in typing errors, and the nature of the Internet itself means that links in books (or on web pages) frequently point to web pages which no longer exist.

Instead, I recommend that readers use a popular search engine and adopt a cautious analytical attitude when reading the search results. I do not recommend a search engine because their long-term longevity is uncertain. At the time of writing the first edition of this book, then Alta Vista was the clear leader. By the time the second edition began to be considered, Alta Vista had seen a change of corporate owner, and *google* had become very widely used. Google's technology produces results which are typically of high quality and 'on topic', but it is just as prone to being replaced if and when a better solution is found.

(The reader might like to consider how this ongoing competitive evolution of web browsers mirrors (or otherwise) the development of sound synthesis and sampling technology.)

The mental approach to dealing with web page contents is almost more important than the source of the links. Whereas most books, magazines and journals have been through a formal publishing process, the informality and ease of creating a web page means that the validity, correctness, lack of bias and truthfulness can all be called into question. Cross-referencing facts is a good tool for checking that web sources of information at least corroborate each other, although checking that one is not a slightly edited copy of the information presented by the other is also a good check. In general, specialist web sites which have been recently (and frequently) updated, and which can be reached via links from more general web sites, and which have high standards of grammar and spelling, are often good candidates.

Ultimately, the best technique for finding information is to use an online selling site which has large volume, broad choice, and provides sophisticated methods for allowing purchasers to feed back their comments to other prospective purchasers. At the time of writing, Amazon was arguably the largest and best organised of the online retailers, although there were also a number of dedicated online book-only retailers with similar recommendation systems. Peer review of books, magazine articles and journals can provide a way of avoiding the book which appears to be perfect, right up until the moment you have paid for it and start reading it properly.

Glossary

AC
Abbreviation for alternating current; current which changes its value cyclically.
See also DC, Chapter 1 Background.

Accuracy
Repeatable measurements.
See also Precision, Chapter 2 Analogue synthesis.

Acoustics
The science of sound.
See also Chapter 1 Background.

Acronym
A word whose letters represent the initial letters of a sentence or phrase. MIDI: Musical Instrument Digital Interface.
See also Synonym, Chapter 1 Background.

ADBDR
Abbreviation for a type of envelope generator which has attack, 2 × decay and release segments.
See also EG, Chapter 2 Analogue synthesis.

ADC
Analogue-to-digital converter. Changes continuous values into discrete digital samples.
See also DAC, Chapter 1 Background.

Additive synthesis
Synthesis using many simple parts combined. Usually lots of sine waves.
See also Subtractive synthesis, Chapter 5 Digital synthesis.

ADR
Abbreviation for a type of envelope generator which has attack, decay and release segments.
See also EG, Chapter 2 Analogue synthesis.

ADS
Abbreviation for a type of envelope generator which has attack, decay and sustain segments.
See also EG, Chapter 2 Analogue synthesis.

ADSR
Abbreviation for a type of envelope generator which has attack, decay, sustain and release segments.
See also EG, Chapter 2 Analogue synthesis.

After-touch
A synonym for channel or poly-pressure: pressing on the key once it has reached the bottom of its travel.
See also Pressure, Chapter 7 Controllers.

Algorithm
A series of operations to carry out a process. Used by Yamaha for the arrangement of modules in an FM synthesizer.
See also FM, Chapter 5 Digital synthesis.

Aliasing
The 'reflection' of frequencies around the half-sampling frequency. If a 750-Hz signal is sampled at 1000 Hz, then a 250-Hz alias frequency is the result.
See also Sampling, Chapter 1 Background.

AM
Amplitude modulation. A change in the amplitude of a signal (usually repetitive).
See also FM, Chapter 2 Analogue synthesis.

Amplitude

The loudness, level, amount or volume of a signal or waveform.
See also Level, Chapter 2 Analogue synthesis.

Analog

American spelling of 'analogue'
See also Analogue, Chapter 2 Analogue synthesis.

Analogue

Continuous values used to represent parameters.
See also Digital, Chapter 2 Analogue synthesis.

Anti-aliasing filter

A low-pass filter which prevents frequencies greater than half the sampling frequency being sampled. It thus prevents aliasing.
See also Reconstruction filter, Chapter 1 Background.

Antinode

The part of a vibrating object which has the largest range of movement.
See also Node, Chapter 1 Background.

AR

Abbreviation for a type of envelope generator which has only attack and release segments.
See also EG, Chapter 2 Analogue synthesis.

Arpeggiator

Cyclic repetition of a group of notes. Often used on the notes which are held down on a keyboard.
See also Sequence, Chapter 7 Controllers.

ASCII

American Standard Code for Information Exchange: computer coding system used for representing character symbols.
See also Keyboard, Chapter 5 Digital synthesis.

Assignment

The process whereby incoming notes are allocated to resources inside a synthesizer using an algorithm.
See also Keyboard, Chapter 2 Analogue synthesis.

Attack

The initial (start) segment of an envelope. Often a rapid rise to the maximum level.
See also ADSR, Chapter 2 Analogue synthesis.

Attack velocity

The speed with which a keyboard note is initially pressed as it travels from the up to the down position.
See also Pressure, Chapter 7 Controllers.

Attenuation

A reduction in the amplitude, level, loudness or volume of a signal or control voltage. Usually expressed as a ratio in dB.
See also Decibel, Chapter 2 Analogue synthesis.

Attenuator

An electronic device or component which reduces the amplitude, level, loudness or volume of a signal or control voltage.
See also Volume, Chapter 2 Analogue synthesis.

Auto-bend

Pitch chirp from AD envelope generator, triggered by start of note.
See also Pitch EG, Chapter 7 Controllers.

Auto-correct

Another word for quantisation. Used specifically for time quantisation, where note values are adjusted to the nearest value within a range set by the quantisation amount.
See also Quantisation, Chapter 6 Using synthesis.

Auto-glide

Synonym for portamento.
See also Portamento, Chapter 7 Controllers.

Auto-tune

Retuning and calibration of voltage controlled oscillators (VCOs), often at power-up.
See also DCOs, Chapter 3 Hybrid synthesis.

Balanced modulator	Also known as a ring modulator. Takes two inputs and produces sum and difference frequencies. *See also Ring modulation, Chapter 2 Analogue synthesis.*
Band	A range of frequencies. *See also Bandwidth, Chapter 2 Analogue synthesis.*
Band-pass	A filter which passes all the frequencies between two values (the pass-band) unchanged, but attenuates all other frequencies (the stop bands). *See also Filter, Chapter 2 Analogue synthesis.*
Band-reject filter	A filter which does not pass the frequencies which are in a band of frequencies. *See also Filter, Chapter 2 Analogue synthesis.*
Band-stop filter	Synonym for band-reject filter. A filter which does not pass the frequencies which are in a band of frequencies. *See also Band-reject filter, Chapter 2 Analogue synthesis.*
Bandwidth	The range of frequencies which are passed by a circuit or device. Normally implies an attenuation of less than 3 dB (the 3-dB points set the bandwidth). *See also Filter, Chapter 2 Analogue synthesis.*
Bank	A set of memories, sounds, patches, programs, etc. *See also Patch, Chapter 1 Background.*
Bar	(1) Unit for the measurment of pressure. (2) A number of beats. *See also Phon, beat, Chapter 1 Background.*
Baud rate	The serial information rate. MIDI operates at 31 250 Baud = 31 250 bits per second. *See also Serial, Chapter 6 Using synthesis.*
Beat	A unit of musical time. Beats are grouped into bars. *See also Bar, Chapter 2 Analogue synthesis.*
Bend	A change of pitch or frequency. As used in the term 'pitch bend' *See also Pitch bend, Chapter 7 Controllers.*
Bender	An alternative to the pitch-bend wheel controller. Benders are normally either side-to-side levers, or joysticks. *See also Pitch bend, Chapter 7 Controllers.*
Bias	High frequency signal applied to tape recorder record heads to improve the frequency response of the head. *See also Tape recorder, Chapter 2 Analogue synthesis.*
billion	One million million. In the UK, one billion used to be one thousand million, but the usage has now commonly come into line with the US usage. *See also Million, Chapter 1 Background.*
Binary	System of counting which uses base two. Only two values: one and zero. Used in computers. *See also Decimal, Chapter 1 Background.*
bit	A binary digit. Bits can have one of two values: on or off = 1 or 0 = true or false. The smallest representation of a number in a computer: roughly equivalent to a unit in the decimal counting system. *See also byte, Chapter 1 Background.*
Block diagram	A diagram used to show high-level relationships between functional elements of a design. *See also Chapter 1 Background.*

BPM
Beats per minute. The tempo of a piece of music is expressed in BPM.
See also Clock, Chapter 6 Using synthesis.

Break-point
A place where a change takes place: applied to scaling and envelopes.
See also Chapter 5 Digital synthesis.

Buffer
A circuit which has a gain of 1, and serves to isolate subsequent circuits from the input.
See also Chapter 3 Hybrid synthesis.

Bus
A carrier of information, usually within a computer, although sometimes used when a set of signals are used outside the computer: the small computer system interface (SCSI) bus is one example.
See also SCSI, Chapter 5 Digital synthesis.

byte
Eight bits. Sometimes arranged in two 4-bit 'nibbles'.
See also kilobyte, Chapter 1 Background.

Card
A plug-in printed circuit board-like device which can be used to store sounds.
See also Cartridge, Chapter 5 Digital synthesis.

Carrier
The output part of an FM synthesis process. The carrier waveform is the one which is frequency modulated by the modulator. Usually related to the fundamental pitch of the resulting sound.
See also FM, Chapter 5 Digital synthesis.

Cartridge
A plug-in device which is used to store sounds.
See also Card, Chapter 5 Digital synthesis.

CD
Compact disk: digital storage medium for up to 75 minutes of stereo digital audio.
See also CD-I, Chapter 1 Background.

CD-I
Compact disk-interactive. A variation on CD-ROM, designed for use with a dedicated stand-alone console rather than a general-purpose computer.
See also CD-ROM, Chapter 1 Background.

CD-ROM
CD-ROM. A compact disk used as a means of storing digital data: not necessarily audio data, but any computer data.
See also CD, Chapter 1 Background.

Cent
Subdivision of pitch. There are 100 cents in a semitone. The ratios of two frequencies which are one cent apart is 1.00057779.
See also Semitone, Chapter 1 Background.

Centre frequency
The middle frequency of a notch or band-pass filter.
See also Filter, Chapter 2 Analogue synthesis.

Chain
Short series of patterns or sequences, arranged into an order. Often used specifically with drum machines.
See also Pattern, Chapter 1 Background.

Channel
The complete path through a recording system.
See also Stereo, Chapter 6 Using synthesis.

Channel (MIDI)
A MIDI channel is a means of splitting messages about musical events into separate streams, numbered from 1 to 16.
See also MIDI, Chapter 7 Controllers.

Chip
Synonym for 'integrated circuit', although often used to mean specifically a microprocessor.
See also IC, Chapter 1 Background.

Chorus	Notch-like cancellations caused by detuning of two audio signals. *See also Phasing, Chapter 6 Using synthesis.*
Clamped sync	Synchronisation of two VCOs where the frequencies are locked to integer multiples. *See also Chapter 2 Analogue synthesis.*
Clock	Basic timing signal used in digital circuitry: rather like a heartbeat in an animal. *See also Chapter 6 Using synthesis.*
CMOS	Complementary metal oxide semiconductor: type of digital logic chip made from field effect transistors (FETs). Features low power consumption. *See also FET, Chapter 1 Background.*
Coarse tune	Semitone steps in the tuning of a VCO. *See also Chapter 2 Analogue synthesis.*
Collages	Assemblies of sounds which overlap. *See also Chapter 1 Background.*
Colour	Usually refers to the tone or timbre of a sound. Noise can be described as white (all frequencies), pink (high frequencies are attenuated) or blue (low frequencies are attentuated). *See also Filter, Chapter 1 Background.*
Companding	The process of compressing an audio signal before storage, and expanding it on retrieval. *See also PCM, Chapter 1 Background.*
Compression	A reduction in the dynamic range of an audio signal. *See also Chapter 6 Using synthesis.*
Computer music	Music produced by, with or from computers. *See also Chapter 1 Background.*
Continuous	Producing a constant output: no gaps or missing values. *See also Discrete, Chapter 5 Digital synthesis.*
Contour	Synonym for the 'amount of envelope applied to a VCF or VCA': Moog-speak. *See also Envelope, Chapter 2 Analogue synthesis.*
Contour generator	Synonym for envelope generator: Moog-speak. *See also Envelope, Chapter 2 Analogue synthesis.*
Control input	The input to a synthesizer module for signals which vary in internal parameters. *See also Chapter 2 Analogue synthesis.*
Control output	The output from a synthesizer module which can be used to vary other parameters. *See also Chapter 2 Analogue synthesis.*
Control sink	Synonym for a control input: used in matrix modulation descriptions. *See also Control input, Chapter 2 Analogue synthesis.*
Control source	Synonym for a control output: used in matrix modulation descriptions. *See also Control output, Chapter 2 Analogue synthesis.*
Control voltage	Signals carried by patch cables in a modular synthesizer: used to vary parameters. *See also Chapter 2 Analogue synthesis.*

Controller	A general-purpose performance control. MIDI controllers are numbered: 1 is the modulation wheel. *See also Wheel, Chapter 7 Controllers.*
Counter	A circuit which increments or decrements each time a specific 'clock' input is received. *See also Chapter 3 Hybrid synthesis.*
Cross-modulation	Connecting the outputs of a pair of oscillators to the other oscillator's control input. *See also Chapter 2 Analogue synthesis.*
Cross-point matrix	A patching system where inputs are presented on one edge of a rectangular matrix of cross-points, and outputs are presented on the other edge. Pins, connectors or jack plugs placed at the cross-point connect the respective inputs and outputs at that joint. *See also Matrix panel, Chapter 2 Analogue synthesis.*
Cut-off frequency	The frequency at which a filter starts to have an effect: the 3-dB point. *See also Chapter 2 Analogue synthesis.*
Cut-off	Synonym for cut-off frequency. *See also Cut-off frequency, Chapter 2 Analogue synthesis.*
Cut-off slope	The rate at which attenuation occurs outside of the filter pass-band. *See also Pass-band, Chapter 2 Analogue synthesis.*
CV	Control voltage. *See also Chapter 2 Analogue synthesis.*
CV-to-MIDI	A device which converts from control voltages to MIDI messages. *See also MIDI-to-CV, Chapter 2 Analogue synthesis.*
DAC	Digital-to-analogue converter. *See also ADC, Chapter 1 Background.*
Damp	Colloquialism for a signal which has been processed by an effects unit (often just an audio signal with some reverb). *See also Chapter 6 Using synthesis.*
Data rate	The number of bits or bytes of information transmitted in a given time period. *See also Baud rate, Chapter 1 Background.*
dB	Unit symbol for decibel. A measure of relative level, using a logarithmic scale for audio power level. *See also Chapter 1 Background.*
DC	Abbreviation for direct current: a voltage which stays at the same value. *See also AC, Chapter 1 Background.*
DCO	Digitally controlled oscillator. *See also VCO, Chapter 3 Hybrid synthesis.*
Decay	A fall in amplitude or level. Often the second segment of an envelope. *See also Attack, Chapter 2 Analogue synthesis.*
Decibel	A measurement unit used in audio. It represents the ratio between a reference level and the measured level. Unit is the dB. *See also Reference, Chapter 2 Analogue synthesis.*
Delay	An event which does not immediately follow another event is said to be delayed in time. *See also Chapter 2 Analogue synthesis.*

Destination	Synonym for control input. *See also Chapter 2 Analogue synthesis.*
Detent	A physical marker used to indicate a position on a rotary control by a change of force required to move the control. *See also Chapter 7 Controllers.*
Deterministic	Synonym for predictable. Many waveforms are deterministic, since each cycle is very similar to the previous cycle and the next one! *See also Stochastic, Chapter 2 Analogue synthesis.*
Detune	Amount of pitch difference between two sounds or oscillators. *See also Chapter 2 Analogue synthesis.*
Detuning	Deliberate pitch difference between two sounds or oscillators. *See also Chapter 2 Analogue synthesis.*
Digital	Using numbers (often binary numbers) to represent values. *See also Logic, Chapter 1 Background.*
DIN	Deutsche Industrie Norm: the German Industry Standard, rather like BSI in the UK. Often used to describe connectors, as in the 5-pin DIN plug used for MIDI. *See also MIDI, Chapter 1 Background.*
Discrete	Split into separate elements: opposite of continuous. *See also Sample, Chapter 5 Digital synthesis.*
Distortion	A non-linear process that changes the waveshape and adds extra frequencies. *See also Non-linear, Chapter 1 Background.*
Dither	Slight variation in the timing of a clock: used to improve the performance of analogue-to-digital conversion systems. *See also Clock, Chapter 1 Background.*
Double	Synonym for stack or layer: playing the same notes using more than one sound at once. *See also Chapter 6 Using synthesis.*
Dry	Unprocessed audio signal. Not processed by an effects unit. Opposite of wet. *See also Wet, Chapter 6 Using synthesis.*
DSP	Abbreviation for digital signal processor. A computer processing chip which is optimised for dealing with computations on numbers. *See also Computer, Chapter 5 Digital synthesis.*
Duophonic	An instrument that can play only two notes at once. Some analogue synthesizers have this feature. *See also Monophonic, Chapter 2 Analogue synthesis.*
DVA	Dynamic voice allocation or dynamic voice assignment. *See also Assignment, Chapter 7 Controllers.*
Dynamic range	The ratio between the loudest and softest parts of a piece of music. *See also Compression, Chapter 1 Background.*
Dynamics	(1) Synonym for MIDI velocity. (2) The ratio between the loudest and softest parts of a piece of music. *See also Compression, Chapter 2 Analogue synthesis.*
Echo	A repeat of an audio signal, delayed in time. Normally at a lower level than the original signal. *See also Delay, Chapter 6 Using synthesis.*

Edit	(v) To change or modify, particularly a parameter in a synthesizer. (n) An edited sound. *See also Chapter 6 Using synthesis.*
Edit buffer	A temporary store, often used to hold the current changes to a sound. Some synthesizers use this as a buffer for incoming MIDI sounds. *See also Chapter 6 Using synthesis.*
Editor	(1) A computer program which assists in editing a sound. (2) Person who edits. *See also Chapter 6 Using synthesis.*
Editor/librarian	Computer program which edits, categorises and stores sounds for later recall. *See also Chapter 6 Using synthesis.*
Effects	Modifications of signals. Audio effects include reverberation, echo, phasing and equalisation. Video effects are also possible. 'Special effects' are techniques used to produce 'the impossible' in films. *See also FX, Chapter 6 Using synthesis.*
Effects unit	An audio processing unit which modifies an audio signal. Multi-effects units can produce more than one type of effect at once. *See also Chapter 6 Using synthesis.*
EG	Envelope generator. *See also ADSR, Chapter 2 Analogue synthesis.*
Electro-acoustic music	Music produced by electronic methods. *See also Chapter 1 Background.*
Electronic music	Music produced by electronic methods. *See also Chapter 1 Background.*
Electronics	The study of devices which use electrons. *See also Chapter 1 Background.*
Emphasis	(1) Performance controller: usually a foot pedal which is used to increase the volume. (2) Boosting the audio level of a specific band of frequencies. *See also Chapter 7 Controllers.*
Enharmonic	Notes which can have two names depending on the context. A flat and G sharp are one example: they have different names, but are often the same pitch. *See also Chapter 1 Background.*
Envelope	The shape of the change of volume or level of an audio signal. *See also ADSR, Chapter 2 Analogue synthesis.*
Envelope follower	A circuit which outputs a voltage which represents the envelope shape of an audio signal. *See also Chapter 2 Analogue synthesis.*
Envelope generator	A circuit which produces a segmented sequence of segments which represent the stages of an envelope. Used as a control voltage source. *See also Chapter 2 Analogue synthesis.*
Envelope tracking	A parameter which controls the variation of the length of envelope segments; often the controller is keyboard note position or key velocity. *See also Chapter 2 Analogue synthesis.*
EPROM	Electrically programmable read-only memory. A ROM memory which can be erased using UV light and reprogrammed. Often used to hold the operating system software in computers and synthesizers. *See also ROM, Chapter 1 Background.*

EQ	Abbreviation of equalisation. Tone control: altering the level of specific bands of frequencies. *See also Filter, Chapter 2 Analogue synthesis.*
Equal temperament	A keyboard scale where the ratios between the frequencies of adjacent notes is the same for all notes. *See also Chapter 1 Background.*
Equalisation	Changes to the relative level of different frequencies (or frequency bands) in an audio signal. *See also EQ, Chapter 1 Background.*
Equaliser	A device which changes the relative level of different frequencies (or frequency bands) in an audio signal. *See also EQ, Chapter 6 Using synthesis.*
Event editing	Part of a sequencer. It allows the time location and type of musical events to be altered. *See also Sequencer, Chapter 6 Using synthesis.*
Excitation	The signal which is connected to the input of a vocoder. *See also Vocoder, Chapter 2 Analogue synthesis.*
Expander	A device which increases the dynamic range of an audio signal. A non-linear amplifier. *See also Chapter 2 Analogue synthesis.*
Exponential	A relationship between an input and an output where the output increases more rapidly (or less rapidly) than the input. A power law relationship: $y = a^x$. *See also Chapter 2 Analogue synthesis.*
Feedback	The connection of the output signal of a system to the input. A total gain of greater than 1 often results in oscillation at the frequency with the highest gain. *See also Chapter 2 Analogue synthesis.*
FET	Field effect transistor. A semiconductor device which can act as a voltage controlled switch. *See also Transistor, Chapter 2 Analogue synthesis.*
FFT	Abbreviation for fast Fourier transform. A mathematical process which calculates the frequency content of a sample or waveform. *See also Chapter 5 Digital synthesis.*
Filter	A circuit which has a gain that depends on the frequency of an audio signal. *See also Chapter 1 Background.*
Filter tracking	Changes to the cut-off frequency of a filter caused by a control signal: often the keyboard note position or the key velocity. *See also Chapter 2 Analogue synthesis.*
Final level	The final (quiescent) voltage level of an envelope generator. *See also EG, Chapter 2 Analogue synthesis.*
Fine tune	Sub-semitone steps in the tuning of a VCO, often providing cent resolution. *See also Chapter 2 Analogue synthesis.*
Flanging	Sound effect produced by mixing together a time delayed audio signal with the original. *See also Effects, Chapter 6 Using synthesis.*

Flutter echo	Echoes with a short time delay: from the point at which separate echoes become apparent (about 10 milliseconds) to the point at which they become 'full' echoes (about 100 milliseconds). *See also Echo, Chapter 6 Using synthesis.*
FM	Frequency modulation. A method of synthesis that uses a modulator waveform to frequency modulate a carrier waveform. Typically uses sine waveforms. *See also Carrier, Chapter 5 Digital synthesis.*
Foot controller	A foot pedal that is used to produce a control voltage. *See also Chapter 7 Controllers.*
Foot pedal	A hinged flat plate which is rotated by the foot of the operator. Often used to control volume. *See also Chapter 7 Controllers.*
Foot switch	A hinged metal plate which is pressed down by the foot of the operator and it provides two positions: on and off. *See also Chapter 7 Controllers.*
Formant	A peak in a frequency spectrum. Formants are responsible for much of the characteristic timbre of a specific instrument. *See also Chapter 1 Background.*
Found sounds	Sounds which are literally 'found' rather than deliberately made; normally recorded in specific locations. *See also Chapter 1 Background.*
Four-pole	A filter with four sections. The combination produces a 'sharper' filter with a narrower pass-band or a sharper cut-off slope. *See also Chapter 2 Analogue synthesis.*
Fourier analysis	Conversion of an audio signal into its constituent sine waves. *See also Chapter 5 Digital synthesis.*
Fourier synthesis	Construction of an audio signal by combining sine waves. *See also Additive synthesis, Chapter 2 Analogue synthesis.*
FPU	Floating point unit. A computer co-processor which is optimised for carrying out mathematical operations on numbers. *See also DSP, Chapter 5 Digital synthesis.*
Frequency	Repetition rate. Can apply to waveforms or events. *See also Hertz, Chapter 2 Analogue synthesis.*
Frequency response	The bandwidth between the two 'half power' points, or 3-dB points. A measure of the frequency range. *See also Chapter 1 Background.*
FSK	Abbreviation for frequency shift keying. *See also Chapter 6 Using synthesis.*
FX	Abbreviation for effects. *See also Effects, Chapter 6 Using synthesis.*
Gain	The ratio between the input and the output of a circuit. Gains of less than 1 are normally called attenuation. *See also Chapter 1 Background.*
Gate	A switching action. A control signal or voltage can be used to open or close a 'gate'. *See also Chapter 2 Analogue synthesis.*

Gate pulse	A control signal which indicates when a key is held down on a synthesizer keyboard. *See also Chapter 2 Analogue synthesis.*
Gate signal	Control signal which indicates when a key is held down on a keyboard. The end of the gate usually initiates the release portion of an envelope. *See also Trigger, envelope, Chapter 2 Analogue synthesis.*
Gated	A signal whose level is controlled by another control signal. *See also Gate, Chapter 2 Analogue synthesis.*
Generic editor	An editor which can be used to change the parameters of a number of devices and whose characteristics can be easily varied to suit any new devices. *See also Universal editor, Chapter 6 Using synthesis.*
Generic librarian	A librarian program which can be used to store and recall information from a number of devices and which can be reconfigured to cope with new devices. *See also Chapter 6 Using synthesis.*
Gigabyte	One billion bytes (1 million million). *See also Chapter 1 Background.*
Glide	Synonym for portamento. *See also Chapter 2 Analogue synthesis.*
Glissando	The playing of all the notes which are present in-between two notes. *See also Portamento, Chapter 3 Hybrid synthesis.*
Global	A control voltage which is available throughout a synthesizer. Often used as a synonym for generic or universal. *See also Generic, Chapter 1 Background.*
Graphic equaliser	An equaliser which is controlled by a number of slider controls arranged to provide an approximation of the frequency response which is produced by the equalisation filters. *See also Chapter 6 Using synthesis.*
Graphical	Shown in pictures rather than words. Many computers have both text and graphical modes. *See also Chapter 1 Background.*
Growl	Low frequency vibrato or vibrato applied to low notes. *See also Chapter 2 Analogue synthesis.*
Hard disk	A storage device for digital information which uses a rigid rotating disk coated with magnetic material. *See also Floppy disk, Chapter 1 Background.*
Hard sync	Synchronisation where the two oscillators are locked together at integer multiples of their frequencies. *See also Chapter 2 Analogue synthesis.*
Hard-wired	Either permanently wired or available via a switch. *See also Chapter 1 Background.*
Harmonic	A frequency which is above the fundamental, and an inverse integer multiple of the fundamental frequency. *See also Overtone, Chapter 2 Analogue synthesis.*
Headroom	The gap between the signal which is being recorded and the maximum allowable signal level. 'dBs below distortion'. *See also Dynamics, Chapter 1 Background.*

Hertz	Unit of measurement for 'how many times an event occurs in a given time (usually a second)'. *See also Chapter 1 Background.*
Hex	Colloquialism for hexadecimal. *See also Hexadecimal, Chapter 1 Background.*
Hexadecimal	Numbers expressed in base 16. The first few hex digits look like decimal digits, whilst the extra digits use alphabetic characters: 0 1 2 3 4 5 6 7 8 9 A B C D E F. Often indicated by a preceding $ or H, or a trailing subscript: 16. *See also Decimal, Chapter 1 Background.*
High-note	The topmost played note on a keyboard. Often used in the context of portamento or key assignment. *See also Portamento, key assignment, Chapter 2 Analogue synthesis.*
High-pass	A filter whose pass band is above the cut-off frequency. Low frequencies are attenuated, whilst high frequencies are passed through. *See also Filter, Chapter 2 Analogue synthesis.*
Hold	(1) Latch. (2) Sustain or delay segment of an EG. *See also Chapter 2 Analogue synthesis.*
Hybrid	A mixture of two technologies. *See also Chapter 3 Hybrid synthesis.*
Hz	Unit symbol for hertz: measure of frequency or repetition rate. *See also Chapter 1 Background.*
IC	Abbreviation for integrated circuit. A semiconductor device which contains several components (transistors, gates, etc.). *See also Transistor, Chapter 1 Background.*
In	Abbreviation for input. One of the three MIDI ports. *See also MIDI, Chapter 1 Background.*
Inharmonic	Frequencies which are not related to the fundamental frequency by whole number ratios. Non-harmonics. *See also Harmonic, Chapter 2 Analogue synthesis.*
Initial level	The start voltage level of an envelope generator output. *See also EG, Chapter 2 Analogue synthesis.*
Input	The port where signals are presented to a circuit. *See also Chapter 2 Analogue synthesis.*
Interface	The input and output ports of a device or system. *See also Chapter 2 Analogue synthesis.*
Interpolate	Calculating in-between values from existing values. *See also Chapter 4 Sampling.*
Intervals	The spacing between notes. Often expressed in semitones. *See also Chapter 2 Analogue synthesis.*
Invert	To turn 'upside down'. In digital terms, to change 1s into 0s and vice versa. In analogue terms, to change the sign of the signal. *See also Chapter 2 Analogue synthesis.*
Inverter	A circuit which inverts anything presented at its input. *See also Chapter 1 Background.*
IRCAM	French experimental music research centre in Paris: Institute for Research into Co-ordination of Acoustics and Music. *See also Bell Labs, Chapter 8 Performance.*

Jack	Abbreviation for Jack connector, based on a telephone exchange connector. Available in mono and stereo versions in a variety of sizes. For synthesizers the quarter-inch jack connector is almost a standard for line level audio interconnections. *See also Connectors, Chapter 1 Background.*
Jargon	Please see separate section.
Jitter	Variation in the timing of an edge in a digital circuit, often clock signals. *See also Chapter 3 Hybrid synthesis.*
Joystick	A two-axis controller consisting of a stick and two orthogonal potentiometers. Normally produces a control voltage or MIDI controllers' messages. *See also Wheels, Chapter 7 Controllers.*
k	Prefix for kilo, meaning one thousand. *See also Chapter 1 Background.*
K	Prefix meaning 1024, as used in computer terminology, for example Kbyte = 1024 bytes. *See also Chapter 1 Background.*
Kb	A kilobyte: 1024 bytes. *See also bit, Chapter 1 Background.*
Kbyte	1024 bytes: abbreviated to Kb (1000 bytes is a kilobyte, abbreviated to kbyte). *See also Kb, Chapter 1 Background.*
Key assignment	Also known as key/note allocation. The process of routing incoming note events to voice resources inside a synthesizer. *See also Chapter 2 Analogue synthesis.*
Key follow	Synonym for key tracking: the amount of the keyboard control voltage (determining the pitch of a note played on the keyboard). *See also Key tracking, Chapter 2 Analogue synthesis.*
Keyboard	(1) Musical keyboard, consisting of a series of black and white notes. (2) Qwerty keyboard, consisting of a number of switches with alpha-numeric legends. *See also Controller, Chapter 1 Background.*
Keyboard amount	Synonym for key tracking: the amount of the keyboard control voltage (determining the pitch of a note played on the keyboard) which is applied to a control sink. *See also Key tracking, Chapter 2 Analogue synthesis.*
Keyboard level scaling	The variation of level with keyboard position. Can be used to split sounds on a keyboard. *See also Chapter 2 Analogue synthesis.*
Keyboard logic	The circuitry which converts key depressions into control voltages, gates and triggers, or digital events. *See also Chapter 2 Analogue synthesis.*
Keyboard rate scaling	The variation in timing of an envelope caused by note position on the keyboard. *See also Chapter 2 Analogue synthesis.*
Keyboard scaling	A change in a parameter due to note position. Can be applied to level, envelope rates and the keyboard note position control voltage. *See also Chapter 2 Analogue synthesis.*

Keyboard tracking	The amount of the keyboard control voltage (determining the pitch of a note played on the keyboard) which is sent to a control sink. *See also Chapter 2 Analogue synthesis.*
Keyboard voltage	The output of the keyboard controller: one or more control voltages which represent any notes which are being held down. *See also Chapter 2 Analogue synthesis.*
kHz	Unit symbol for thousands of hertz. The 'k' is not capitalised, since this would mean 1024ths of hertz! *See also Chapter 1 Background.*
Knob	The front panel appearance of a rotary control. Normally a small cyclinder with some means of indicating the rotation position. *See also Chapter 7 Controllers.*
Lag processor	A low-pass filter with a cut-off frequency of a fraction of 1 Hz. Intended to 'smooth' out rapid changes in voltages, and so can be used to extract the envelope from a sound. Portamento can be produced by using a lag processor on the keyboard voltage. *See also Chapter 2 Analogue synthesis.*
Last note	The last note played on a keyboard so far. Often used to control the mode of portamento or note assignment. *See also Chapter 2 Analogue synthesis.*
Latch	A simple memory circuit. The state at the input is stored when a trigger input is activated. *See also Chapter 2 Analogue synthesis.*
Latching	A latching control is one where it maintains a state until changed. Some switches are latching and some are momentary. *See also Momentary, Chapter 7 Controllers.*
Layer	A composite sound made up from two or more separate sounds. *See also Stack, Chapter 6 Using synthesis.*
Layering	The process of making a composite sound from two or more separate sounds. *See also Stacking, Chapter 6 Using synthesis.*
LCD	Abbreviation for liquid crystal display. A type of display used on electronic equipment. LCDs require only low power because they use electric fields to alter the rotation of light as it passes through the LCD. Polarisation effects either let the light through or stop it, depending on the rotation, and so the dots on the LCD can appear as either transparent or opaque. *See also LED, Chapter 1 Background.*
LED	Abbreviation for light-emitting diode. A semiconducting device which produces light when a current is passed through it. *See also LCD, Chapter 1 Background.*
Level	The amplitude, level, loudness or volume of a signal or waveform. *See also Volume, Chapter 2 Analogue synthesis.*
Lever	Method of controlling pitch on some synthesizers. A performance controller, often fabricated as a 'paddle' connected to rotary control. *See also Chapter 7 Controllers.*
LFO	Low frequency oscillator. *See also VCO, Chapter 2 Analogue synthesis.*
Librarian	Computer application program which stores data, usually sounds from synthesizers. *See also MIDI, Chapter 6 Using synthesis.*

Limit	The maximum or minimum value which can be reached. *See also Chapter 1 Background.*
Linear	A transfer function which is a straight line. Circuits which are linear do not cause significant distortion. *See also Chapter 1 Background.*
Lock	Synonym for 'phase-lock'. *See also Chapter 2 Analogue synthesis.*
Log	Colloquialism for logarithmic. Derived from logarithms, which are a means of carrying out multiplication and division using just addition and subtraction operations. *See also Logarithmic, Chapter 2 Analogue synthesis.*
Logarithmic	A relationship where the output is the power of a number that equals the input number. If $y = e^x$, then x is the log of y. Logs are normally to base e or base 10. *See also Chapter 1 Background.*
Look-up	A mapping between two variables. One is used as an index and points to the other number. *See also Chapter 3 Hybrid synthesis.*
Loop	A repeated series of events. Can be audio, MIDI messages or a physical loop of tape. *See also Chapter 2 Analogue synthesis.*
Loudness	(1) The volume of an audio signal. (2) Compensation of the response of the ear at low volumes. *See also Chapter 1 Background.*
Low-note	The lowest note which is being pressed on a keyboard. Low-note priority is a type of key assignment where the lowest note has precedence over higher notes. *See also High-note, Chapter 2 Analogue synthesis.*
Low-pass	Low-pass filter: a circuit which passes frequencies below a special 'cut-off frequency'. *See also Filter, Chapter 2 Analogue synthesis.*
LSB	Least significant byte or least significant bit. *See also MSB, Chapter 1 Background.*
μ	Unit prefix for 'one millionth'. *See also 'm', Chapter 1 Background.*
M	Unit prefix for 'million'. *See also 'm', Chapter 1 Background.*
m	Unit prefix for 'one thousandth'. *See also 'M', Chapter 1 Background.*
Map	A relationship between two variables. It can be one-to-one, or one-to-many. A 'MIDI program change' table is a map which connects program change messages to stored sounds. *See also Look-up, Chapter 6 Using synthesis.*
Mapper	A device which carries out mapping, especially changing one MIDI message to another. *See also Chapter 6 Using synthesis.*
Mapping	The process of noting down the map which connects two variables. *See also Chapter 6 Using synthesis.*

Master	A keyboard which is used as a source for MIDI messages is called a 'master' keyboard. *See also MIDI, Chapter 6 Using synthesis.*
Matrix	A two-axis table which can be used to show the relationship between an input and an output. *See also Chapter 2 Analogue synthesis.*
Matrix modulation	A method of connecting control sources and sinks so that any source can be connected to any sink. *See also Chapter 2 Analogue.*
Matrix panel	A flexible patching system found on some modular synthesizers: ARP, EMS, Dewtron, etc. Not restricted to rectangular matrices like cross-point matrices. *See also Cross-point matrix, Chapter 2 Analogue synthesis.*
Matrix switching	Synonym for patching systems which use a matrix of switches. *See also Chapter 2 Analogue synthesis.*
Mb	Unit symbol for megabyte: just over one million bytes. *See also Byte, Chapter 1 Background.*
Mbyte	Abbreviation for megabyte. *See also Chapter 1 Background.*
Megabyte	Just over 1 million bytes. Abbreviated to MB. *See also Byte, Chapter 1 Background.*
Memories	The sounds that a synthesizer makes are stored in memories. Different manufacturers use different names for memories. *See also Jargon, Chapter 1 Background.*
Memory	A store for analogue or digital information. *See also Chapter 1 Background.*
Messages	MIDI commands: sent as one or more bytes of MIDI data. *See also Chapter 1 Background.*
MG	Acronym for modulation generator: an LFO in Korg-speak. *See also LFO, Chapter 2 Analogue synthesis.*
MIDI	Acronym for musical instrument digital interface. An international standard for the interchange of information about musical events, and the intercommunication of electronic musical instruments. *See also Serial, Chapter 1 Background.*
MIDI byte	8 bits from the MIDI serial stream. Also used for the 7 bits which are used in a byte for carrying information. *See also MIDI, Chapter 1 Background.*
MIDI clock	A MIDI message which indicates timing. Sent at the rate of 24 clocks per quarter note. *See also MIDI, Chapter 1 Background.*
MIDI message	One or more MIDI bytes with a specific meaning as defined in the MIDI specification. *See also MIDI, Chapter 1 Background.*
MIDI mode	One of the four ways of operating a MIDI system: mono, 'multi', omni and poly. *See also Chapter 1 Background.*
MIDI-to-CV	A conversion between control voltages and MIDI messages. *See also Chapter 2 Analogue synthesis.*

millisecond	Unit of time measurement: one thousandth of a second. *See also 'ms', Chapter 1 Background.*
Mixer	A device which mixes together two or more audio signals. *See also Chapter 2 Analogue synthesis.*
Mod	Abbreviation for modulation or modifier. *See also Chapter 2 Analogue synthesis.*
Mode	Colloquialism for one of the four ways of operating a MIDI system: mono, 'multi', omni and poly. *See also Chapter 1 Background.*
Modifier	A circuit or device which changes a control voltage, audio signal or digital information. *See also Chapter 2 Analogue synthesis.*
Modular	Built-up from simpler units. Modular synthesizers have discrete VCOs, VCFs, etc. *See also Chapter 2 Analogue synthesis.*
Modulation	Changing or varying a parameter. Often used to mean a cyclic variation, but not exclusively. *See also Chapter 2 Analogue synthesis.*
Modulation generator	Term used by Korg for an LFO. *See also LFO, Chapter 2 Analogue synthesis.*
Modulator	A device which imposes modulation. *See also FM, Chapter 5 Digital synthesis.*
Module	One of the simpler component units which form a larger and more complex complete assembly. *See also Modular, Chapter 2 Analogue synthesis.*
Momentary	A momentary control is one which is normally in one state, and can be temporarily put into another state. Some switches are momentary, others latch in one of several states. *See also Latching, Chapter 7 Controllers.*
Monitor	(1) To note down or keep track of. (2) A television display. *See also Chapter 6 Using synthesis.*
Mono	Single channel of audio. Many synthesizers have mono synthesis functions, but pan controls and stereo effects processors are used to provide a 'pseudo-stereo' output. Some samplers provide 'true stereo' samples. *See also Stereo, Chapter 6 Using synthesis.*
Mono mode	The monophonic MIDI mode. Provides up to 16 separate individual mono-phonic channels. *See also Chapter 1 Background.*
Monochord	A stringed instrument with only one string. *See also Chapter 1 Background.*
Monophonic	An instrument that can only play one note at once. Most wind and brass instruments are naturally monophonic. *See also Polyphonic, Chapter 2 Analogue synthesis.*
Monostable	A circuit which has a default state, and which can be forced into an alternative state which remains in for a fixed time interval, after which it reverts to the default state. *See also Chapter 1 Background.*

Mother	Synonym for 'master', as in a master keyboard. *See also Chapter 7 Controllers.*
Mouse	An *x–y*-positioning device, often used as a pointing device for a computer. *See also Chapter 7 Controllers.*
ms	Unit symbol for millisecond: one thousandth of a second. *See also μs, Chapter 1 Background.*
MSB	Most significant byte or most significant bit. *See also LSB, Chapter 1 Background.*
MTC	MIDI time code. A simple form of SMPTE-like time code, used to provide synchronisation in MIDI systems. *See also SMPTE, Chapter 6 Using synthesis.*
Multi-mode	An 'unofficial' MIDI mode. When several channels can be used in polymodes, then the resulting multi-timbral mode is often called 'multi'-mode. *See also Chapter 1 Background.*
Multi-trigger	(1) An envelope generator which can restart the envelope. (2) A monophonic keyboard which can produce additional triggers, if a note is pressed whilst another is held down. *See also Chapter 2 Analogue synthesis.*
Multi-mode filter	A filter whose design allows it to operate in low-pass, high-pass and band-pass modes. Sometimes also provides band-reject modes, and sometimes multiple outputs. *See also Chapter 2 Analogue synthesis.*
Multi-sample	Assigning several samples to notes (velocity layering) or groups of notes (keyboard layering). *See also Chapter 4 Sampling.*
Multi-timbral	An instrument which is able to produce several different timbres simultaneously. A guitar can produce noises from the strings and from the soundboard by slapping it. *See also Polyphony, Chapter 2 Analogue synthesis.*
Multiple	A passive splitting module which is found in modular synthesizers. Can also be used as a passive mixer. *See also Chapter 2 Analogue synthesis.*
Multiple trigger	Synonym for multi-trigger. *See also Chapter 2 Analogue synthesis.*
Musique concrète	Music produced using pre-prepared audio fragments. *See also Tape, Chapter 1 Background.*
μs	Unit of time measurement, a microsecond: one millionth of a second. *See also 's', 'ms', Chapter 1 Background.*
Natural harmonic series	A sequence of harmonics produced by taking a fundamental frequency and multiplying it by integers: for example, 55, 110, 165, 220, 275 Hz, etc. *See also Chapter 1 Background.*
nibble	Half of a byte: 4 bits. *See also bit, byte, Chapter 1 Background.*
Node	A position on a vibrating string where no movement occurs. *See also Antinode, Chapter 1 Background.*

Noise	A random mix of all frequencies. *See also Chapter 2 Analogue synthesis.*
Noise generator	A circuit which produces noise. *See also Chapter 2 Analogue synthesis.*
Non-linear	A transfer function which is not a straight line. Non-linear circuits normally distort a signal passed through them. *See also Chapter 1 Background.*
Non-realtime	Not happening in real time. Either with a delay, or taking longer to process than the length of the information being processed. *See also Chapter 1 Background.*
Non-sinusoidal	Not a sine wave waveform or waveshape. *See also Sine, Chapter 2 Analogue synthesis.*
Non-volatile	Permanently stored. *See also Chapter 1 Background.*
Notch	Synonym for a notch filter. *See also Chapter 2 Analogue synthesis.*
Notch filter	A filter which rejects one frequency. *See also Chapter 2 Analogue synthesis.*
Note stealing	The process of reassigning playing notes when the available polyphony has all been allocated. *See also Chapter 2 Analogue synthesis.*
Nyquist frequency	The highest frequency which can be coded and restored in a digital codec system which is sampled at twice the Nyquist frequency. *See also Chapter 1 Background.*
Octal	Numbers expressed in base 8. The first eight digits are used: 0 1 2 3 4 5 6 7. *See also Chapter 1 Background.*
Omni mode	A MIDI mode where messages on any channel are played by the receiving instrument. *See also Chapter 1 Background.*
One-shot	Synonym for monostable. A circuit which produces a short pulse when it is triggered. *See also Monostable, Chapter 1 Background.*
Operating system	The overall underlying software which runs a computer. *See also Chapter 5 Digital synthesis.*
Operator	Yamaha FM-speak for an oscillator, VCA and envelope generator. *See also FM, Chapter 5 Digital synthesis.*
Orthogonal	(1) Two axes or planes which are at 90 degrees to each other. (2) Two quantities which are unrelated. *See also Chapter 6 Using synthesis.*
OS	Operating system. The overall underlying control program of a computer. *See also Computer, Chapter 5 Digital synthesis.*
Oscillator	A circuit which produces a repetitive output. *See also Chapter 2 Analogue synthesis.*
Oscillator sync	Locking together of two oscillators so that the two frequencies have some relationship. *See also Hard sync, Chapter 2 Analogue synthesis.*

Out
Abbreviation for output. One of the three MIDI ports.
See also MIDI, Chapter 1 Background.

Overdrive
Synonym for distortion.
See also Distortion, Chapter 2 Analogue synthesis.

Overflow
When a digital circuit exceeds the largest number it can cope with, it has overflowed.
See also Chapter 1 Background.

Overtone
Similar to harmonic: whole number multiples of the fundamental. The second harmonic is the *first* overtone.
See also Harmonic, partial, Chapter 2 Analogue synthesis.

Paddle
Method of controlling pitch on some synthesizers. A performance controller, often made from a flat-textured vertical piece of plastic, with a rectangular hole in the front panel, and a fulcrum below the surface of the panel.
See also Chapter 7 Controllers.

Page
User interface term, meaning a new set of controls in the same screen or panel area.
See also Chapter 7 Controllers.

Parallel
In a parallel communication channel the data is sent with several bits simultaneously, using at least as many wires as there as bits.
See also Serial, Chapter 4 Sampling.

Parameter
An abstract value or variable quantity. Often used to represent the controls of a synthesizer.
See also Value, Chapter 5 Digital synthesis.

Part
(1) One line in a piece of music. (2) One channel in a multi-timbral arrangement.
See also Chapter 2 Analogue synthesis.

Partial
(1) Synonym for harmonic, although can be used for inharmonics too.
(2) Roland-speak for a 'voice'.
See also Voice, harmonic, Chapter 2 Analogue synthesis.

Patch
A sound, originally referred to the layout of the patch cords and control settings, but now generic.
See also Chapter 2 Analogue synthesis.

Patch cord
A cable with connectors at each end. Used for connecting modules together in a modular synthesizer.
See also Modular, Chapter 2 Analogue synthesis.

Patching
The process of connecting together modules using patch cords.
See also Chapter 2 Analogue synthesis.

PCM
Pulse code modulation. A method of encoding audio signals into digital form; the audio is sampled and the level is converted into a number.
See also ADC, DAC, sampling, Chapter 4 Sampling.

Peak level
(1) The highest voltage present in an audio signal. (2) The voltage level reached by the attack segment of an envelope generator.
See also EG, Chapter 2 Analogue synthesis.

Performance
Many meanings: live performance. User controls. Complete front panel settings.
See also Memories, Chapter 6 Using synthesis.

Phase
The position in a waveform or waveshape. Normally given in degrees. One cycle contains 360 degrees.
See also Chapter 1 Background.

Phase-shifter	A circuit which alters the relative phase of an audio signal. *See also Chapter 6 Using synthesis.*
Phase sync	A method of locking two oscillators together so that they have a fixed phase relationship between their outputs. *See also Chapter 2 Analogue synthesis.*
Phase-lock	A method of locking two oscillators together so that they have a fixed phase relationship between their outputs. *See also Chapter 2 Analogue synthesis.*
Phasing	Sound effect produced by mixing together a phase changed audio signal with the original. *See also Effects, Chapter 6 Using synthesis.*
Phon	Unit of perceived loudness, which takes into account the response of the human ear. *See also Loudness, Chapter 1 Background.*
Pink noise	Noise passed through a low-pass filter so that the level decreases with frequency. Produces equal energy for any bandwidth in the audio spectrum. *See also Chapter 2 Analogue synthesis.*
Pitch	Synonym for the frequency of a note. Pitch is a musical term for frequency. *See also Frequency, Chapter 1 Background.*
Pitch bend	Changes of pitch or frequency. *See also Chapter 7 Controllers.*
Pitch-to-MIDI	The conversion of pitched audio signals into the equivalent MIDI note numbers. Normally only monophonic conversions are possible. *See also Chapter 5 Digital synthesis.*
Pitch-to-voltage	The conversion of pitched audio signals into a control voltage. Normally only monophonic conversions are possible. *See also Chapter 2 Analogue synthesis.*
Pivot point	Synonym for a break-point in scaling. *See also Chapter 2 Analogue synthesis.*
Pole	One section of a complex filter. Two- and four-pole VCFs are commonly found in analogue synthesizers. The number of poles affects the cut-off slope: two-pole = 12 dB/octave, four-pole = 24 dB/octave. Poles are normally associated with peaks in the frequency response. *See also Cut-off slope, Chapter 2 Analogue synthesis.*
Poly-mode	MIDI mode where polyphonic notes can be produced from a single MIDI channel. *See also Mode, Chapter 1 Background.*
Poly-mod	Synonym for cross-mod. Modulation of an oscillator by another oscillator. *See also Cross-mod, Chapter 2 Analogue synthesis.*
Polyphonic	An instrument that can play more than one note at once. *See also Monophonic, Chapter 2 Analogue synthesis.*
Polyphony	The total number of simultaneous 'notes' that an instrument can produce at once. A piano is fully polyphonic, with enough hands you can play all of the notes simultaneously! *See also Timbrality, Chapter 2 Analogue synthesis.*
Port A	MIDI port. *See also Chapter 1 Background.*

Portamento	A smooth change of frequency between two notes instead of the normal abrupt transition. *See also Glissando, Chapter 2 Analogue synthesis.*
Pot	Abbreviation for potentiometer: a rotary or slider control whose output is proportional to the position of the indicator knob. *See also Knob, Chapter 7 Controllers.*
PPQ	Acronym for pulses per quarter-note. *See also MIDI clock, Chapter 2 Analogue synthesis.*
PPQN	Pulses per quarter-note. The number of pulses used to represent a quarter note length of time in a sequencer. *See also Sequencer, Chapter 7 Controllers.*
Pre-emphasis	The pre-processing of an audio signal before digitally coding it. A slight gain at high frequencies is intended to improve the dynamic range for high frequency components. *See also Chapter 1 Background.*
Precision	The number of significant digits. *See also Accuracy, Chapter 2 Analogue synthesis.*
Preset	Synonym for patch. *See also Chapter 1 Background.*
Pressure	Pressing on the key with the finger, once it has reached the bottom of its travel. *See also After-touch, Chapter 7 Controllers.*
Pressure sensitivity	The size of the control signal produced by key pressure depends on the 'pressure sensitivity'. *See also Chapter 2 Analogue synthesis.*
Priority	Part of an assignment strategy. *See also DVA, Chapter 7 Controllers.*
Program	(1) Synonym for sound or patch. (2) A sequence of instructions to a computer. *See also Chapter 1 Background.*
Program change	A MIDI message which is used to select a sound. *See also Sound, Chapter 1 Background.*
Programmable	The behaviour can be controlled using a parameter or a script. *See also Chapter 2 Analogue synthesis.*
Programmer	(1) A person who writes computer programs. (2) A remote front panel for expander modules which do not have a complete front panel. *See also Chapter 2 Analogue synthesis.*
Programming	The process of writing a computer program or the process of producing a sound on a synthesizer. *See also Chapter 2 Analogue synthesis.*
PROM	Programmable read-only memory, a fixed one-time-programmable storage medium for computer programs. *See also EPROM, Chapter 1 Background.*
Pulse	A digital signal which consists of a change to one state for a fixed period of time, followed by a reversion to the original state. *See also Chapter 2 Analogue synthesis.*
Pulse width	The ratio between the two parts of a pulse waveform. Normally expressed as a percentage. *See also Chapter 2 Analogue synthesis.*

PW	Abbreviation for pulse width. *See also Pulse width, Chapter 2 Analogue synthesis.*
Q	Synonym for resonance. A measure of the selectivity/sharpness/resonance of the filter. *See also Resonance, Chapter 2 Analogue synthesis.*
Quantisation	Converting a continuous value into a series of discrete values. *See also Chapter 6 Using synthesis.*
Quantisation noise	Noise produced by a digital coding system because of the limitations of the representation of small signals. *See also Chapter 1 Background.*
Quantise	The process of converting a continuous value into discrete values. *See also Chapter 1 Background.*
Quantised	A value which has been converted to a quantised form. *See also Chapter 1 Background.*
Quantiser	A circuit or device which quantises. *See also Chapter 1 Background.*
RAM	Random-access memory. Memory which can be written to, and read from. *See also ROM, Chapter 1 Background.*
Ramp	Synonym for sawtooth. Description for a waveform shape which rises linearly and then abruptly returns to the start level. *See also Chapter 2 Analogue synthesis.*
Random	Unpredictable. *See also Noise, Chapter 2 Analogue synthesis.*
Rate/level EG	A type of envelope generator, where the shape is defined by a number of rate and level values. *See also Chapter 2 Analogue synthesis.*
Rate scaling	Variation of envelope generator times by a control voltage, usually derived from the keyboard control voltage. *See also Chapter 2 Analogue synthesis.*
RCA	Abbreviation for Radio Corporation of America, an electronics innovation company. *See also Chapter 1 Background.*
Real time	Happens at the same rate as normal time. Implies a short time delay, almost instantaneous. *See also Chapter 1 Background.*
Reconstruction filter	Filter which removes all frequencies above the Nyquist frequency from the output of a DAC. *See also Nyquist, Chapter 1 Background.*
Rectangle	A waveshape which is equivalent to a pulse waveshape. The 50% pulse is also known as a square wave. *See also Square, Chapter 2 Analogue synthesis.*
Rectangular	Shaped like a rectangle. *See also Rectangle, Chapter 2 Analogue synthesis.*
Reference	A standard tone or level which is used to make comparisons against. *See also Chapter 1 Background.*
Regeneration	Synonym for feedback or resonance. *See also Chapter 2 Analogue synthesis.*

Release	The final segment of an envelope generator. *See also EG, Chapter 2 Analogue synthesis.*
Release velocity	The speed with which a key is released. *See also Attack velocity, Chapter 2 Analogue synthesis.*
Remote keyboard	A controller keyboard which is remote from the expander module (or modules) which it is controlling. Often limited in functionality with respect to a master keyboard. *See also Master keyboard, Chapter 7 Controllers.*
Resolution	The smallest change which can be represented. In integer mathematics, the resolution is one. *See also Precision, Chapter 1 Background.*
Resonance	A measure of the way that a system responds to an external stimulus. A tuning fork is resonant because it vibrates when struck. Filters can 'ring' or self-oscillate when the resonance is high. *See also Q, Chapter 2 Analogue synthesis.*
Resynthesis	Analysing an audio signal and extracting parameters which are then used as the basis for synthesising the sound with modifications. *See also Chapter 5 Digital synthesis.*
Reverb	Abbreviation for reverberation. *See also Chapter 6 Using synthesis.*
Reverberation	A series of echoes caused by reflections from the acoustic environment. *See also Echo, Chapter 6 Using synthesis.*
Ribbon	(1) Type of microphone. (2) Colloquialism for a performance controller using a flexible strip which is activated by pressing with the finger. *See also Chapter 7 Controllers.*
Ribbon controller	A performance controller using a flexible strip which is activated by pressing with the finger. *See also Chapter 7 Controllers.*
Ring modulation	Balanced modulation with two inputs and one output. Output is sum and difference of input frequencies. *See also AM, Chapter 2 Analogue synthesis.*
Ringing	Resonant circuits will oscillate momentarily when there is a sudden change at their input. This is called *ringing*. Some VCFs can be forced to ring by using high resonance settings. *See also Oscillator, Chapter 2 Analogue synthesis.*
Roll-off	Synonym for filter cut-off slope. *See also Slope, Chapter 2 Analogue synthesis.*
Roll-off slope	Synonym for filter cut-off slope. *See also Slope, Chapter 2 Analogue synthesis.*
ROM	Read-only memory. Memory which can only be read, the contents are fixed and so cannot be altered. *See also EPROM, Chapter 1 Background.*
Routing	The path which an audio or control signal takes through a device or circuit. *See also Patching, Chapter 2 Analogue synthesis.*
Running status	A method of removing redundant information in repeated MIDI messages to reduce the required data bandwidth. *See also Chapter 1 Background.*

s

Unit of time measurement: second. One sixtieth of a minute.
See also Chapter 1 Background.

S&H

Abbreviation for sample and hold. Also known as S/H.
See also Sample and hold, Chapter 2 Analogue synthesis.

S/H

Abbreviation for sample and hold. Also known as S&H.
See also Sample and hold, Chapter 2 Analogue synthesis.

Sample

Data which represents an audio signal. Samples can be analogue or digital, although they are usually digital.
See also Sampling, Chapter 4 Sampling.

Sample and hold

An analogue latch which stores an analogue voltage when triggered by a clock or trigger signal.
See also Chapter 2 Analogue synthesis.

Sample rate

The frequency at which analogue samples are converted to digital representations.
See also Sampling, Chapter 4 Sampling.

Sampler

A device which can sample and replay sounds. Often used to mean a sample-replay device.
See also Chapter 4 Sampling.

Samples

Analogue or digital values which make up the information contained within an audio signal.
See also Chapter 4 Sampling.

Sampling

The process of capturing part of an audio signal. Usually refers to digital sampling.
See also Chapter 4 Sampling.

Sawtooth

A waveshape found in analogue synthesizer VCOs. Consists of a rising slope followed by a sudden return to the start value.
See also Ramp, Chapter 2 Analogue synthesis.

Scale

The relationship between the keyboard control voltage and the VCO frequency. Normally set so that 1 volt of control voltage change produces an octave change in pitch.
See also Chapter 2 Analogue synthesis.

Scaling curve

The shape of the curve used to modify the rate or level scaling with keyboard note position.
See also Chapter 2 Analogue synthesis.

Schmitt trigger

A circuit which can be used to convert analogue signals into digital signals with two values.
See also Chapter 2 Analogue synthesis.

Scrub

Moving backwards and forwards through events in a sequencer (audio, video or MIDI) or on a tape (audio or video).
See also Chapter 6 Using synthesis.

SCSI

Acronym for small computer systems interface. A method of connecting peripherals (disks, CD-ROM drives, etc.) to a computer.
See also Bus, Chapter 4 Sampling.

SDS

Acronym for sample dump standard, the method defined in the MIDI specification for transferring samples using MIDI messages.
See also Chapter 1 Background.

Second touch

Synonym for after-touch or pressure.
See also After-touch, Chapter 7 Controllers.

Segment	One part of an envelope cycle. An AD envelope has two segments: attack and decay, whilst an ADSR envelope has four segments. In a rate and level envelope, one of the pairs of rates and levels. *See also Envelope, Chapter 2 Analogue.*
Self-oscillation	VCOs and clock generators are designed to maintain self-oscillation, where a continuous repetitive output is produced. Often associated with resonant circuits. *See also Oscillator, Chapter 2 Analogue synthesis.*
Semiconductor	A material whose electrical properties can be changed by applying voltages or currents. The basis of electronic devices. *See also Electronics, Chapter 1 Background.*
Semitone	The smallest change of pitch on a conventional music keyboard. The ratios of two frequencies which are one semitone apart is 1.059463. *See also Cent, Chapter 1 Background.*
Sequence	A series of musical events. *See also Chapter 6 Using synthesis.*
Sequencer	A computer or hardware device which is designed to produce a series of musical events. *See also Chapter 6 Using synthesis.*
Serial	In a serial communication channel the data is sent 1 bit at a time along a single piece of wire. *See also Parallel, Chapter 4 Sampling.*
Sideband	A frequency produced above or below the carrier frequency by an FM, AM or RM modulation technique. *See also Chapter 5 Digital synthesis.*
Sine	A smooth-curved waveform which contains only one frequency component. *See also Harmonic, Chapter 2 Analogue synthesis.*
Single trigger	A keyboard which produces only one trigger when a key is held down and other keys are subsequently pressed. *See also Multi-trigger, Chapter 2 Analogue synthesis.*
Single-step	Method of recording notes and other musical events into a sequencer. Each event is entered individually, in non-realtime. *See also Real time, Chapter 6 Using synthesis.*
Sinusoidal	Shaped like a sine wave. *See also Sine, Chapter 2 Analogue synthesis.*
Slave	A 'slave' is a keyboard or expander synthesizer which receives MIDI messages. *See also MIDI, Chapter 6 Using synthesis.*
Slope	The ratio between the horizontal and vertical parts of a line. A slope of 1 : 1 is at an angle of 45 degrees. *See also Chapter 2 Analogue synthesis.*
SMPTE	Society of Motion Picture and Television Engineers. Short-hand for the time code that is used to synchronize video, television and audio. *See also MTC, Chapter 7 Controllers.*
SNR	Signal-to-noise ratio. A measure of the signal level against the background noise. *See also THD, Chapter 1 Background.*

Song

A collection of phrases of music. Typically starting with an introduction, then a verse, followed by a chorus, repeats of verse and chorus, a break for 8 bars where a second theme is introduced, then more verse and chorus pairs, a key change and the closing section.
See also Chapter 6 Using synthesis.

Sostenuto pedal

A performance controller which mimics the behaviour of the piano pedal: only notes which are being held down on the keyboard are sustained when the sostenuto function is activated.
See also Sustain, Chapter 7 Controllers.

Source

A control voltage or control signal output.
See also Chapter 2 Analogue synthesis.

Spectrum

A plot of the frequency content of an audio signal. The horizontal axis is the frequency axis, whilst the vertical axis shows the level of the frequency components.
See also Harmonic, Chapter 2 Analogue synthesis.

Speed

The tape speed of a tape recorder is the speed at which the magnetic tape passes through the machine.
See also Chapter 1 Background.

Spillover

Synonym for overflow.
See also Chapter 6 Using synthesis.

Splice

A join between 2 bits of magnetic tape. This has also come to mean a join between any 2 bits of audio, even in software-based audio editors and samplers.
See also Tape, Chapter 1 Background.

Split

The allocation of two sounds to different parts of the keyboard. Named from the major usage, which is to play different parts with each hand on the same keyboard.
See also Chapter 6 Using synthesis.

Split point

The point on the keyboard at which a split between two sounds occurs.
See also Split, Chapter 6 Using synthesis.

Square

A waveshape which has equal time periods in two states. A special case of a rectangular or pulse waveshape.
See also Rectangle, Chapter 2 Analogue synthesis.

Stack

A composite sound made up of two or more sounds.
See also Layer, Chapter 6 Using synthesis.

Stacking

The process of making a composite sound from two or more sounds.
See also Layering, Chapter 6 Using synthesis.

Stage

Synonym for a segment in an envelope. One of the pairs of rates and levels, or one of the attack, decay or release periods.
See also Chapter 2 Analogue synthesis.

State-variable

A multi-mode filter made from a loop of operational amplifiers (op-amps), which produces several types of filter response outputs simultaneously.
See also Chapter 2 Analogue synthesis.

Step input

Sequencer input mode: one note/event at a time. Non-realtime.
See also Single step, Chapter 6 Using synthesis.

Stereo

Two audio channels (left and right), which can provide the illusion of a complete sound stage.
See also Mono, Chapter 6 Using synthesis.

Stochastic	Synonym for unpredictable. Noise is an example of a stochastic waveform, since no two cycles are the same. *See also Deterministic, Chapter 2 Analogue synthesis.*
Storage	Often implies 'data storage media'; somewhere to hold information. *See also Chapter 1 Background.*
Sub-octave	A note one or two octaves down from the fundamental pitch. Often produced using digital divider circuits. *See also Chapter 3 Hybrid synthesis.*
Subtractive	Synthesis technique which uses harmonically rich waveforms and filters them to produce sounds. *See also Additive, Chapter 2 Analogue synthesis.*
Subtractive synthesis	Synthesis technique which uses harmonically rich waveforms and filters them to produce sounds. *See also Additive, Chapter 2 Analogue synthesis.*
Sustain	One of the segments or stages in an ADSR envelope. This is the level which the envelope will decay to and stay at whilst the key is held down. *See also Chapter 2 Analogue synthesis.*
Switch trigger	Envelope trigger signal which uses a brief pulse to initiate an envelope. *See also Chapter 2 Analogue synthesis.*
Sync	Abbreviation for synchronisation: (1) Locking a sequencer or tape recorder to another timing source. (2) Locking two VCOs together. *See also Chapter 2 Analogue synthesis.*
Sync track	A track on a tape recorder which is reserved for the recording and playback of synchronisation signals. *See also Chapter 6 Using synthesis.*
Synchronisation	(1) Locking a sequencer or tape recorder to another timing source. (2) Locking the frequency of two VCOs together. *See also Chapter 2 Analogue synthesis.*
Synonym	A word having almost exactly the same meaning as another word. *See also Acronym, Chapter 1 Background.*
Synthesiser	Alternative UK spelling for synthesizer. A musical instrument which produces sound from simple resources. *See also Chapter 1 Background.*
Synthesizer	US spelling of synthesiser. A musical instrument which produces sound from simple resources. *See also Chapter 1 Background.*
Sysex	Abbreviation for system exclusive: instrument and manufacturer specific information. *See also System exclusive, Chapter 7 Controllers.*
System	MIDI-speak for global MIDI messages. *See also MIDI, Chapter 7 Controllers.*
System exclusive	Instrument and manufacturer specific information; transmitted as MIDI messages with a distinctive header and footer ($F0 and $F7). *See also MIDI, Chapter 7 Controllers.*

Tape	Recording tape. Polyester plastic backing with an iron oxide magnetic coating. *See also Speed, Chapter 4 Sampling.*
Tape loops	A piece of magnetic tape looped around and the end spliced to the beginning. Also used for looped audio in digital equipment. *See also Chapter 4 Sampling.*
THD	Abbreviation for total harmonic distortion. A measure of the distortion produced by an audio system. Measures the amount of additional harmonics which are added to a signal by a process. Usually given in per cent. *See also Distortion, Chapter 1 Background.*
Theremin	A non-contact controller for pitch and volume. Invented by Lev Termen (aka Leon Theremin) in 1919, the Theremin uses capacitive coupling between two antennas and the operator's hands as the control mechanism, the loop antenna controls the volume of the controller. *See also Chapter 7 Controllers.*
Thru	Abbreviation for 'through' (US). One of the three MIDI ports. *See also MIDI, Chapter 1 Background.*
Timbrality	The number of different 'timbres' or 'sounds' that an instrument can produce at once. Different to polyphony, which is the total number of notes which can be played at once, regardless of the 'timbre' or 'sound'. *See also Polyphony, Chapter 6 Using synthesis.*
Timbre	(1) The 'tone colour' of a sound. (2) Roland-speak for a sound. *See also Chapter 1 Background.*
Time code	A method of attaching information about timing to an audio or video recording. *See also Chapter 6 Using synthesis.*
Tone colour	The timbre of a sound. *See also Timbre, Chapter 1 Background.*
Tone control	Synonym for equalisation. *See also EQ, Chapter 1 Background.*
Tone poems	A piece of music which attempts to describe a place, event, emotion or person using just notes and timbre. *See also Timbre, Chapter 1 Background.*
Top-note priority	A key assignment algorithm which gives priority to the highest note played on a monophonic keyboard. *See also Low-note, Chapter 6 Using synthesis.*
Total harmonic distortion	Audio measurement which is used to determine the amount of additional harmonics which are added to a signal by a process. *See also THD, distortion, Chapter 1 Background.*
Touch pad	A performance controller. *See also Chapter 7 Controllers.*
Touch sensitive	Some thing which is changed by pressure. Normally additional pressure once a key has been depressed. *See also After-touch, Chapter 7 Controllers.*
Track	One of several paths on a piece of magnetic recording tape. This is also used in the context of digital sequencers. *See also Chapter 4 Sampling.*

Tracking generator	A module which is concerned with controlling the rate and level scaling functions. *See also Chapter 2 Analogue synthesis.*
Transient	Brief, non-sustaining, events. Can refer to sounds, harmonics or one-off occurrences. *See also Chapter 2 Analogue synthesis.*
Transient generator	Synonym for envelope generator. *See also EG, Chapter 2 Analogue synthesis.*
Transistor	Abbreviation for transfer resistor: a simple semiconducting device which can switch or amplify signals. *See also Semiconductor, FET, Chapter 1 Background.*
Tremolo	Periodic variation in the volume, level or amplitude of a signal (audio or control voltage). The LFO speed is usually between 5 and 25 Hz. *See also LFO, AM, Chapter 2 Analogue synthesis.*
Triangle	A waveform which has linear slopes which rise and fall alternately. *See also Sawtooth, Chapter 2 Analogue synthesis.*
Trigger	Signal which starts an event, usually an envelope. *See also Chapter 2 Analogue synthesis.*
Trigger pulse	(1) The initiation signal for an EG. (2) The output of a keyboard controller note detector, indicates that a key has been pressed. *See also Chapter 2 Analogue synthesis.*
TTL	Transistor–transistor logic: type of digital logic chip made from transistors. *See also CMOS, Chapter 1 Background.*
TVA	Abbreviation for time-variant amplifier. Roland-speak for a VCA in its digital synthesizers. *See also VCA, Chapter 5 Digital synthesis.*
TVF	Abbreviation for time-variant filter. Roland-speak for a VCF in its digital synthesizers. *See also VCF, Chapter 5 Digital synthesis.*
Two-pole	A filter with two sections, normally with a cut-off slope of 12 dB/octave. *See also Chapter 2 Analogue synthesis.*
UART	Universal asynchronous receiver/transmitter: converts data into serial format (as used in a MIDI port).
Unison	Several events happening as one. Often used for a stack of sounds which all sound when a key is played on a keyboard. *See also Chapter 6 Using synthesis.*
Universal	Synonym for generic, as in the sense of globally available or usable. Often used in the context of editor/librarian software. *See also Generic, Chapter 6 Using synthesis.*
VCA	Abbreviation for voltage controlled amplifier. An amplifier whose gain can be controlled by a control voltage. *See also VCO, Chapter 2 Analogue synthesis.*
VCF	Abbreviation for voltage controlled filter. A filter whose cut-off frequency can be controlled by a control voltage. *See also VCA, Chapter 2 Analogue synthesis.*

VCO	Abbreviation for voltage controlled oscillator. An oscillator whose frequency can be controlled by a control voltage. *See also VCF, Chapter 2 Analogue synthesis.*
Vector synthesis	A synthesis technique where two or four separate sounds can be mixed in real time using a joystick, and the mix recorded for subsequent replay as part of the sound. *See also Chapter 2 Analogue synthesis.*
Velocity	The speed with which a key is pressed. Most synthesizers merely measure the time from just after the key is depressed to just before it stops moving at the bottom of its travel. *See also Chapter 2 Analogue synthesis.*
Velocity curve	The relationship between the speed of depressing a key and the control voltage or control signal which is produced. This can often be mapped. *See also Chapter 2 Analogue synthesis.*
Velocity sensitivity	Synonym for velocity curve. *See also Chapter 2 Analogue synthesis.*
Vernier	A method of allowing precise measurement by having a small scale attached to a rotary or linear control. Often used for the frequency control knobs of analogue synthesizers. *See also Chapter 2 Analogue synthesis.*
Vibrato	Periodic variation in pitch or frequency of an audio signal. (The LFO speed is usually between 5 and 25 Hz.) *See also LFO, FM, Chapter 2 Analogue synthesis.*
Vocoder	A modifier which splits an audio signal into separate frequency bands for processing. *See also Chapter 2 Analogue synthesis.*
Voice	A complete sound producing system for one note. *See also Chapter 2 Analogue synthesis.*
Voice channel	Synonym for voice. *See also Voice, Chapter 2 Analogue synthesis.*
Voice stealing	Synonym for note stealing. *See also Note stealing, Chapter 2 Analogue synthesis.*
Volatile	Memory which loses its contents when the power is removed. *See also Non-volatile, Chapter 1 Background.*
Voltage control	Using a voltage as a means of controlling parameters. *See also CV, Chapter 2 Analogue synthesis.*
Voltage pedal	A performance controller which produces a control voltage as the output of a foot pedal. *See also Chapter 7 Controllers.*
Voltage trigger	Envelope triggered signal which uses a brief pulse to initiate an envelope. *See also Chapter 2 Analogue synthesis.*
Volume	The amplitude, loudness or level of a signal or waveform. *See also Loudness, Chapter 2 Analogue synthesis.*

Volume pedal   A performance controller which varies the attenuation (or gain) of an audio signal as it passes through the foot pedal.
See also Chapter 7 Controllers.

Wave sequencing   The concatenation of two or more wavecycles, multi-cycle waveforms, or samples.
See also Chapter 3 Hybrid synthesis.

Wavecycle   A single cycle of a waveform.
See also Waveform, Chapter 3 Hybrid synthesis.

Waveform   The shape of a wave. Some waves only repeat over several cycles.
See also Wavecycle, Chapter 2 Analogue synthesis.

Waveform modulation   Changes in the shape of a waveform under the control of a control voltage or a control signal. PWM is one example.
See also PWM, Chapter 2 Analogue synthesis.

Wavelength   An alternative name for a cycle of a waveform: the time for one complete cycle.
See also Cycle, Chapter 1 Background.

Waveshape   Synonym for waveform.
See also Waveform, Chapter 2 Analogue synthesis.

Wavetable   An area of memory which is used to hold wavecycle and multi-cycle waveforms.
See also Chapter 3 Hybrid synthesis.

Wet   A processed audio signal. Often refers to reverberation.
See also Chapter 6 Using synthesis.

Wheel   Performance control where a vertical wheel is moved to change a controlled parameter. Often used for pitch or modulation parameter.
See also Parameter, Chapter 7 Controllers.

Wheels   Wheels are performance controllers used to control pitch or modulation functions. Usually take the form of a 50-cm disk set next to the left-hand side of the keyboard.
See also Joystick, Chapter 7 Controllers.

White noise   Noise which has an equal mix of all frequencies.
See also Chapter 2 Analogue synthesis.

Wobble   Low frequency vibrato is sometimes called wobble.
See also Wow, Chapter 2 Analogue synthesis.

Word   A set of bits or bytes containing a single sample or block of data. Often 8, 12 or 16 bits.
See also Chapter 1 Background.

Workstation   (1) A powerful networked personal computer. (2) A synthesizer with a built-in sequencer and disk storage.
See also Chapter 6 Using synthesis.

Wow   Low frequency variation in the speed of a tape recorder, which results in wobble or vibrato.
See also Vibrato, Chapter 2 Analogue synthesis.

XLR   Identifier name for a specific type of audio connector; also called a Cannon connector.
See also DIN, Chapter 2 Analogue synthesis.

Y leads   An audio adapter which has one input and two outputs, or two inputs and one output. Cannot be used with MIDI cables.
See also Chapter 1 Background.

Zero   Part of the mathematical background to filter design. Zeroes are associated with notches and dips in the filter response.
See also Pole, Chapter 1 Background.

Zero crossing   The points at which the audio signal crosses the zero axis at zero volts.
See also Chapter 1 Background.

Zone   Part of a keyboard. Used in describing splits. Can be a single note, or a contiguous range of notes.
See also Split, Chapter 6 Using synthesis.

Jargon

Jargon

Specialist	Used only in a given subject
Particular	Words used in one context
Subject	Synthesizers is the current subject
Alternative	Jargon often replaces more common words
Glossary	Also in the *Glossary* section
Acronym	Jargon frequently uses acronyms

Jargon is a descriptive term for those specialist words which are used only in a particular subject. This entry shows how they are dealt with here. Look up the word which is used throughout this book, and the alternatives will be shown. This is somewhat like a thesaurus, but for jargon. The entries include both synonyms and antonyms. Most of these words will also be in the *Glossary* section.

After-touch

Touch	Additional key pressure used as a controller
Pressure	Alternative name for touch
Sustain pressure	After-touch is often used in the sustain segment
Second touch	Alternative name for touch (initial touch = velocity)
Expression	Alternative name for touch
After-touch	Alternative name for touch

After-touch is the name for the additional pressure that some keyboards allow to be used as a controller after the key has travelled from the up to the down position. After-touch sensors vary from soft rubber, which provides several millimetres of movement, to ones which are hard rubber and hardly move at all. After-touch sensing can be global for the entire keyboard (the hardest pressure is normally the output) or polyphonic (much rarer).

Amplitude

Level	Alternative name. How big the signal is
Volume	Alternative name. How big the signal is
Size	Alternative name. How big the signal is
Peak	The highest amplitude reached by the waveform
Normalised	Using all of the available volume range
Compressed	Increasing the volume of quiet audio, reducing loud

Low-pass filters are good for making tonal changes to the brightness of a sound, but for making more major changes a band-pass filter is required. By adding feedback to a low-pass filter, it is possible to turn it into a resonant band-pass filter which has a strong peak at the cut-off frequency.

Sink

Destination	The input to the circuit = destination of control signal/voltage
Input	The input of the circuit
In	Abbreviation for input
Sink	Alternative name for destination

Controls can be divided into sources and destinations. Destinations take control voltages and signals and use them to make changes to parameters. Matrix modulation schemes use the words source and destination for the outputs and inputs to a matrix of connection points.

Source

Output	The source of the control voltage = output of circuit
Out	Abbreviation for output
Source	The output of the circuit
Origin	Alternative name for source

Controls can be divided into sources and destinations. Sources provide control voltages and control signals. Matrix modulation schemes use the words source and destination for the outputs and inputs to a matrix of connection points.

Storage

Disk	Hard or floppy: rotating magnetic/optical data memory
Disc	Vinyl record or album: holds audio information
RAM	Random-access memory: R/W solid-state data chips
ROM	Read-only memory: permanent solid-state data chips
Media	The physical carrier/holder of the data: disk, tape, card, etc.
Floppy	Flexible circle of magnetic material in a rigid casing
DAT	Tape-based data media using rotating helical scan heads
Card	Generic term for memory card (using ROM or RAM)
PCMCIA	Specific type of standardised card and interface
Network	A means of connecting computers together: non-local
Server	A remote computer used to hold data, accessed via a network
Optical	Disk using optical rather than magnetic storage for data

Virtual	RAM used as a hard disk replacement (faster)
Tape	Flexible plastic strip with magnetic coating for data
Off-line	Offline storage needs to be physically loaded by hand
Flash	Rewriteable memory technology used in memory cards
Removable	Opposite to built-in (memory cards are removable)
Built-in	Flash, ROM or RAM memory inside the device.
CD/DVD	Compact disk/digital versatile disk, optical storage

Computers use many forms of storage for holding data (information). Storage is really a shortened form of the compound word 'mass storage'.

Vibrato

FM	Frequency modulation: cyclic pitch changes
AM	Amplitude modulation: cyclic volume changes (not vibrato)
Tremolo	Cyclic changes in volume (not vibrato)
Flutter	Rapid random changes in pitch (usually on a tape recorder)
Wow	Slow cyclic pitch changes (usually on a tape recorder)
Wobble	Cyclic pitch changes
Modulation	Can be any pitch, volume or timbre change, often vibrato
Leslie effect	A complex combination of vibrato and tremolo effects

Vibrato is the name applied to cyclic changes in frequency. Tremolo is the term for cyclic changes in volume. The terms vibrato and tremolo are often mistakenly used interchangeably. For example, the 'tremolo' arm on a guitar produces vibrato and pitch bending, whilst a violinist's 'vibrato' is a mixture of vibrato and tremolo.

Volatile

RAM	Random-access memory = read/write memory
Battery-backed	The battery maintains the data when the power is down
Eraseable	The memory contents can be changed
Non-volatile	The device holds its data without any power present
Permanent	The device holds its data without any power present
EPROM	Erasable programmable read-only memory
EEPROM	Electrically erasable ROM
EAROM	Electrically alterable ROM
Flash EPROM	EPROM that behaves like non-volatile RAM
Flash chips	Alternative name for flash EPROM

Flash memory	Alternative name for flash EPROM
Flash	Alternative name for flash EPROM

'Volatile' means that the storage medium depends on a supply of electrical power to hold data. If the power fails then the data is lost. RAM is usually volatile. 'Non-volatile' means that the storage does not depend on an external power supply. It may have its own local power supply, or use a technology which requires no power to maintain the data.

Index